A Concise Encyclopedia of Additive Manufacturing

Sanjay Kumar

A Concise Encyclopedia of Additive Manufacturing

Springer

Sanjay Kumar
Swarup Niwas
Gumla, Jharkhand, India

ISBN 978-3-031-98393-1 ISBN 978-3-031-98394-8 (eBook)
https://doi.org/10.1007/978-3-031-98394-8

Professional book

This Springer imprint is published by the registered company Springer Nature Switzerland AG
The registered company address is: Gewerbestrasse 11, 6330 Cham, Switzerland

Preface

This book is an attempt to present the subject of additive manufacturing (AM), alphabetically. It is not claimed that everything present in AM can be found in this book. But it is claimed that everything present in AM can be found if it is started from this book, for which ample references are given. I acknowledge everyone whose work is listed in the references, as without their work, I would not have been in a position to make this claim.

The entries (going to be published as chapters) given in this book do not follow any guidelines, as entries in a multi-author editied encylopedia follow a guideline. It was hoped that the absence of guidelines would allow the natural evolution of the content and improve readability. The absence of guidelines and its consequences mean there are some entries that have a few lines and they cannot be removed, and there are some entries that span some pages and they cannot be divided into multiple entries.

This book was composed with the help of my earlier books. Therefore, in addition to new content, this book consists of updated versions of selected topics from my previous books. To maintain alphabetic order and keep the same topics together, the names of the topics needed to be adjusted, which might have caused not an exact match between the topic and the name of the entry.

I hope that this book will be useful for everyone.

Gumla, Jharkhand, India
May 29, 2025

Sanjay Kumar

Competing Interests The author has no competing interests to declare that are relevant to the content of this manuscript.

Contents

List of Figures

List of Table

About the Author

Sanjay Kumar is the author of five books on additive manufacturing, published by Springer, Cham.

Chapter 1
Additive Friction Stir Deposition

A rotating pin attached to a shoulder (friction stir tool) is used to stir a solid material in friction stir process. The pin is an extended tip of a solid cylinder (shoulder). The role of the shoulder is to prevent stirred materials from being extruded out of plane.

In additive friction stir deposition (AFSD), there is no pin while the shoulder has a thorough hole (Fig. 1.1a). The purpose of the hole is to pass the material onto a substrate from above. The material can be a rod, powder [1], or small blocks of any shape.

A rod cannot be deposited onto a substrate from above in the same way as powder through the hole, unless the rod is converted into small pieces. Therefore, a consumable rod, which is rotating and moving, is abraded against a substrate so that the rod loses material from its surface to the substrate. The material getting lost is deposited in the form of a paste (Fig. 1.1b) [2].

A rotating and moving cylindrical tool deposits powder vertically through its hole. The cross-sectional area of the hole is smaller than the non-hole cross-sectional area of the tool. The non-hole section of the tool is called the shoulder. When the tool rotates, part of the shoulder encounters already deposited material, which deforms the material before it can escape and become free. Depending on its linear speed, the shoulder encounters the same material times and again, which deforms, mixes, and plasticizes it—this is stirring.

A deposited material will be free from this cycle of processing if it is spherical ceramic powder deposited on a ceramic substrate so that the rotating shoulder will be able to neither deform and flatten nor embed the powder onto the substrate. The powder will then roll over through the gap between the shoulder and the substrate and will eject.

Friction between the shoulder and the substrate raises the temperature that increases the deformation and stirring. The temperature does not exceed the melting point of the material, so the benefit gained through solid-state processing is not lost [3]. High rotation speed of the shoulder may induce melting, which will limit the maximum permissible speed, which in turn will limit the fabrication speed.

S. Kumar, *A Concise Encyclopedia of Additive Manufacturing*,
https://doi.org/10.1007/978-3-031-98394-8_1

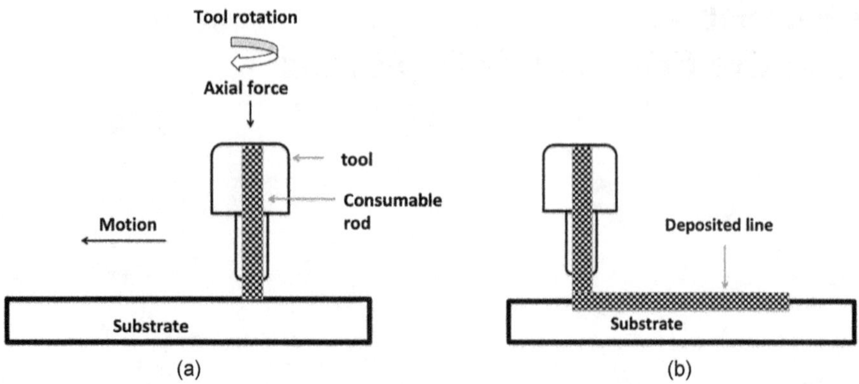

Fig. 1.1 Schematic diagram of AFSD. (**a**) Deposition. (**b**) Deposited line partially in the substrate due to stirring

In the absence of high temperature, there is no driving force for grains to increase in size, resulting in small grains responsible for high strength and ductility. A grain refinement up to 82% is possible for aluminium alloy [4]. Though if many layers are fabricated, long exposure of the initial layers to friction-generated temperature causes an increase in the grain size. Stirring and mixing do not let grains settle, resulting in high-angle grain boundaries that are responsible for a further increase in the strength. However, if the temperature is low, it needs to be increased by using an additional heat source to get better properties [5].

An axial force is applied in AFSD process, which is an essential component of the process for either transforming or joining the material. The applied force may deform a complex structure. As for making the complex structure, the tool that applies force needs to tread over from a thick to a thin support, pass through a gap, go over a bridge, or move in a corner. These may require the application of less force to prevent a delicate feature of the structure from collapsing.

Less force implies no material transformation—this brings a situation where either the material needs to be transformed or the structure needs to be built. This situation can be eliminated if no complex structure is built, i.e., simple structures such as rectangular, circular, annular, cylindrical, or any such blocks are made. This situation can be partially eliminated if the process is optimized for a range of tool sizes so small blocks can be made with small tools and big blocks with big tools, and the design of a structure is chosen that is the superset of a number of such blocks.

References

1. Zhang, M., Jiang, T., Sun, Z., et al. (2025). Screw-fed powder-based additive friction stir deposition: A study on pure aluminum. *Journal of Materials Processing Technology, 337*, 118730.
2. Schultz, J. P., & Creehan, K. D. (2014). Friction stir fabrication. In *US Patent (US 8893954 B2)*.

3. Yu, H. Z. (2022). Emerging processes—Friction stir based. In F. G. Caballero (Ed.), *Encyclopedia of materials: Metals and alloys* (pp. 153–161). Elsevier.
4. Razmjoo, H., Mirsalehi, S. E., Abbasi-Nahr, M., & Navidi-Helan, M. (2025). Microstructure, corrosion, and tribological performance of multilayered cylindrical AA5083 parts fabricated by modified friction stir deposition—Additive manufacturing. *Materials Characterization, 221*, 114777.
5. Qiao, Q., Tam, C. W., Lam, W. I., et al. (2025). Hybrid heat-source solid-state additive manufacturing: A method to fabricate high performance AA6061 deposition. *Journal of Materials Science and Technology, 228*, 107–124.

Chapter 2
Additive Layer Manufacturing

In additive layer manufacturing, a part is visualized as an assembly of layers, and, therefore, effort is done to make or have layers and then assemble them. When a part needs to be fabricated, its CAD model needs to be transformed in the form of layers (Fig. 2.1a). Before it is transformed (sliced), the model needs to be converted into an assembly of many small interconnected triangles, which is called tessellation, that generates STL (standard triangle language or stereolithography) files. To slice the STL model, a horizontal plane intersects it at various points and collects data at these points. The separation between two consecutive points is equal to the layer thickness.

These data determine the tool path—information is collected by the machine in the form of G-code, which is required by the machine to use its tool accordingly and convert these data (virtual layers) into physical layers. The sequence of steps is given in Fig. 2.1b. The lowest layer is fabricated first, followed by the fabrication of successive layers over it till the topmost layer is fabricated [1, 2].

S. Kumar, *A Concise Encyclopedia of Additive Manufacturing*,
https://doi.org/10.1007/978-3-031-98394-8_2

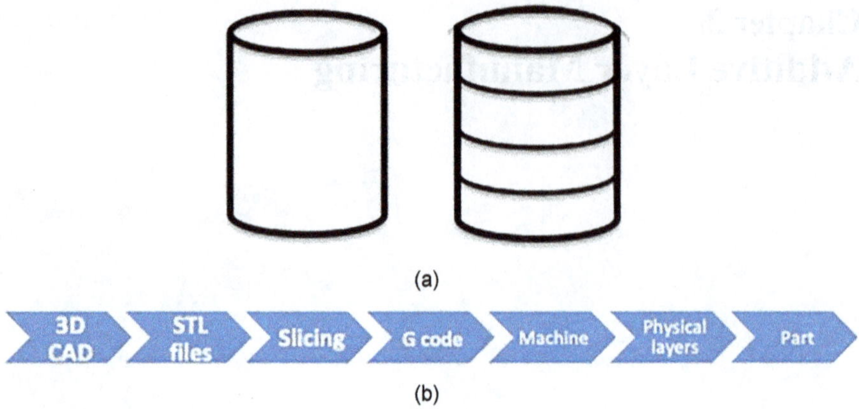

Fig. 2.1 Conversion of a CAD model into a part: (**a**) slicing of a cylindrical model into layers, (**b**) sequence of transformation

References

1. Youssef, H., El-Hofy, H., & Ahmed, M. (2024). *Fundamentals of additive manufacturing: Principles, technologies, and applications*. CRC Press.
2. Zhang, X., & Liou, F. (2021). Introduction to additive manufacturing. In *Additive manufacturing* (pp. 1–31). Elsevier.

Chapter 3
Additive Layerless Manufacturing

The following are examples of additive layerless manufacturing:

- CNC accumulation [1].
- Two-photon photopolymerization [2].
- Layerless fused deposition modeling [3].
- Continuous liquid interface production [4].
- Continuous three-dimensional printing [5].
- Volumetric additive manufacturing [6].
- Immiscible-interface assisted direct metal drawing [7].
- Linear volumetric 3D printing [8].

Additive layerless manufacturing is part of AM [9].

References

1. Zhao, X., Pan, Y., Zhou, C., et al. (2013). An integrated CNC accumulation system for automatic building-around-inserts. *Journal of Manufacturing Processes, 15*(4), 432–443.
2. Cho, H., Lee, W. S., & Chang, W. S. (2024). Three-dimensional photopolymerization additive manufacturing technology based on two-photon polymerization. *JMST Advances, 6*(9), 1–7.
3. Kanada, Y. (2015). Support-less horizontal filament-stacking by layer-less FDM.
4. He, H., Yang, Y., & Pan, Y. (2019). Machine learning for continuous liquid interface production: Printing speed modelling. *Journal of Manufacturing Systems, 50*, 236–246.
5. Jiang, Y., Wang, Y., He, H., et al. (2020). Constrained window design in projection stereolithography for continuous three-dimensional printing. *3D Printing and Additive Manufacturing, 7*(4), 163–169.
6. Whyte, D. J., Doeven, E. H., Sutti, A., et al. (2024). Volumetric additive manufacturing: A new frontier in layer-less 3D printing. *Additive Manufacturing, 84*, 104094.

S. Kumar, *A Concise Encyclopedia of Additive Manufacturing*,
https://doi.org/10.1007/978-3-031-98394-8_3

7. He, L., Fei, F., Wang, W., & Song, X. (2019). Layerless additive manufacturing of metal alloy lattices using immiscible-interface assisted direct metal drawing. *Procedia Manufacturing, 34*, 647–654.
8. Regehly, M., Garmshausen, Y., Reuter, M., et al. (2020). Xolography for linear volumetric 3D printing. *Nature, 588*(7839), 620–624.
9. Kumar, S. (2020). Introduction. In *Additive manufacturing processes* (pp. 1–19). Springer.

Chapter 4
Advantage

All advantages of AM are derived from its fundamental characteristic, i.e., freedom to add material layer by layer (Fig. 4.1). Though the freedom is restricted to adding in a plane before moving to add in the next plane, the freedom left after the restriction is enough to lead to advantages.

The advantages are.

1. Freedom to design—ability to make products of various designs. It gives three sub-advantages: freedom to save material, freedom to save energy, and freedom to make multi-material parts [1–3].
2. Freedom to move—ability to be used at various (including nonconventional) places [4–6] due to its versatile and compact systems.
3. Fast process—ability to manufacture fast, as the manufacturing is direct and not through design-specific tools [7].

S. Kumar, *A Concise Encyclopedia of Additive Manufacturing*,
https://doi.org/10.1007/978-3-031-98394-8_4

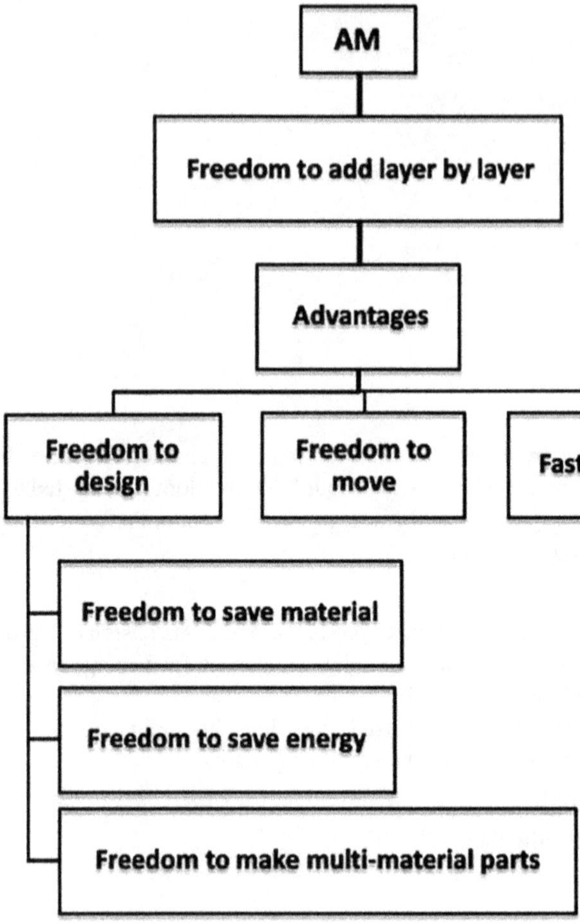

Fig. 4.1 Advantages

References

1. Teng, T., Zhi, Y., & Akbarzadeh, M. (2025). Prototyping high-fidelity multifunctional objects using single-nozzle multi-filament additive manufacturing system with active mixing. *Materials & Design, 249*, 113479.
2. Zhu, G., Wang, L., Wang, B., et al. (2023). Multi-materials additive manufacturing of Ti64/cu/316L by electron beam freeform fabrication. *Journal of Materials Research and Technology, 26*, 8388–8405.
3. Oel, M., Rossmann, J., Bode, B., et al. (2023). Multi-material laser powder bed fusion additive manufacturing of concentrated wound stator teeth. *Additive Manufacturing Letters, 7*, 100165.
4. Alogla, A. A., Baumers, M., Tuck, C., & Elmadih, W. (2021). The impact of additive manufacturing on the flexibility of a manufacturing supply chain. *Applied Sciences, 11*(8), 3707.
5. Verboeket, V., & Krikke, H. (2019). Additive manufacturing: A game changer in supply chain design. *Logistics, 3*(2), 13.

6. Rama Murthy, S., Evans, S., & Sarkis, J. (2019). The four freedoms-of-movement and distributed manufacturing. In *Sustainable development goals and sustainable supply chains in the post-global economy* (pp. 47–66).
7. Go, J., & Hart, A. J. (2017). Fast desktop-scale extrusion additive manufacturing. *Additive Manufacturing, 18*, 276–284.

Chapter 5
Advantage: Cause and Perspective

What if a composite material is developed by AM? Is this an advantage? Consider, for example, that two nozzles are used; one supplies steel powder, and the other supplies ceramic. When they meet at a substrate during and after processing by a beam, they become a steel-ceramic composite.

The development of a composite is an advantage, but this advantage is not exclusive to AM. The development does not come from AM's ability, i.e., layer making. It is developed because the beam can convert a mixture of different powders into a composite. The beam is part of an AM system, but it can be part of any system, such as welding, cutting, and hardening. Therefore, what a beam does cannot be the advantage of a system unless all the advantages of the beam give the exclusive advantage of the system.

For example, when a beam is used in AM, it is used to make a layer. Beam qualities that count are as follows: how well a beam helps make a layer; whether it gives a well-defined dimension, melts precisely, makes various shapes on a layer, processes different sizes of a layer, etc. The advantage of AM comes from its ability to form layers. If the beam does something that is not essential for the layer formation, that will not be the contribution of AM but of the beam. If the beam makes a composite during layer formation, this is an advantage of the beam. But it is not the advantage of AM because the composite formation does not contribute to a better layer formation.

If the composite formation gives a better layer formation, even then, the composite formation is not an advantage of AM. Because any material that is processed is required to give better layer information. If the material does not contribute to a better layer formation, it is a disadvantage that needs to be overcome. If the material contributes to a better layer formation, it helps AM realize its advantages. Thus, the ability of a material to be processed well by AM is not required to be called an advantage.

But this will not help, because this does not distinguish between two materials: one processed well by AM and the other not able to be processed. To take this into

S. Kumar, *A Concise Encyclopedia of Additive Manufacturing*,
https://doi.org/10.1007/978-3-031-98394-8_5

account, materials that cannot be processed well are kept under the disadvantage "lack of processable materials." It implies the materials that are processed well will not come under this disadvantage, which acknowledges the materials that are processed well.

This helps distinguish between the advantage of AM and the advantage of the tool of AM. If a laser beam, as a tool of AM, improves material properties, and if this advantage of the beam is considered an advantage of AM, it will not help find why AM is better than those techniques that also use the same beam and improve material properties.

Thus, if the ability of a beam to convert powders into a composite is not considered an advantage of AM, it is not because this ability of the beam is of no use for AM, but because this ability of the beam has no ability to distinguish between AM and those non-AM that also possess the same ability.

But not acknowledging this ability as an advantage can lead to concluding it is a disadvantage; therefore, a category "lack of processable materials" is given, which is not the same as "lack of the ability of beams"; but when the ability of a beam is evaluated by its ability to process materials, the category is sufficient to represent the lack of the ability of the beam.

When a beam converts two powders coming from two nozzles into a composite, both the ability of the beam and the role of the nozzles enable the conversion. If the role is not acknowledged, there will be no difference between the same conversion that takes place due to two powders (from two nozzles) and two powders (from a single nozzle). If the result is the same, i.e., formation of a composite, there is no difference between the two. Besides, if the result is the fabrication of the same design, there is again no difference between the two. But, if two nozzles increase the processing speed, this contribution has a category of advantage (i.e., fast process) to acknowledge.

However, if two nozzles make a multi-material structure, this advantage is acknowledged as a sub-advantage. How does a multi-material structure form? The ratio of powders coming from nozzles is changed, which gives different composites comprising different ratios of ceramics, resulting in a multi-composite structure.

If using two nozzles instead of one nozzle does not increase the processing speed, does not improve a design, or does not make a multi-material structure, then using the two is neither an advantage nor a disadvantage. However, if it decreases the processing speed, there is a category (i.e., slow process) for this disadvantage.

1 Perspective

If AM uses steel-ceramic powder to make a complex composite part, and the same material is used to make a part by sintering, and if the tensile strength of the AM part is higher, how can this result be reported?

If it is reported that AM gives better properties than sintering, it means AM has an advantage over sintering. This advantage is not exclusively due to AM but due to its tool (laser beam). The role of the tool is not unverifiable, as different beams (laser, electron beam) give different material properties. The role of layers cannot be underestimated, but in the presence of a beam, the material property advantage is not exclusively due to layers.

If it is reported that AM has an advantage over sintering because AM has a material property advantage over sintering, it will not work because the material property advantage is not considered an advantage of AM, as it is not exclusively due to layers. It means AM has no advantage over sintering because the reason for which AM does have an advantage is not acknowledged as an advantage.

Therefore, it should be reported that AM and sintering are equal. If it needs to be reported in terms of advantages and disadvantages, it should be reported that AM has neither an advantage nor a disadvantage over sintering. But this does not provide the information that AM gives better properties than sintering. There would not have been any problem in reporting if AM had not given better properties. Then it could have been reported that AM had a disadvantage over sintering because AM did not give better properties than sintering. It is because the material property disadvantage is acknowledged, while the material property advantage is not acknowledged.

There shall not be any problem in the reporting if both the material property disadvantage and advantage are acknowledged. It means the list of advantages and disadvantages needs to be revisited, and "better material property" should be added as an advantage category. It will give a balance, as the list of advantages is not complete. This brings up a question of whether a balance is lacking.

If it is reported that sintering has a disadvantage over AM because sintering does not provide better properties than AM, it will work because no information is missing. Thus, without revisiting the list of advantages and disadvantages, reporting can be done well. Reporting has become right not because there was a trick missing in the earlier reporting, but because perspective was missing.

The list seems incomplete because whatever other techniques do is considered normal, while whatever AM does is considered better or inferior. If whatever AM does is considered normal, then, without listing "better material property" as an advantage, the reporting can work well. If not listing this advantage as an advantage helps bring the right perspective, why not other advantages from the list be removed as well? Other advantages are not removed because they are exclusively due to layer making, while this advantage is not exclusively due to layer making.

Chapter 6
Advantage: Fast Process and Freedom to Design

Freedom to design (FD) comes from an ability to add material without using design-specific tools. Fast process (FP) arises because design-specific tools are not used. Both are due to the absence of design-specific tools. Then, why should there be two advantages?

FD means AM can make a complex part, which conventional manufacturing (CM) can and cannot. If AM cannot make it fast, AM does not have the advantage of being fast, but it still has an advantage that it can make. Thus, the absence of FP does not mean the absence of FD. It implies that both are required.

If AM makes a part fast, AM is called fast because it is compared with other processes, and its speed is found to be faster. If there is no comparison, AM is neither fast nor slow. If AM makes fast, this only concludes that AM makes faster than CM. If a design is considered complex, it is not necessarily that it will not be fast; vice versa is also true. For example, a half-hollow sphere is more complex than a half-solid sphere, as a hollow structure is more difficult to make than a solid structure [1, 2]. When powder bed fusion (PBF) makes both spheres, it is faster to make the hollow one; it does not conclude that a complex design is slow to make.

Comparing two parts made by AM is not the comparison of AM with CM. The comparison of the two parts is important, but knowing about the comparison will be important when AM is either better than or similar to CM. For example, if AM takes 1 week for a hollow sphere and two weeks for a solid sphere to make, it will not be interesting to know how fast AM makes the hollow sphere compared to the solid one when CM makes them within hours.

The comparison can take place only when things are comparable. If AM makes a complex design, CM has to make that so there will be a comparison. Because the comparison is about the speed to make one design and not about the ability to make a better design. If AM has the advantage of making a complex design, this advantage has no ability to inform whether AM is faster than CM in the absence of any comparison. When AM has the advantage to make even higher levels of complex designs, there will be even less chance to have them made by CM to enable

S. Kumar, *A Concise Encyclopedia of Additive Manufacturing*,
https://doi.org/10.1007/978-3-031-98394-8_6

comparison, which provides no ability to inform whether AM is faster than CM. Hence, design advantage is not sufficient to describe how AM compares with CM. One more advantage is required.

References

1. Noronha, J., Rogers, J., Leary, M., et al. (2023). Ti-6Al-4V hollow-strut lattice materials by laser powder bed fusion. *Additive Manufacturing, 72*, 103637.
2. Veeraiahgari, V., Regalla, S. P., & Kurra, S. (2023). Geometric form errors in hollow cylindrical parts manufactured by powder bed fusion. In *International conference on processing and fabrication of advanced materials* (pp. 71–83). Springer Nature Singapore.

Chapter 7
Advantage of Simple Products

Making simple products by AM will be advantageous if making them is economic, fast, or if they are made at such places where other techniques are not suitable to be used. These advantages can be summarized as economic, fast, and places to fabricate. The first advantage, i.e., economic, is temporary. Therefore, it is not included. The second advantage is included as a fast process. The third advantage is included as freedom to move.

Why is the first advantage temporary? If AM makes a simple product economically, it is due to two reasons:

first reason—the conditions of the market: when the product is made and when it is out for sale.

second reason—the technical aspect of a manufacturing technique.

Therefore, the second reason, i.e., the technique, is not the only reason for a product to be economic. The technique cannot claim that it solely provides the advantage. Even if it claims, the claim will be time-dependent because the market conditions will not always be the same. Therefore, even if AM makes a simple product economically, this advantage has no guarantee to last long.

The second advantage, i.e., the fastness of AM, is already recognized when AM is called rapid prototyping or rapid manufacturing. If the product is simple, the product does not benefit from the design advantage of AM. Nonetheless, if the product is required to be made fast, its fabrication is an advantage.

S. Kumar, *A Concise Encyclopedia of Additive Manufacturing*, https://doi.org/10.1007/978-3-031-98394-8_7

Chapter 8
Air Deposition

Air deposition, also known as aerosol jet printing or aerosol jetting (AJ), is used to fabricate small features. Its applications are the formation of electronic circuits [1], drug printing [2], conformal films [3], optical fibers [4], electrodes [5], etc. Aerosol implies that a liquid or solid particle is suspended in air. In AJ, liquid particles are used, which become constituents of a part made, while the air acts as a carrier to transport particles. If particles are suspended in the air, they are not big but are the size of a nanometer or micrometer. If such small particles are transported, there will be a small number of particles per second ready to be deposited to make a part, which is suitable to make only small parts from ~10 μm up to 1 mm (~100 nm resolution).

Most of the parts are used along with substrates on which they are made. The process is thus used more for modifying a substrate and adding value to it than for making a part that is used after being removed from the substrate. Its applications are making structures such as antennas, interconnects, electronic circuits, and conductive lines on a planar or non-planar surface.

For making a structure, AJ requires aerosol and its subsequent deposition (Fig. 8.1). The creation of the aerosol requires the creation of liquid droplets and their mixing with air. The creation of the droplets requires the agitation of the liquid so that the separation of the droplets from the bulk liquid can take place. The agitation is possible by vibrating the liquid placed in a container so that some liquid particles on the surface of the liquid overcome surface tension and get separated from the bulk to form droplets. This is also the concept of ultrasonic atomization, where ultrasonic vibration is transferred to the liquid through another liquid medium.

The agitation is also possible by impacting the liquid with a gas moving at a high velocity, generally coming from a compressed source. Such gas moving parallel to a liquid surface and brushing the surface will detach some liquid from the surface and turn it into droplets. For a high-viscosity liquid, a higher velocity is required. The requirement of the higher velocity can be overcome by heating the liquid to make it internally agitated. Because a gas moving not at the higher velocity can still

S. Kumar, *A Concise Encyclopedia of Additive Manufacturing*, https://doi.org/10.1007/978-3-031-98394-8_8

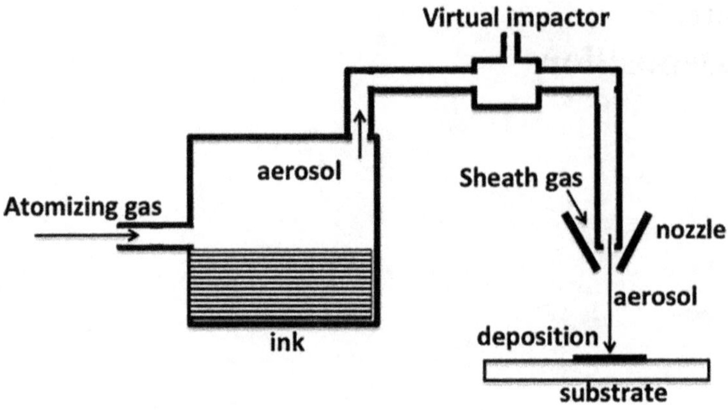

Fig. 8.1 Schematic diagram of aerosol jetting

give an increased number of droplets when impacted on internally agitated liquid. Heating as a method to facilitate droplet formation is generally used when the liquid kept still in a container needs to be atomized. If a high-velocity gas moves perpendicular to a liquid stream coming from a nozzle, the collision of the gas with the liquid stream creates droplets. This method is used for atomizing liquid streams.

Liquid droplets thus created will fall and get lost if they are not carried away. The gas that creates them, called an atomizing gas, can also carry them. The gas can then be called a carrier gas. After atomizing, droplets of various sizes are created. Bigger-sized (or of higher inertia) droplets will require a higher velocity carrier gas to be transported. If the atomizing gas is a carrier gas, the option for increasing or decreasing the velocity is limited. After the creation and transportation of aerosols and before their deposition, some mechanism is required to exclude droplets of extreme (high and low) sizes. Such droplets are not suitable if a high resolution is required.

When aerosol moves in a pipe, the presence of an exhaust or outlet in the pipe may let some gas (containing aerosol) move out to the atmosphere. The amount of gas moving out depends on the exhaust. If the size of the exhaust is big or the exhaust flow rate is high, more gas will move out, resulting in a loss of smaller-sized droplets. Besides causing the loss of droplets, the exhaust causes a decrease in the gas velocity. If the gas velocity decreases, it will not be able to carry large droplets as they will drop from the gas. Thus, using an exhaust eliminates both the smallest and the biggest droplets from the gas. This mechanism or controlling device is called a virtual impactor that eliminates extreme-size droplets because of their lowest or highest inertia.

Carrier gas deprived of droplets of extreme sizes, while containing medium-sized ones, is directed toward a nozzle to be deposited through it. When the gas passes through the nozzle, its flow diameter increases constantly as it approaches the substrate, which causes the deposited patterns not to have a high definition and density. Therefore, collimating the carrier gas is required.

To collimate the carrier gas, another gas called sheath gas is used [6]. The sheath gas flows in the direction of the carrier gas flow and covers it with an aim to converge at some point. If the shape of the carrier gas is just like a solid cylinder, the sheath gas is a hollow-converging cone surrounding the cylinder, which collimates the carrier gas. By collimating, the sheath gas helps constrict the deposition.

If there are several carrier gases (containing different materials) flowing at the same velocity and originating from different atomizers, it is possible to merge these flows into one and make a multi-material deposition. Changing the amount of material in one gas flow by changing the setting of atomizer will provide variations in the deposited material.

References

1. Zhang, H., Kim, Y., Cui, L., et al. (2025). AI-driven process optimization framework for enhancing print quality in aerosol jet printing. *International Journal of Precision Engineering and Manufacturing-Green Technology*, 1–15.
2. Turner, A. J., Prasad, E., Florence, A. J., & Halbert, G. W. (2025). Investigation of aerosol jet printing for the preparation of solid dosage forms. *International Journal of Pharmaceutics, 671*, 125288.
3. Niu, Y., Wang, Z., Li, Y., et al. (2025). Ultrathin MXene/ag-ag nanocomposite films for 3D-conformal electromagnetic shielding via aerosol jet printing. *Chemical Engineering Journal, 506*, 160122.
4. Wang, K., Liu, Z., Zhou, C., et al. (2025). A novel structure luminous flexible fiber was prepared via aerosol jet printing. *Scientific Reports, 15*(1), 3440.
5. Arango-Marín, V., Rocha-Ortiz, J. S., Osterrieder, T., et al. (2025). Aerosol-jet-printed silver nanowires as top electrodes in organic photovoltaic devices. *Solar RRL, 9*, 2400874.
6. Binder, S., Glatthaar, M., & Rädlein, E. (2014). Analytical investigation of aerosol jet printing. *Aerosol Science and Technology, 48*(9), 924–929.

Chapter 9
AM and 3D Printing

3D printing is another name for AM, and both names are used interchangeably. However, in practice, 3D printing is used differently.

The term 3D printing is used mostly in the following types of cases:

- When application of AM is more important than knowing the technical details of AM. For example, in a primary or secondary school or in an educational field where it is more important to convey how various important pieces of items can be produced from a small gadget than what the various methods of manufacturing are. If the term additive manufacturing is used, it may not be possible to convey the idea because students may not be aware of any types of manufacturing. But if 3D printing is used, conveying will be easier as students may be aware of both the words—3D and printing [1–5].
- In nontechnical environments, e.g., politics and policy, where the term printing is more well-known than the term additive manufacturing. In those cases, 3D printing can be intuitively understood as the next higher level of printing. If the term additive manufacturing is used, it will require an explanation of why additive manufacturing is not the manufacturing of additives [6–10].
- In those technical environments where the technical field is other than manufacturing, such as the medical field or food industry, the term 3D printing is understood by the majority of practitioners [11–15].
- For low-value AM products when quality is not critical [16–18]. For example, 3D printing is used in the gaming industry since the quality of a model in the gaming industry is not as important as the quality of a part in the aerospace industry.

S. Kumar, *A Concise Encyclopedia of Additive Manufacturing*,
https://doi.org/10.1007/978-3-031-98394-8_9

References

1. Pearson, H. A., & Dubé, A. K. (2022). 3D printing as an educational technology: Theoretical perspectives, learning outcomes, and recommendations for practice. *Education and Information Technologies, 27*, 1–28.
2. Salazar, D., Thompson, M., Rosen, A., & Zuniga, J. (2022). Using 3D printing to improve student education of complex anatomy: A systematic review and meta-analysis. *Medical Science Educator, 32*(5), 1209–1218.
3. Anđić, B., Lavicza, Z., Ulbrich, E., et al. (2024). Contribution of 3D modelling and printing to learning in primary schools: A case study with visually impaired students from an inclusive biology classroom. *Journal of Biological Education, 58*(4), 795–811.
4. Chen, J., & Cheng, L. (2021). The influence of 3D printing on the education of primary and secondary school students. *Journal of Physics: Conference Series, 1976*(1), 012072. IOP Publishing.
5. Lin, K. Y., Lu, S. C., Hsiao, H. H., et al. (2023). Developing student imagination and career interest through a STEM project using 3D printing with repetitive modeling. *Interactive Learning Environments, 31*(5), 2884–2898.
6. Thierer, A., & Marcus, A. (2016). Guns, limbs, and toys: What future for 3D printing? *Minnesota Journal of Law, Science & Technology, 17*, 805.
7. Stein, J. A. (2017). The political imaginaries of 3D printing: Prompting mainstream awareness of design and making. *Design and Culture, 9*(1), 3–27.
8. Fordyce, R. D. E. (2015). Manufacturing imaginaries: Neo-nazis, men's rights activists, and 3D printing. *Journal of Peer Production, 6*, 1–11.
9. Daly, A., Mann, M., Squires, P., & Walters, R. (2021). 3D printing, policing and crime. *Policing and Society, 31*(1), 37–51.
10. Volpe, T. A. (2021). Dual-use distinguishability: How 3D-printing shapes the security dilemma for nuclear programs. In *Emerging technologies and international stability* (pp. 220–246). Routledge.
11. Va, A., Udayarajan, C. T., Goksen, et al. (2023). A brief review on 3D printing of chocolate. *International Journal of Food Science and Technology, 58*(6), 2811–2828.
12. Parid, D. M., Talib, A. T., Baharuddin, A. S., et al. (2025). Mechanics of 3D printing process of white chocolate. *Journal of Food Engineering, 391*, 112429.
13. Tomašević, I., Putnik, P., Valjak, F., et al. (2021). 3D printing as novel tool for fruit-based functional food production. *Current Opinion in Food Science, 41*, 138–145.
14. Oyinloye, T. M., & Yoon, W. B. (2021). Application of computational fluid dynamics (CFD) simulation for the effective design of food 3D printing (a review). *Processes, 9*(11), 1867.
15. Mitsouras, D., Liacouras, P., Imanzadeh, A., et al. (2015). Medical 3D printing for the radiologist. *Radiographics, 35*(7), 1965–1988.
16. Petersen, E. E., Kidd, R. W., & Pearce, J. M. (2017). Impact of DIY home manufacturing with 3D printing on the toy and game market. *Technologies, 5*(3), 45.
17. Eisenberg, M. (2013). 3D printing for children: What to build next? *International Journal of Child-Computer Interaction, 1*(1), 7–13.
18. Derossi, A., Caporizzi, R., Azzollini, D., & Severini, C. (2018). Application of 3D printing for customized food. A case on the development of a fruit-based snack for children. *Journal of Food Engineering, 220*, 65–75.

Chapter 10
AM and CM (Conventional Manufacturing)

1 Rationale for Considering AM as a Fast Process

Powder bed fusion (PBF) is taken as an example to represent AM because it makes the most complex metallic parts AM can make [1]. Therefore, how fast PBF can make a complex part shows how fast AM can make it. AM is used for all materials, but why is ceramic not taken? Because AM makes metal parts better than ceramic parts [2]. Why not plastics? Because AM makes most high-value products in metals rather than in plastics, although the amount of plastics used is far more than that of metals [3].

When PBF makes one part (half hollow sphere) faster than another (half solid sphere), it gives no indication of what will happen when PBF is compared with conventional manufacturing (CM). If one CM process (i.e., forming) makes one part (a half-hollow sphere) faster than PBF, and the other CM process (i.e., machining) makes another part (a solid sphere) faster than PBF, it does not conclude that AM is faster than CM. If forming makes the solid sphere slower than PBF and machining makes the hollow one slower than PBF, it again concludes that AM is not faster. For AM to be faster, any of the CMs must not be faster. AM needs to compete with all CM, which again means the design advantage of AM is not enough to check how AM fares compared to CM. One more attribute is required.

The above shows AM is not faster. Then, why should the attribute be an advantage, not a disadvantage?

Compared to AM, when forming makes a hollow sphere faster and machining makes a solid sphere faster, AM is slower than CM. But the problem is not with comparison; it is about the type of comparison. The above comparison is about which process will make simple geometry, i.e., a solid or hollow sphere, fast. This is a problem. The promise of AM is to make a complex part [4]; the comparison should be to make a complex one.

S. Kumar, *A Concise Encyclopedia of Additive Manufacturing*,
https://doi.org/10.1007/978-3-031-98394-8_10

If a hollow sphere is more complex than a solid one, why should both be called simple geometry when both are not the same? When they are not the same, they are neither equally simple nor equally complex. This difference leads one of them to be more complex than the other. That's why the hollow one is called more complex than the solid one when both are made by PBF, as it is more difficult for PBF to make a hollow one than a solid one. If another technique replaces PBF, the solid one can be more complex. But the difference between these two does not change the fact that both are basic shapes. When they need to be clubbed together, they can only be clubbed together as a simple geometry.

Consider a geometry consisting of both spheres, i.e., a half hollow sphere is at the top of a solid sphere (Fig. 10.1). Both forming and machining will have problems making the other half (solid or hollow), for which they are less comfortable. While for AM, the difficulty in making any half will not change. Therefore, compared to AM, CM will not be as fast as it was when it only had to make either of the spheres. With an increase in complexity, AM starts to compete with any CM, as machining and forming start to show their limitations due to the tools.

Machining has a limitation that its tool is not as at ease to remove material from the inside of a section (for a hollow sphere) as from the outside of a section (for a solid sphere). Forming has a limitation that its tool is not at ease to bend a thick section to make a solid sphere as it is to bend a thin section to make a hollow one. Though the forming can be scaled up to work with a thick section (to make a solid

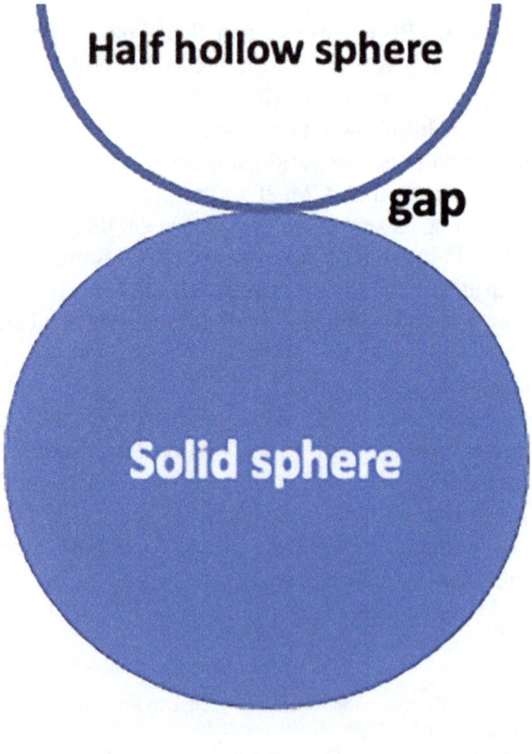

Fig. 10.1 A design where a half-hollow sphere is on a solid sphere

sphere) or scaled down to work with a thin section (to make a hollow sphere). But the geometry, depending upon its size, may allow it to work in only one machine (either large or small). It means one half will be made using one machine and the other half in the other. Or tools of different sizes will be used to make each half. Moving from one machine to another or selecting different tools will delay fabrication, making the forming process slower. This delay is absent when the geometry is simple; with an increase in complexity, the delay comes.

In AM, whether a geometry is simple or not, there always remains an absence of design-specific tools. Therefore, when CM gets delayed because its reliance on tools increases, it is not faster than AM. Therefore, advantage rather than disadvantage should be the attribute of AM.

2 Effect of Shape and Size on Fastness

In the above example (Fig. 10.1), why did an increase in complexity not delay AM as much as it delayed CM? In PBF, if a beam makes the thin wall of a hollow sphere, it implies that it has the ability to make both spheres, because the thin wall is of the highest complexity, as making the thinnest wall is difficult [5, 6]. For making the solid sphere, the beam has to melt powder within the periphery of a sphere, using some scanning strategy [7], which is simpler than making the thin wall. The solid sphere requires one outer boundary to make, while the hollow one requires two boundaries (outer and inner). The requirement of one less boundary for the solid sphere makes the job simpler. Though PBF takes more time to make a half-solid sphere than a half-hollow sphere of the same size, it is not because PBF faces manufacturing problems while making the solid sphere. It is because the quantity of work increases as beam-material interaction has to fill up the spherical space, while it does not have to fill up to make the hollow sphere. Unless the wall is very thin, the beam-material interaction for making either of the spheres is not different.

In machining, if a solid sphere takes a different amount of time, it is because the tool faces different problems for making both spheres. For the solid one, how the tool moves without colliding with a workpiece is not as much of a concern as in the hollow one. If the spheres are of different sizes, tools of different sizes are required, which changes the material removal rate. The rate changes with the change of the sphere size, but the reason is not that the size change causes machining to become different, but because it causes two different tools to be required. For example, when 20 and 40 cm diameter metal spheres are processed with tools of different sizes, the removal rates are different, but the machining problems are not. This is not a preferred method to make spheres, but this shows the difference between PBF and machining. In PBF, the material adding rate does not change with a change in the sphere size (unless one size is far smaller, needing extra attention).

References

1. Rao, J., Sing, S. L., Liu, P., et al. (2023). Non-destructive testing of metal-based additively manufactured parts and processes: A review. *Virtual and Physical Prototyping, 18*(1), e2266658.
2. Dadkhah, M., Tulliani, J. M., Saboori, A., & Iuliano, L. (2023). Additive manufacturing of ceramics: Advances, challenges, and outlook. *Journal of the European Ceramic Society, 43*(15), 6635–6664.
3. Cotteleer, M., & Joyce, J. (2014). 3D opportunity: Additive manufacturing paths to performance, innovation, and growth. *Deloitte Review, 14*(1), 3–19.
4. Mecheter, A., Tarlochan, F., & Kucukvar, M. (2023). A review of conventional versus additive manufacturing for metals: Life-cycle environmental and economic analysis. *Sustainability, 15*(16), 12299.
5. Chakraborty, A., Tangestani, R., Batmaz, R., et al. (2022). In-process failure analysis of thin-wall structures made by laser powder bed fusion additive manufacturing. *Journal of Materials Science and Technology, 98*, 233–243.
6. Lee, J., Hussain, A., Ha, J., et al. (2024). Mechanical properties of Ti-6Al-4V thin walls fabricated by laser powder bed fusion. *Additive Manufacturing, 94*, 104484.
7. Doğu, M. N., Ozer, S., Yalçın, M. A., et al. (2024). A comprehensive study of the effect of scanning strategy on IN939 fabricated by powder bed fusion-laser beam. *Journal of Materials Research and Technology, 33*, 5457–5481.

Chapter 11
AM and CM: Comparison

1 Basic Requirement to Compare Speed

In machining, the number of tools will increase if the tools that make both spheres are not sufficient to make gaps between the two spheres. When two spheres meet, there remain gaps between them because their peripheries do not coincide (Fig. 11.1). To make a fine gap, a fine tool is required. In powder bed fusion (PBF), the beam makes the gap. Whether the beam can accurately make the gap depends on the requirement. If the requirement is high, the beam will not provide accuracy. Thus, two cases can happen—PBF will make it, or PBF will not make it. When PBF makes it, it happens by the same beam that makes every section of the geometry. When PBF does not make it, the requirement should be made lower so PBF can make it, which will allow the comparison to take place.

A comparison between AM and conventional manufacturing (CM) will happen only when PBF is able to make the structure, because the comparison is not about the ability to make a better structure but about the speed to make a structure that both can make. When machining uses an additional tool to make the interface, the fabrication gets further delayed compared to AM. It brings further questions of why the requirement should not be made further lower so CM will be free from using the additional tool for making the interface, and CM will not be further delayed. The requirement is not made further lower to help CM because the comparison is about what will happen when both make a structure to their best ability—CM makes it using all its tools, and AM makes it using a beam.

What if PBF uses an additional beam? Then, PBF is further delayed. It will be delayed not because additional beams do not increase the fabrication rate [1, 2] but because the additional beam is used for another purpose. What is the need for using an additional beam? Since, without using an additional beam, PBF makes a structure for which machining needs to take the help of many tools. If PBF uses the additional beam, the structure will be too fine to be made by machining, or PBF will

S. Kumar, *A Concise Encyclopedia of Additive Manufacturing*,
https://doi.org/10.1007/978-3-031-98394-8_11

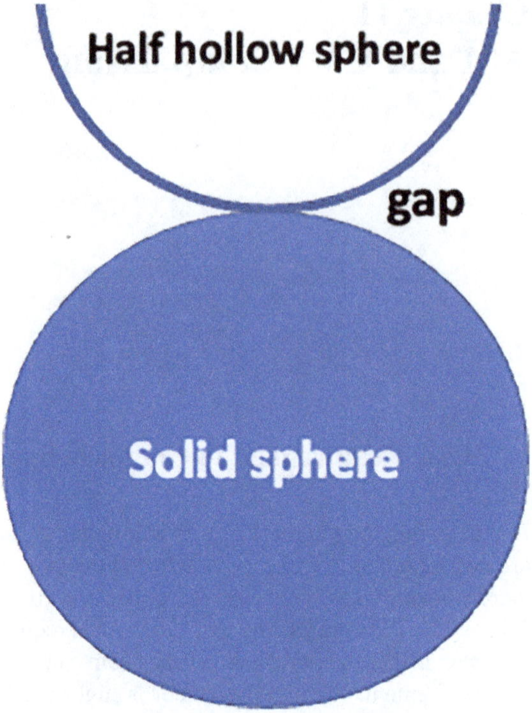

Fig. 11.1 A design where a half-hollow sphere is on a solid sphere

make a still finer gap that machining may or may not make. If machining is not able to make that structure, the comparison will not be able to take place.

Whether it is assumed that PBF is more advanced than machining, so what PBF does with one beam, machining needs more than one tool to do the same? It is not assumed that PBF is more advanced than machining; this comparison shows the basic difference between PBF and machining. PBF can make various structures using one beam, while machining cannot make them using one tool.

2 Geometrical Complexity and Problems

From the above example, the complexity in manufacturing problems increases in machining with a change in shape and size, while this does not happen in PBF. It infers that when manufacturing problems increase, manufacturing will not be faster than when the problems do not increase. AM becomes faster than CM when geometrical complexity causes the problems to increase. But it leaves several questions to be answered.

What if an increase in geometrical complexity does not cause manufacturing problems to increase? What is geometrical complexity? What if the complexity increases problems in AM but not in CM? Why will a geometry be considered complex when it does not increase problems in AM?

A shape is simple when it is basic. For example, cylinder, cube, sphere, and cone are basic shapes and therefore are simple. When these are added to or subtracted from each other, they become more complex than the basic shapes. For example, a cylinder plus a cube is more complex than either a cylinder or a cube (Fig. 11.2). A shape in which a cylinder is at the top of a cube has more geometrical complexity than a cylinder (or cube). This increase in the complexity will be important to know if it increases complexity in other aspects, such as making a 3D model, modeling, or manufacturing.

If an increase in geometrical complexity does not increase manufacturing complexity, the increase will not be important from the perspective of manufacturing. For example, machining can make a cube from a workpiece using one tool. It can further make a geometry consisting of a cube and a cylinder using one tool without having an increase in problems. Then, complexity in manufacturing for the machining process has not increased. Because this has not increased the number of tools, the difficulty in creating tool paths, the number of workpiece orientations to complete the machining, the number of fixtures, or problems of surface finish or accuracy at different areas of the geometry. If manufacturing time has still increased, it is not due to an increase in manufacturing complexity but to the quantity of machining. This increase in the machining quantity is not different from what happens due to an increase in the size of the same shape. The increase in manufacturing time due to an increase in the machining quantity cannot be attributed to an increase in the manufacturing complexity.

A cube is machined without any problem because the cube is simple in geometry. A cube plus cylinder is again machined without any problem because the cube plus cylinder is simple in geometry, because if it were complex, the machining could have detected it. If machining is the only process involved in making two shapes and machining does not find any difference in problems in making them, there is no method left to know if they are different, even if one of them is named more complex than the other.

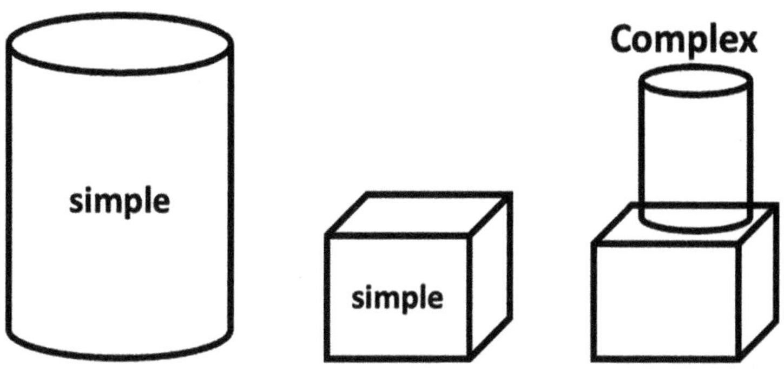

Fig. 11.2 Simple and complex shapes

References

1. Zhang, W., Tong, M., & Harrison, N. M. (2020). Scanning strategies effect on temperature, residual stress and deformation by multi-laser beam powder bed fusion manufacturing. *Additive Manufacturing, 36*, 101507.
2. Stump, B. C., Gibson, B. T., Reynolds, J. T., et al. (2023). Load balancing for multi-beam additive manufacturing systems. *Additive Manufacturing, 74*, 103708.

Chapter 12
AM and CM: Perspective

1 Perspective of Machining

If an increase in geometrical complexity increases a machining problem, the cause of the increase in the machining problem can be ascribed to the geometry of a part. Thus, the increase in machining problems will help find a geometry that is complex. This is what the problem is. The problem is that it is the machining that decides which geometry is complex. If the geometry brings a problem for machining, the geometry is complex. Otherwise, it is not complex. This is what is seen from the perspective of machining. It is seen only from this perspective because the perspective of AM was not present until recently.

When a geometry is already decided by machining that it is complex because this geometry increases the machining problem, it is a foregone conclusion that machining needs to overcome the problem to be fast. If AM makes the same geometry and is faster than conventional manufacturing (CM), it is what is required from AM. But it does not conclude that the geometry AM made is complex. Whether a geometry is complex or not is decided by a technique. This geometry was proved complex by machining and then brought before AM to be tested. It is not possible to find whether a geometry is complex or not for AM by going through the route of machining. Machining can only say which geometry is complex or not for machining. Machining cannot say the same on behalf of AM. It does not mean that the geometry that is complex for machining will not be complex for AM. It only means there is no guarantee that a given geometry will be complex for both.

S. Kumar, *A Concise Encyclopedia of Additive Manufacturing*,
https://doi.org/10.1007/978-3-031-98394-8_12

2 Perspective of AM

If a geometry that is complex for machining is tested for AM and it is concluded that AM is faster than machining, the conclusion can be erroneous. Because it is based on the assumption that what is complex for machining is complex for AM. Therefore, the conclusion can hide the fact that the geometry was simple instead of being complex for AM.

A geometry can be checked to verify it. Consider a geometry that consists of a rectangular block, two pillars, and a plate. On the top surface of the block, the pillars stand not far away from the center of the surface. The plate is on top of two pillars; the plate is an overhang structure (Fig. 12.1). The geometry needs to be made by both PBF and machining to check the fabrication time for each technique.

The thickness of the block is assumed to be small. Because if it is thick, it will take more time for PBF to fill up the volume, which will increase the quantity of work and delay fabrication. While the thickness will not delay the machining because a thicker workpiece will be taken for the machining process to start. Thus, with an increase in thickness, the quantity of work will increase in PBF but not in machining, which will delay PBF but not machining. As a consequence, it will not help identify whether the total delay in fabrication is due to only complexity or not.

Why is the thickness of the block not selected at random? If the thickness is selected at a lower value beforehand, it means there is an awareness of what will happen when a certain thickness is selected in PBF. This awareness leads to selecting the thickness in such a way that there will be no bias against PBF, so there will not be any delay in PBF; nevertheless, the selection of the thickness is also not biased against machining. But why should there be an awareness about what will happen in PBF? The comparison could have been the best when any one of many

Fig. 12.1 A geometry of a block, pillars, and a plate

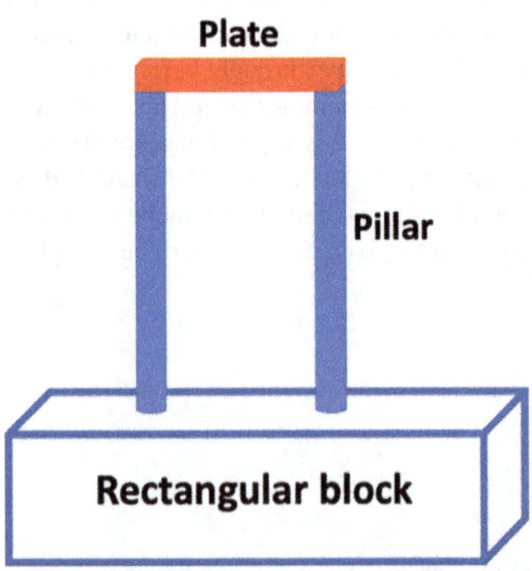

geometries could have been selected for testing. Then the comparison could have given an unbiased result.

It is because geometry is not selected for a random reason; it is selected for a purpose. In this case, the purpose is to compare. When the comparison is going to be made, there remains an awareness of what is supposed to be done. If there is already an awareness beforehand, why is the comparison made? The comparison is made because the detail is unknown, the results need to be verified, and a confirmation is sought. The problem is not about the awareness; the problem is about the type of awareness. When there was machining and not AM, the awareness was only about the machining. In that awareness, when the geometries were selected for comparison, there was no awareness to check what would happen in PBF when the thickness of the block increases. There was no awareness of AM to advise either for or against the selection. When AM is present, if again the awareness of AM is ignored at the design stage, the comparison will not be total, as it will not include the perspective of AM.

3 Advanced: AM or Machining?

In the above geometry, the block thickness should be thick enough to sustain the construction of a structure made up of two pillars and an overhang. If it is made by PBF, the fabrication will be delayed because the overhang will require extra work. The overhang may require a support structure (SS). Changing the orientation is one way to avoid SS, but this geometry does not allow avoiding SS by an orientation change. The geometry can be turned by 90° in any of four directions (left, right, front, back) to change the orientation. In any of the four directions, the two pillars will become overhangs. Out of the four directions, two directions (left or right, front or back) will give symmetrical results, and there will be no change in overhang problems. If the geometry is turned upside down, the block will become an overhang structure (Fig. 12.2). This geometry thus does not allow speeding up the fabrication by changing orientations.

If this geometry is made in machining, one rectangular block of the same area but of higher thickness will be taken as a workpiece. One milling tool will be sufficient to make the geometry if a sharp corner at the intersection of two pillars or at that of a pillar and the block is not required.

Thus, for this geometry, machining is faster than PBF, as PBF is delayed because it consumes surplus time to make overhangs. Is this geometry simple or complex for machining? It is unknown. Why? It is because the method by which a geometry is known, whether it is simple or complex, is absent. The method is comparison. There must be at least two geometries for machining, which are absent. If there is only one geometry, it is neither simple nor complex. If it is called simple, the question is which one is complex. If it is called complex, the question is where the simple one is. But the geometry, whatever it is, shows that it poses fewer manufacturing problems to machining than to PBF; therefore, machining is faster than PBF. Checking

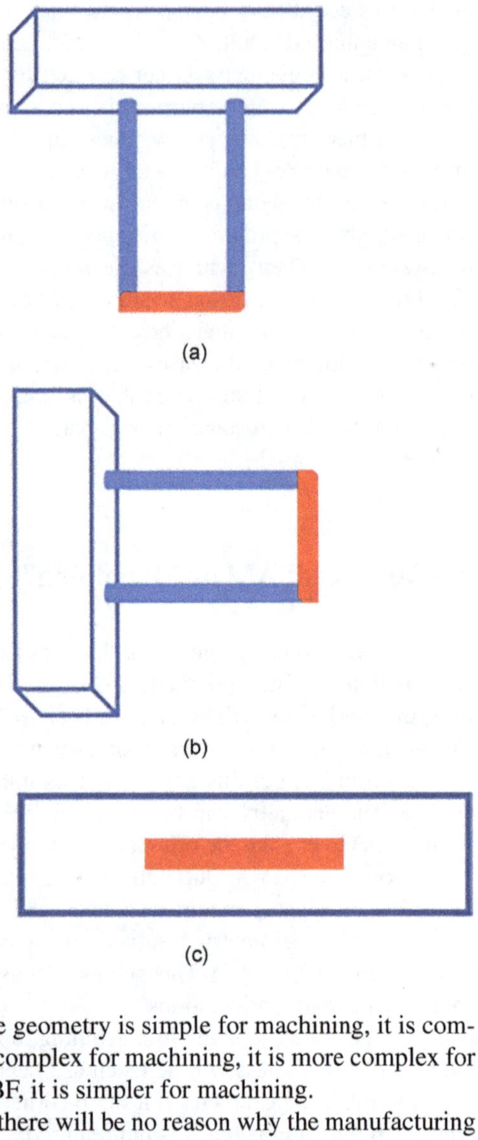

Fig. 12.2 Orientations of the geometry: (**a**) upside down, (**b**) right, and (**c**) front

the speed, it can be inferred that if the geometry is simple for machining, it is complex for PBF. And if the geometry is complex for machining, it is more complex for PBF. If the geometry is simple for PBF, it is simpler for machining.

Besides manufacturing problems, there will be no reason why the manufacturing speed will be slow when there is a need to check the maximum speed. It can be slowed down in the anticipation of manufacturing problems when there is no need to check the maximum speed, but this is not the case.

This gives an inference that a geometry that is difficult to make in PBF should be made in machining because that geometry is simple to manufacture through machining. It is because machining is an advanced technique than PBF. If machining were not an advanced technique, it would not have been found that manufacturing a part is simple. It was the same part, manufacturing of which was found complex by PBF.

But this inference does not conform to the general opinion. The general opinion says AM is more advanced because AM can make complex geometry that machining cannot. Which one is true: AM is more advanced than machining, or vice versa?

AM solves those problems that machining cannot; therefore, AM is a more advanced technique. But the narrative of why AM is better is wrong. The narrative that AM is better because it can make complex geometry that machining cannot is wrong. Because there is no guarantee that what a complex geometry is for one technique is also complex for the other. Therefore, when AM makes the complex geometry, it is better. But AM is not better because the geometry it makes is also complex for it.

Then, why is AM considered more advanced than machining? Why is it not inferred that, since it is not clear which geometry is complex, it is also not clear which technique is better? Why is it not said why it matters which technique is better when the problem is already solved?

AM is better than machining because AM solves the problems that machining could not. There were no problems waiting to be solved by machining when machining came, which could not be solved by AM. It is because machining came before AM. In the presence of only machining, it will be the machining that will define which one is complex or simple, which one is the problem to be solved, and which one is the problem already solved. It is the perspective of machining.

Does it matter that machining came before AM? It matters because the technique, which came first and stayed, created its own set of expectations that a coming technique has to fulfill. It is not the other way around, that a coming technique will promise to fulfill those expectations that were not perceived as expectations. If a new technique comes and proposes to solve all those problems that were not considered problems, the technique will not fulfill any expectations. The new technique cannot claim that what it considers a problem should also be considered a problem. And therefore, its claim should not be dismissed for the only reason that the claim was made later, as the technique could not arrive earlier than the old technique.

Did AM not exist earlier? AM has been present since antiquity [1], but as a modern engineering technique, it is not more than three decades old.

When there is a problem, a new technique is sought to solve the problem. When the technique solves the problem, the technique is better. Otherwise, it is not better. When there is a problem that machining cannot solve but AM can, AM is better. But when there is a problem that is already solved by machining, the problem is no longer available to be solved by AM. For example, machining has no problem in making a cube from a workpiece; therefore, the expertise of AM is not sought to make the cube. If AM tries to make the cube and cannot make it, it will not prove that AM is not a better technique when a decision is taken on the basis of a part that machining cannot make. Thus, when AM makes a geometry consisting of a rectangular block and pillars that are already made well by machining, the result will not be interesting for machining, even if the result proves that AM is not better. The interest of machining is what AM can do that it cannot do, not what AM cannot do that it can do well.

But this is not different from what the perspective of machining says, or what the general opinion says, or what is already known. What is then the use of the perspective of AM? Checking through the perspective of AM conveys that what the general opinion says is not wrong, but what the general opinion means is wrong.

Reference

1. Kumar, S. (2020). Introduction. In *Additive manufacturing processes* (pp. 1–19). Springer.

Chapter 13
AM and CM: Technique and Complexity

The perspective of AM brings forth that it is necessary to define a complex part. In the absence of a definition, a part will be called complex when a technique will be having a problem making it. Thus, depending on the number of problems faced and techniques, there will be varying perceptions of complexity for the same part. For example, if a part consists of a section that requires a high surface finish, the part will be complex for AM but not for machining. Or if a part consists of a section that has a well-defined cylindrical cavity, the part will be complex for AM but not for drilling.

There is a need for a complex geometry that will not favor one technique over another. And it will be equally complex for all techniques, so it will pose equal manufacturing problems to all. Is it possible to find this geometry? If it poses equal problems to all techniques, the techniques are not different. They are different because they find different degrees of difficulty in making the same feature of a geometry. Though in one case, it can pose equal problems to all when the geometry is too difficult to be made by any technique. In this case, the geometry will never be made. But what will be the use of this geometry? A geometry is required that one technique can make and another cannot, so the difference between the two can be known. If there is a geometry that no technique can make, this will not help create a difference between the two. Therefore, a geometry that poses equal problems to all techniques will not work, as techniques are unequal.

Finding a geometry that will be complex for all techniques implies that there should be previous knowledge about them. If the complete knowledge of a technique is unknown, at least its disadvantages should be known. Due to the disadvantage, it will be known which type of geometry will be complex or pose a problem for a technique. So a geometry can be defined, which will be complex for it. It means the process of defining a complex geometry is dependent on all techniques. And expecting that defining will be independent of techniques will not work.

If a complex geometry is defined from the knowledge of a technique, this geometry will not be useful, as it will not give the information of a technique that is not

S. Kumar, *A Concise Encyclopedia of Additive Manufacturing*,
https://doi.org/10.1007/978-3-031-98394-8_13

known. If the knowledge about a technique is not available, it will not lead to defining a geometry.

Thus, there does not exist a unique complex geometry that can be used to test the performance of a technique compared to other techniques. However, it is possible to find a geometry that will help identify two similar techniques, such as two types of AM, two types of machining, or two types of casting, which is usually done in a benchmark study. In a benchmark study, requirements are known, and the geometry is made accordingly. For example, if a technique can make thin walls or not is the requirement, then the geometry consists of various thicknesses of thin walls. When requirements are not known, the geometry is not made. When requirements are not known, why would a technique be developed?

The reason why a unique geometry is not found is because a technique is not developed at random. But it is developed because an old technique cannot make a geometry, which leads to finding a new technique that will make it. The development of a technique is therefore not free from the knowledge of a geometry that it needs to make, or other techniques could not make. Therefore, expecting a technique to make a unique geometry is not what the technique is developed for.

It means it is difficult to find a geometry that is universally accepted as complex by all techniques when there is at least one technique available that can convert the geometry into a part [1].

Reference

1. Kumar, S. (2023). Advantage and disadvantage. In *Additive manufacturing advantage* (pp. 1–60). Springer.

Chapter 14
AM and Machining: Approach

Machining is a big-small approach where a big block becomes a small product, while AM is a small-big approach where small blocks become a big product. These are two different approaches. This brings up a question of whether these two approaches are still different if observed through the movement of tools. Machining can be a top-down approach, while AM can be a bottom-up approach—these approaches can be due to the direction of the tool movement—the tool moves from top to down in machining while from bottom to up in AM (Fig. 14.1).

In machining, an object is made by machining a block from the top surface. The object is continuously carved out from the block so that it is continuously getting visible from the top surface to its bottom surface with the progress of machining. This is shown from Fig. 14.1a–c through 14.1b, where a block is cut by a machining tool.

This definition does not imply that the machining has limitations and cannot be performed from a side or bottom surface. If machining is performed from the side or bottom surface, these surfaces will act as a top surface for its tool. The top-down approach implies that the tool will go toward the material (block) so that it can access it to machine, while the bottom-up approach implies that the tool will go away from the material (substrate) so that it can have a free space to build.

In AM, a build progresses from the bottom surface or the first layer and reaches the last layer or the top surface. The bottom-up approach conveys that it starts from the first layer, adds layers continuously, and finishes at the last layer. If the orientation of a CAD model changes by 180° so that the top surface becomes the bottom surface and vice versa, the approach does not change—it again needs to start from the first layer that then becomes the bottom surface of the build. If the orientation of the AM system changes by 180° (e.g., inverse stereolithography), so that the approach may look similar to the top-down approach of the machining, the approach remains unchanged—it again needs to start from the first layer that then becomes the bottom surface of the build.

S. Kumar, *A Concise Encyclopedia of Additive Manufacturing*,
https://doi.org/10.1007/978-3-031-98394-8_14

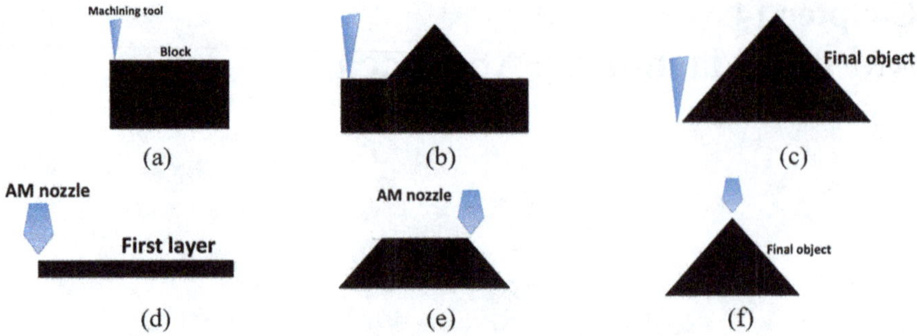

Fig. 14.1 Top-down approach in machining: (**a**) starting point, (**b**) middle stage, (**c**) final stage. Bottom-up approach in AM: (**d**) starting point, (**e**) middle stage, (**f**) final stage

In AM, irrespective of the orientation of the final object or the system, the build starts from nothing and adds building blocks until it gets everything in the form of a final object—in a bottom-up approach, fabrication happens through addition. In machining, the build starts from everything, removing blocks until there is nothing left to be removed except a shaped block in the form of a final object—in a top-down approach, fabrication happens through disintegration.

Figure 14.1d–f shows the first stage, middle stage, and last stage of the build. These figures show an AM nozzle that is not common to all AM processes, but the approach is common. For processes that do not use an AM nozzle, the direction of a build increment replaces the direction of the movement of a tool.

The name "bottom-up" does not intend to imply the limitation of AM. Considering a case when fabrication happens on the side surface of a vertical wall [1], then the nozzle will not move from bottom to up but either from left to right or from right to left—the name of the approach will be "left-right" or "right-left" approach. These new names will not represent anything differently than what is represented by the "bottom-up" approach. If AM is "left-right" instead of "bottom-up," machining will be "right-left" instead of: "top-bottom," and vice versa, as the purpose of the name is to show the relative position between AM and machining.

If a vertical wall is standing in the left corner on a horizontal table, the fabrication on its side surface will be toward the right direction. Thus, the AM tool moves from left to right in a "left-right" approach. If instead the machining is performed on the same surface, the tool will move further left so that the tool will pierce inside the surface, and the movement of the machining tool will move from left to further left in a "right-left" approach.

This brings a question: if AM tool movement can be explained by "bottom-up," "left-right," or "right-left" approach, then which name should be preferred? The answer to this question lies in another question—which is more important, symbol or symbolism? If a symbol is more important, "bottom-up" is preferred as it represents the working of most of the AM systems. If symbolism is more important, any name will suffice.

The top-down or bottom-up approach is about the tool movement and does not indicate the orientation of AM systems or final AM products formed [2, 3].

References

1. Kumar, S. (2022). Fabrication strategy. In *Additive manufacturing solutions* (pp. 111–144). Springer.
2. Hafkamp, T., Baars, G. V., Jager, B. D., & Etman, P. (2017). A trade-off analysis of recoating methods for vat photopolymerization of ceramics. *Proceedings of the 28th Annual International Solid Freeform Fabrication Symposium, 28*, 687–711.
3. Santoliquido, O., Colombo, P., & Ortona, A. (2019). Additive manufacturing of ceramic components by digital light processing: A comparison between the "bottom-up" and the "top-down" approaches. *Journal of the European Ceramic Society, 39*(6), 2140–2148.

Chapter 15
AM and Machining: Build Direction

In machining, all surfaces (A, B, C, D in Fig. 15.1a) can be processed. While AM, in practice, has one build direction, only top surface can be processed (Fig. 15.1b).

If a design needs to be selected, knowing the number of its possible orientations helps find the right manufacturing type. In AM, the orientation of a design determines the speed, cost, quality, anisotropicity, and fabricability [1]. In machining, the orientation of a block can be changed many times (depending upon the geometry). Besides, the orientation does not give rise to anisotropicity as it gives in AM [2]. In some AM techniques, orientation can be changed [3].

S. Kumar, *A Concise Encyclopedia of Additive Manufacturing*,
https://doi.org/10.1007/978-3-031-98394-8_15

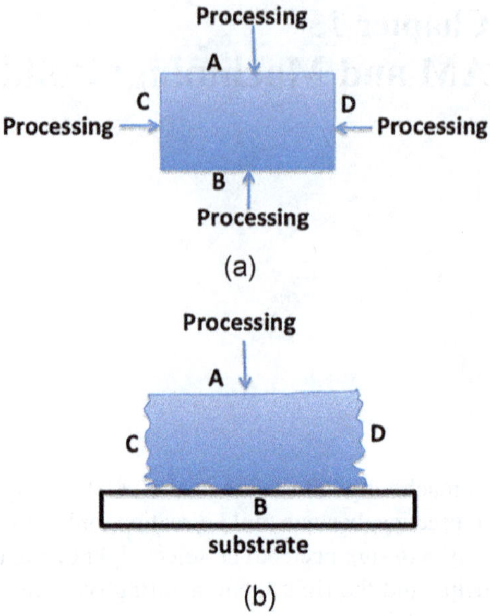

Fig. 15.1 Possibilities of processing in (**a**) machining and (**b**) AM

References

1. Zhang, Y., Bernard, A., Harik, R., et al. (2017). Build orientation optimization for multi-part production in additive manufacturing. *Journal of Intelligent Manufacturing, 28*, 1393–1407.
2. Kok, Y., Tan, X. P., Wang, P., et al. (2018). Anisotropy and heterogeneity of microstructure and mechanical properties in metal additive manufacturing: A critical review. *Materials & Design, 139*, 565–586.
3. Tang, P., Zhao, X., Shi, H., et al. (2024). A review of multi-axis additive manufacturing: Potential, opportunity and challenge. *Additive Manufacturing, 83*, 104075.

Chapter 16
AM and Machining: Lattice Structure

1 In Machining

To know the restriction of machining, a rectangular hollow space having two lattice boundary walls and one lattice internal wall needs to be fabricated (Fig. 16.1c). This is made by hollowing a rectangular solid (Fig. 16.1a) into two sections (Fig. 16.1b), which makes four boundary walls of the hollow section and one internal wall. Two of the boundary walls and the internal wall need to be made of lattice structures. There are methods, such as foaming, investment casting (IC), and coating [1], to make lattices, but machining is selected to show its limitations.

Hollowing is the easiest part of making the desired structure. Developing a toolpath is required to make a lattice on the walls. If the structure has variable cell densities, one milling tool will not be sufficient. If the cell has sharp corners, a fine finish is required. If all walls have different lattices, programming and machining will take more time. If the struts of unit cells are thin, they will break under cutting force; besides, the tool will have no space to access the wall. If the rectangle is big and the machine is robust, the internal wall can be accessed, but it will require changing the rectangle's orientation. Thus, difficulty in accessing an internal wall, fabricating sharp or thin lattices, process planning with tools, and path programming are restrictions.

2 In AM

With an increase in lattice cells, there is no increase in parameter optimization as long as the change in the cell does not reach the process limit, while there is an increase in programming in machining. This advantage of AM allows for working with various lattices [2]. Besides, the access of a tool to a workpiece to make a

S. Kumar, *A Concise Encyclopedia of Additive Manufacturing*,
https://doi.org/10.1007/978-3-031-98394-8_16

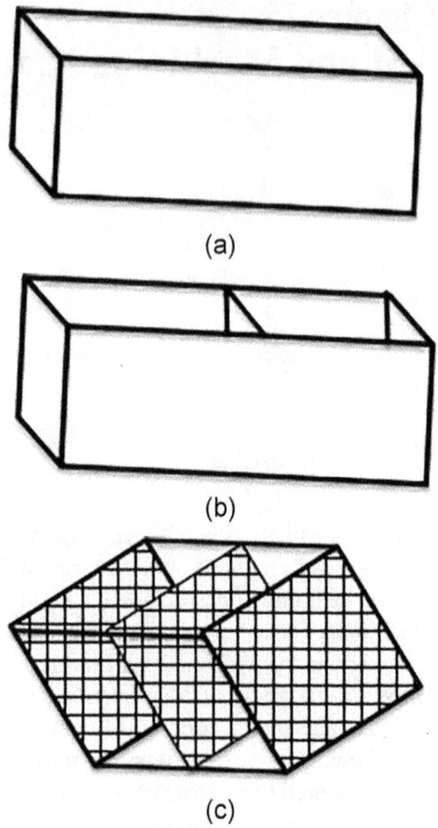

Fig. 16.1 Schematic diagrams of (**a**) rectangular solid, (**b**) hollow sections, (**c**) lattice walls

feature is not a restriction in AM. Because after adding material, the next addition happens in a building direction that has free space. If the next addition were in the reverse building direction, which does not happen, the problem could have occurred due to the lack of free space. If various build orientations are supported by AM systems, various lattice designs can be made [3].

Machining can prefer a circular rather than a rectangular lattice cell, as tools can have difficulty making the corners of the rectangular cell. In PBF, the size of a lattice cell is restricted [4], though there is no difference in making different shapes if the residual stress is optimized [5]. The smallest size will be mainly decided by the size of the beam spot. If the spot size is big, it will be difficult to control the melting of powder to a small area. Machining has restrictions when the shape of a cell changes, while AM has restrictions when the size of a cell decreases.

3 Not Enough

In PBF, the lattice should be in the horizontal section of a part. If the section is in a vertical direction, the lattice can bend or have surface defects. Thus, for a part having different sections, there will always be some sections that will not be in the horizontal direction, which will result in different properties in different sections. Hence, with an orientation change, there will be different toughness and propensity to crack [6], which is not an advantage. In machining, making different sections does not cause them to have different properties.

4 Restriction as Advantage

To overcome the problem, PBF does not have an option to stop ongoing fabrication and change the build direction to make the horizontal section vertical. In machining, this is not a restriction but a strategy to fabricate, which can be a disadvantage if compared with PBF in terms of automation. Because changing the workpiece orientation decreases the manufacturing speed and brings difficulty in automation. What is thus a restriction in PBF is also an advantage from the speed and automation points of view.

References

1. Khoda, B., Ahsan, A. M. M. N., Shovon, A. N., et al. (2021). 3D metal lattice structure manufacturing with continuous rods. *Scientific Reports, 11*, 434.
2. Jin, Q. Y., Yu, J. H., Ha, K. S., et al. (2021). Multi-dimensional lattices design for ultrahigh specific strength metallic structure in additive manufacturing. *Materials & Design, 201*, 109479.
3. Liu, B., Feng, J., Lin, Z., et al. (2024). Spherical path planning for multi axis support free additive manufacturing of truss structures. *Journal of Manufacturing Processes, 109*, 198–212.
4. Defanti, S., Giacalone, M., Mantovani, S., & Tognoli, E. (2024). *Dimensional and mechanical assessment of gyroid lattices produced in aluminum by laser powder bed fusion* (pp. 1–14). Meccanica.
5. Chen, C., Chang, S., Zhu, J., et al. (2020). Residual stress of typical parts in laser powder bed fusion. *Journal of Manufacturing Processes, 59*, 621–628.
6. Gu, H., Li, S., Pavier, M., et al. (2019). Fracture of three-dimensional lattices manufactured by selective laser melting. *International Journal of Solids and Structures, 180-181*, 147–159.

Chapter 17
AM and Machining: Material Property Perspective

Machining does not change mechanical properties except for some possible changes on a surface due to tool-induced heat [1], and, therefore, the properties of a material before and after machining remain the same. AM does not come with such material blocks and their properties. It comes with feedstock, experimental parameters, and noise—these all have influences on properties. Machining can be interesting because material properties will not deteriorate due to the lack of skill in the machining operation, or it cannot be interesting because material properties will not improve even by adopting the best practice.

AM can be interesting because properties at various locations of a part can be changed by optimizing parameters to fulfill different functional requirements—this advantage is not available in machining. AM cannot be interesting because material property is an aggregate of contributions provided by particles, pores, voxels, or volumes, and if any entity is not controlled, the property may vary—machining is free from this challenge. But this challenge has given rise to opportunities—if the addition of tiny volumes of materials is controlled during manufacturing, the properties can be predicted, and the properties can be known without measuring them. Then the desired property can be introduced at a design stage [2].

References

1. Yadav, S. P., & Pawade, R. S. (2023). Manufacturing methods induced property variations in Ti6Al4V using high-speed machining and additive manufacturing (AM). *Metals, 13*(2), 287.
2. Roach, R. A., Bishop, J. E., Johnson, K., et al. (2018). Using additive manufacturing as a pathway to change the qualification paradigm. In *29th annual international solid freeform fabrication symposium - an additive manufacturing conference, SFF 2018* (pp. 3–13).

S. Kumar, *A Concise Encyclopedia of Additive Manufacturing*,
https://doi.org/10.1007/978-3-031-98394-8_17

Chapter 18
AM and Machining: Side Surface

In AM, only the top surface is engineered, while the side surface is the result of making the top surface. The bottom surface is the result of how it is detached from a substrate if it was attached, or the bottom surface is the result of the condition of the platform surface if it was not attached.

Though if side surfaces are given attention, they improve. It happens by usual methods (parameter optimization, orientation changing), but only by applying the methods on the top surface. It gives no guarantee that the top surface and side surfaces will have the same quality (roughness, planarity, etc.). As the basic requirement of AM—that whenever the top surface is made, the side surface results—does not change. As the state of a surface depends upon its orientation with the build direction, there are artifacts comprising surfaces at various orientations to determine the surface quality at various orientations [1–3].

The basic requirement can be negated. For example, the ongoing build of a block can be stopped, and its orientation can be changed by 90°, so the top surface is no longer the top surface. The side surface becomes the top surface. Since the problem is with the side surface, when it becomes the top surface, it can be engineered. The problem can be minimized if geometry allows such an orientation change. But the problem cannot be completely removed because the action of creating a surface will create a side surface. Again, changing the block by 90° to catch and engineer the side surface is time-taking.

If the deposition process improves so that it changes the ongoing build [4], the difference between the side and the top surface can be minimized. But making a new type of system to change the build direction will not make it free from the basic requirement.

A top surface has a boundary. This is the boundary that becomes a side surface when the number of top surfaces increases. If the boundary is managed, the side surface can be managed. Thus, two types of parameters can be used for processing: one for the boundary and the other for the non-boundary area. While in machining, surfaces are already got before machining starts. The surfaces remain as they are

S. Kumar, *A Concise Encyclopedia of Additive Manufacturing*,
https://doi.org/10.1007/978-3-031-98394-8_18

until disturbed by machining. There is no interconnection between the side surface and the top surface as in AM.

For example, if milling is done on the top surface of a block to make a rectangular cavity, action through the top surface creates four side surfaces, i.e., four internal walls of the cavity. But the creation of these four side surfaces does not directly depend on the working of the top surface. Once a smaller cavity than the required cavity size is made by working on the top surface, the milling tool does not need to work on the top surface anymore to create the required cavity size.

The tool instead works on the side surface directly to make a bigger cavity. The surface quality of the side surface depends on what the milling tool does on the side surface. It does not depend upon what the milling tool did earlier on the top surface when the milling tool was making a smaller cavity as a precursor to making a big cavity.

If the tool works on the whole top surface of the block instead of partially working on the top surface to make a cavity, the top layer of the block is removed. With the removal of the top layer, its side surface is also gone (Fig. 18.1a). Thus, if the milling tool reaches the side surface of the block, the side surface is no longer present. If the tool does not reach the side surface of the block, the side surface remains as it is.

This is the difference between AM and machining. In AM, while the top surface is processed, the side surface forms (Fig. 18.1b). In machining, while the top surface

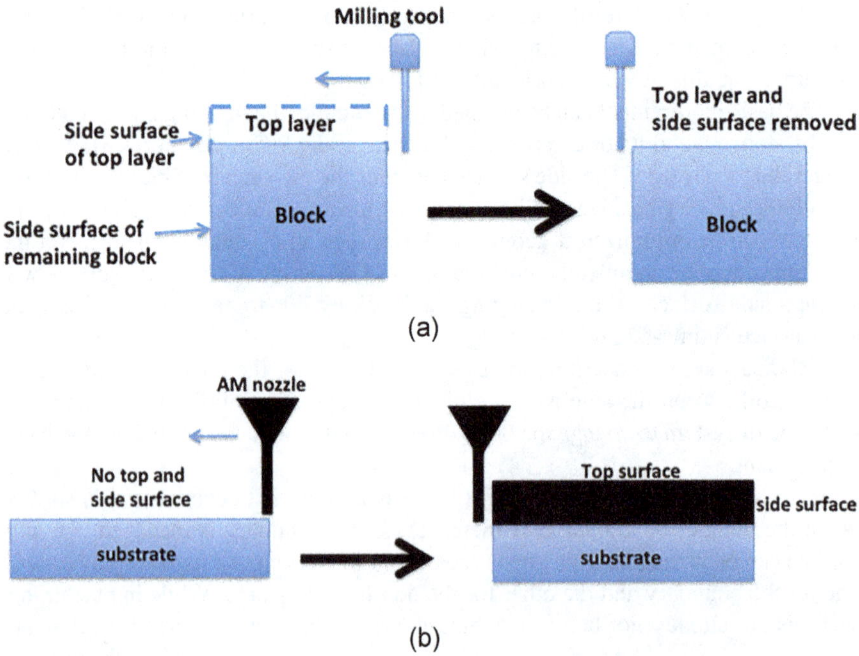

Fig. 18.1 Processing in machining (a) and AM (b)

is processed, the side surface is removed. This is the advantage of machining over AM. In AM, the side surface needs to be accepted as it is, as the problem on the side surface cannot be removed by AM anymore (a secondary process is required). In machining, the side surface is already gone.

When a milling tool removes the top layer of a block, the side surface of the remaining portion of the block (Fig. 18.1a) can get damaged by the tool. The damage can be in the form of a crack or scratch due to the tool, which can be due to the lack of the right tool, the right tool speed, or other conditions. But in an ideal case, there will be no damage, and the surface quality will not depend on the tool. In AM, even in an ideal case, when the side surface is the best, its formation will not be free from its dependence on the top surface. This leads to the conclusion that machining and AM are different.

References

1. de Pastre, M. A., Tagne, S. C. T., & Anwer, N. (2020). Test artefacts for additive manufacturing: A design methodology review. *CIRP Journal of Manufacturing Science and Technology, 31*, 14–24.
2. Rizzuti, S., De Napoli, L., & Ventra, S. (2019). The influence of build orientation on the flatness error in artifact produced by direct metal laser sintering (DMLS) process. In *Advances on mechanics, design engineering and manufacturing II: Proceedings of the international JCM 2018* (pp. 463–472). Springer International Publishing.
3. Strano, G., Hao, L., Everson, R. M., & Evans, K. E. (2013). Surface roughness analysis, modelling and prediction in selective laser melting. *Journal of Materials Processing Technology, 213*(4), 589–597.
4. Xiao, X., & Joshi, S. (2020). Process planning for five-axis support free additive manufacturing. *Additive Manufacturing, 36*, 101569.

Chapter 19
AM and Machining: Strategy

A product (Fig. 19.1) needs to be fabricated with the help of PBF and machining. If the whole product is made by PBF, it is expensive and takes considerable time. If it is made by machining, a lot of materials need to be removed to make thin features 1, 2, 3, and 4. These features have a propensity to break during machining, and if any breaks occur, machining needs to be restarted. What should be the strategy to make it?

If PBF is tried to make it in the same orientation as shown in Fig. 19.1, the product will more likely be as shown in Fig. 19.2, which will show overhang problems in features 3 and 4, as their downfaces will show the dimensional inaccuracy. Overhang is a recurring problem in PBF [1, 2]. Thus, the fabrication will not succeed.

If it is tried in the other orientation possible (Fig. 19.3), it will show the same problem but on different features 1, 2, and the downface of the cavity. Thus, the fabrication will not succeed again. There is no more orientation left that can be fabricated without having the problem.

Fabrication failures (Figs. 19.2 and 19.3) demonstrate that no orientation will work in PBF, as PBF is not developed enough to make this product. But Figs. 19.2 and 19.3 show the partial success. Figure 19.2 shows it can succeed in making all other features except features 3 and 4, while Fig. 19.3 shows it can succeed in making features 3 and 4. Though the process is not successful, a strategy can be successful. The process has already shown partial successes—one partial success at one orientation and another one at another orientation.

If there is a strategy that captures one partial success that happened in one orientation and another partial success that happened at the other orientation, the fabrication will be successful.

The strategy is to make only those parts of the product that can be fabricated. Figure 19.4 shows a part that can be made by PBF in which the fabrication of the problematic features 3 and 4 is excluded. The next attempt can be done to work on the same part or pre-product using the same process, i.e., PBF, but by changing the

S. Kumar, *A Concise Encyclopedia of Additive Manufacturing*,
https://doi.org/10.1007/978-3-031-98394-8_19

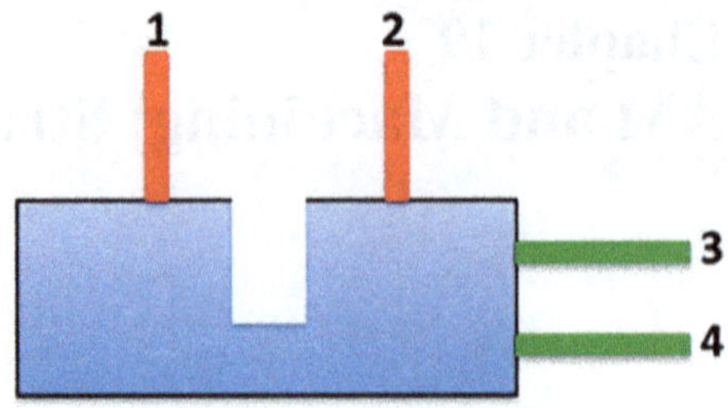

Fig. 19.1 Product design with features 1, 2, 3, 4, and a rectangular cavity

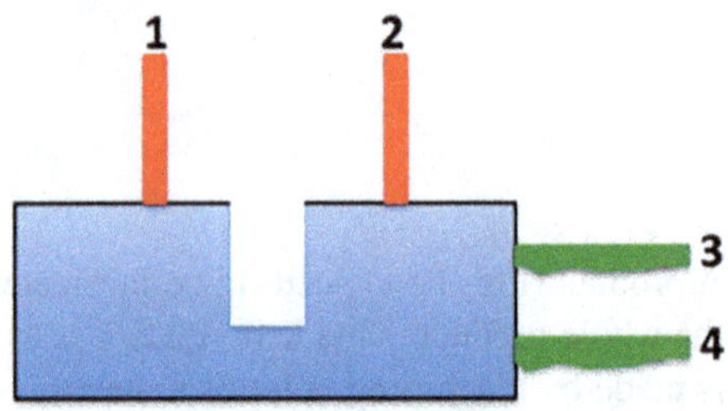

Fig. 19.2 Fabrication by PBF, problems in features 3 and 4

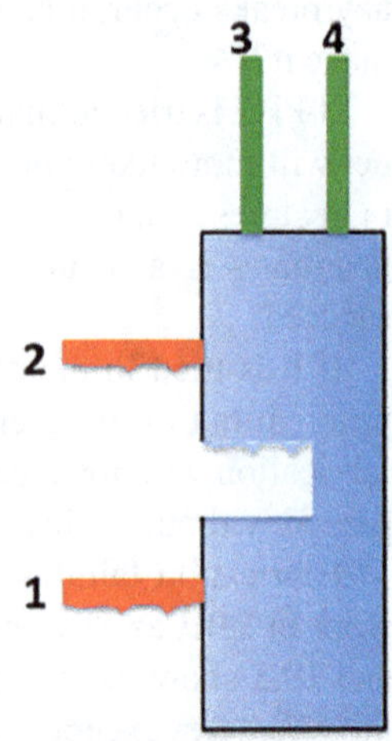

Fig. 19.3 Fabrication by PBF, problems in features 1, 2, and cavity

orientation (Fig. 19.5a). The remaining features, 3 and 4, can be built on it afterwards (Fig. 19.5b).

Changing the orientation is possible because both upper and lower surfaces in this orientation are parallel and normal to each other. Therefore, when the pre-product is fitted in the platform (to remain under the powder bed plane), extra effort is not required to position the pre-product to make its upper surface parallel to the powder bed plane. In the absence of such fitting, it will not be possible to make features 3 and 4.

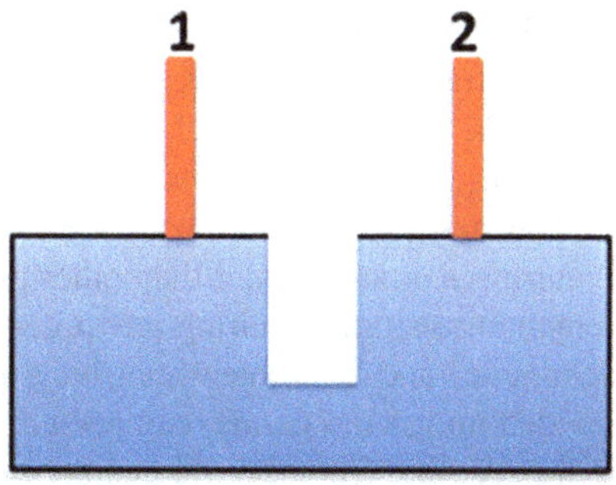

Fig. 19.4 Pre-product fabrication strategy by PBF in the absence of features 3, 4

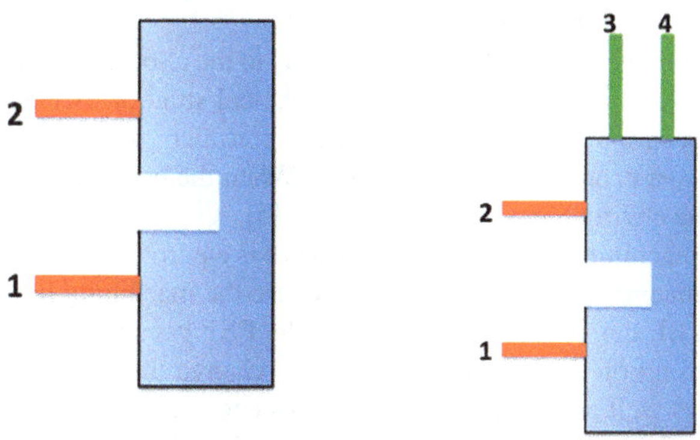

(a) changing orientation (b) remaining features fabrication

Fig. 19.5 Fabrication strategy by PBF by changing the pre-product orientation. (**a**) changing orientation. (**b**) remaining features fabrication

Thus, a product that cannot be built by a process can be built by the same process with the application of a strategy. But the process needs to be tried twice, which may not be a good strategy from the sustainability point of view, as it can consume energy and material twice.

In this strategy, the whole object can be fabricated by a single process without considering other processes. Thus, a bigger picture is absent. This does not mean that if the process is developed enough and the system is used once instead, energy consumption and material wastage occurring once, then it is better. This only means it is also not better.

There are two cases. In the former case, a process is not developed, a strategy is made, and the bigger picture is forgotten. In the latter case, a process is developed; therefore, no strategy is made, and the bigger picture is forgotten. Neither of the two cases is appreciable. Because, in the former case, just a technical strategy is made, while in the latter case, there is a lack of acknowledgement of a strategy.

Fabricating the whole object (Fig. 19.2) by a single process (PBF) is itself a wrong strategy, irrespective of whether the process is used twice or can be used once. The design (Fig. 19.1) shows that it consists of two types of sections—one type consists of fine features 1, 2, 3, and 4, while another type does not.

The design can thus be divided into two sections—one having fine features requiring a process better than other processes to make it, and the other having the remaining section requiring any process. If it is any process, a process that is fast, economic, and convenient can be a process of choice. Thus, the design that is divided into two sections requires a division of labor—one that is advanced (PBF) will make fine features, and another that is economic (machining) will do the bulk work. Machining is chosen for the bulk work because machining is more suitable than AM for making large-sized, noncomplex parts [3, 4].

Thus, the investigation of the design (Fig. 19.1) can lead to the allocation of tasks to different processes. This allocation is more than a technical strategy. Because it takes into account the economic consideration when it considers machining for doing the bulk work, it is an economic strategy as well. When the bulk work is not done by PBF, it makes the environment more sustainable [5].

What if selecting the machining for doing the bulk work is environment-friendly but not economic? The machining is not economic because the machining has not become expensive, but PBF has become inexpensive. The PBF becomes inexpensive or expensive depending on the market and does not change the fact that the production of powders, as well as making a part, requires energy. In this case, the selection of PBF instead of machining for doing the bulk work will be a better strategy from an economic point but not from a sustainability point. Again, the selection of machining instead will be a better strategy from the sustainability point but not from the economic point.

This brings up a question. Which strategy is better: PBF + machining or PBF + PBF from an agility point of view, assuming that both strategies make equally satisfactory products?

PBF + machining is better than PBF + PBF because PBF + machining is faster than PBF + PBF due to PBF being a slower process. Since PBF + PBF does not mean working in a PBF system once but twice (Figs. 19.4 and 19.5), this strategy does not gain a time advantage because the system does not make product in one go—the system still needs to shut down to change the orientation of the pre-product (Fig. 19.5). If the meaning of PBF + PBF was to use PBF once, PBF + PBF would have some more time to outpace PBF + machining.

Figure 19.6a shows the bulk part of the design (Fig. 19.1) that is fabricated by machining. On this pre-product, two features, 1 and 2, are built using PBF (Fig. 19.6b). This part needs to be used again in PBF by changing the orientation (Fig. 19.6c). Using a strategy of PBF + machining, the product can be built in three stages.

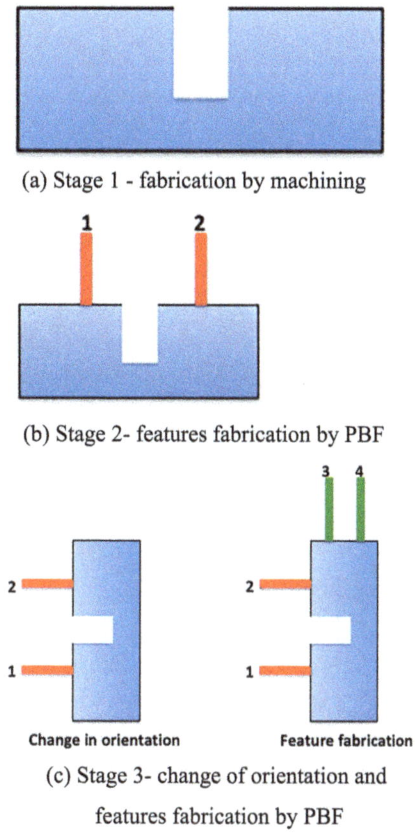

Fig. 19.6 Fabrication by machining and PBF. (**a**) Stage 1 - fabrication by machining. (**b**) Stage 2- features fabrication by PBF. (**c**) Stage 3- change of orientation and features fabrication by PBF

References

1. Bugatti, M., Semeraro, Q., & Colosimo, B. M. (2022). Effect of overhanging surfaces on the evolution of substrate topography and internal defects formation in laser powder bed fusion. *Journal of Manufacturing Processes, 77*, 588–606.
2. Mele, M., Bergmann, A., Campana, G., & Pilz, T. (2021). Experimental investigation into the effect of supports and overhangs on accuracy and roughness in laser powder bed fusion. *Optics & Laser Technology, 140*, 107024.
3. Mecheter, A., Tarlochan, F., & Kucukvar, M. (2023). A review of conventional versus additive manufacturing for metals: Life-cycle environmental and economic analysis. *Sustainability, 15*(16), 12299.
4. Pusateri, V., Hauschild, M. Z., Kara, S., et al. (2024). Quantitative sustainability assessment of metal additive manufacturing: A systematic review. *CIRP Journal of Manufacturing Science and Technology, 49*, 95–110.
5. Van Sice, C., & Faludi, J. (2021). Comparing environmental impacts of metal additive manufacturing to conventional manufacturing. *Proceedings of the Design Society, 1*, 671–680.

Chapter 20
AM and Machining: Wall and Cavity

1 Wall

In machining, there are many tools to choose from to machine features of different sizes. For example, if a design consists of a thin and a thick wall, they are machined using a small and a big milling tool, respectively (Fig. 20.1a).

In AM, an object can be found that can be the counterpart of a machining tool. For example, this object can be a beam in a beam-based AM. It can be a laser beam in laser powder bed fusion (LPBF), an electron beam in electron beam powder bed fusion (EPBF), and a nozzle in jetting or extrusion deposition. This object in AM has to do all the jobs that many machining tools do. The laser beam thus makes both big and small walls, or both wide and narrow cavities, in LPBF, while the nozzle makes different walls (Fig. 20.1b) and cavities in deposition process.

If AM makes a design, it has to be taken into account that a single tool has to make various features (Fig. 20.1b). Therefore, the accuracy and finish of a feature depend upon how the tool performs after parameters are optimized. In machining, the accuracy of a feature depends first on the selection of a tool and then on the optimization of parameters, as having the option to select the right tool takes some burden away from optimizing the parameters.

Consider two beam sizes—a small spot size to make small features and a big spot size to make big features, or two nozzle sizes—a small outlet diameter and a big outlet diameter. Using the two sizes will provide advantages in making small features that cannot be made with one size. For example, if a 400 μm thick wall is to be fabricated in LPBF, a beam of 100 μm rather than 500 μm diameter will be more suitable. It is possible to make all thick walls using a smaller diameter beam, but wall properties will not be the same [1].

To fabricate a 20 mm thick wall, a beam of 500 μm will be suitable to speed up the fabrication, while a 100 μm diameter will be suitable when higher accuracy is required. Thus, for speeding up without having to make small features, a big

S. Kumar, *A Concise Encyclopedia of Additive Manufacturing*,
https://doi.org/10.1007/978-3-031-98394-8_20

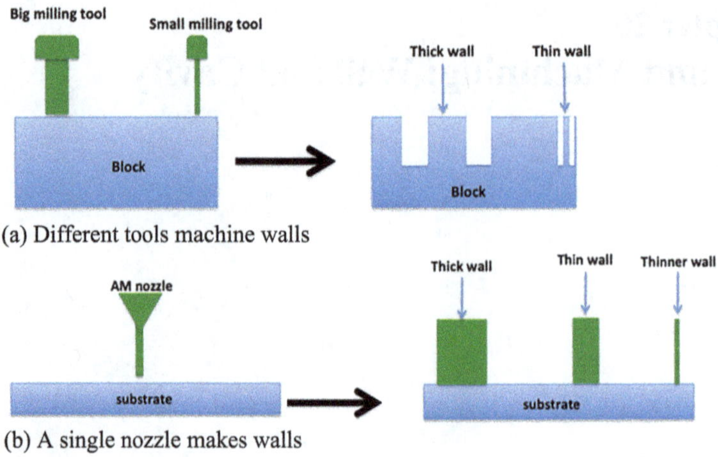

Fig. 20.1 Walls (side view) made by (**a**) milling, and (**b**) AM

diameter is suitable. However, the speed advantage achieved with a big diameter is limited, as layer thickness or scan speed has more influence on speed than the diameter does [2]. Therefore, speeding up using a big-diameter beam in AM is not as good a strategy as using a big milling tool in machining.

Similarly, a big-diameter nozzle will accelerate the fabrication, while a small one will improve accuracy (Fig. 20.2).

If two beam sizes are used, it will increase the complexity in the algorithm for processing a layer; besides, it will require the following:

- Hardware adjustment,
- Planning to determine the sequence of processing,
- Decision to find which intermediate-size feature comes under a small or a big feature,
- To determine an overlap between two features when either is processed by one beam size.

Using many beams increases productivity [3] and decreases residual stress [4], but in the absence of the right adjustment between scanning, the quality deteriorates. If two beams are used, the plume (gas plus small particles) generated by one beam attenuates the other beam. Hence, a part made using two beams results in a lower density. A big-diameter beam can give residual stress, while a small one can vaporize the powder; there is therefore a limitation to the maximum difference in diameters that can be used.

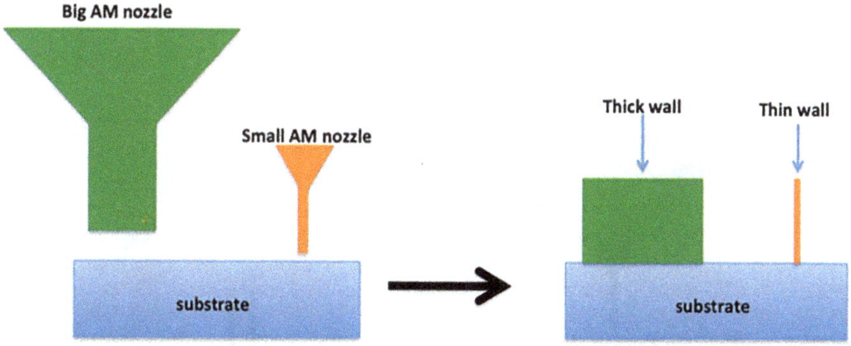

(a) Thick wall made by big nozzle and thin wall by small nozzle (side view)

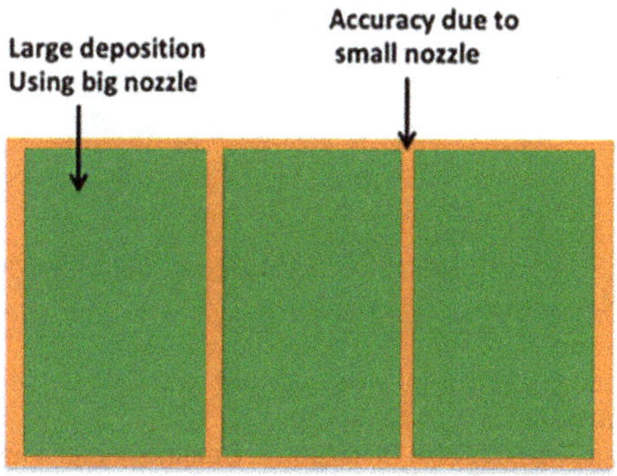

(b) Structure consisting of thick and thin walls (top view)

Fig. 20.2 Deposition due to multi-nozzle AM. (**a**) Thick wall made by big nozzle and thin wall by small nozzle (side view). (**b**) Structure consisting of thick and thin walls (top view)

2 Cavity

In machining, if a 100 μm diameter drilling tool is to process a design consisting of cavities of various diameters, the tool will not be able to make cavities of any diameter other than 100 μm. Because the tool makes a cavity having an internal diameter similar to its diameter, while in AM, a beam can make cavities of various diameters.

In AM, a beam does not dig out a cavity as a drilling tool does (Fig. 20.3a) but processes adjacent areas (Fig. 20.3b) to make a circular wall for the cavity. How small a cavity can be, as long as the solidification at diametrically opposite points at the cavity does not overlap, causing the desired space within the cavity to shrink. If it shrinks, the size of the solidified material needs to be made smaller [5]. This can be done if the feedstock (powder, wire) is smaller, or a photopolymer of low curing capacity is used [6], or the beam diameter is made smaller to restrict its effect to a smaller area.

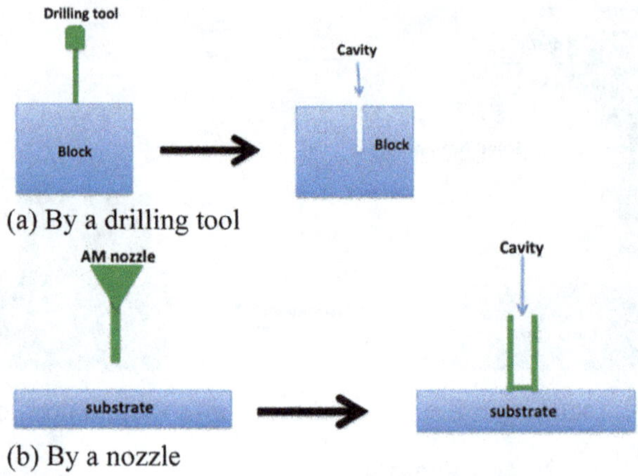

Fig. 20.3 Cavity by (**a**) drilling and (**b**) AM

References

1. Antonysamy, A., Prangnell, P., & Meyer, J. (2012). Effect of wall thickness transitions on texture and grain structure in additive layer manufacturing (ALM) of Ti-6Al-4V. *Materials Science Forum, 706-709*, 205–210.
2. Sow, M. C., Terris, T. D., Castelnau, O., et al. (2020). Influence of beam diameter on laser powder bed fusion (L-PBF) process. *Additive Manufacturing, 36*, 101532.
3. Tsai, C. Y., Cheng, C. W., Lee, A. C., & Tsai, M. C. (2019). Synchronized multi-spot scanning strategies for the laser powder bed fusion process. *Additive Manufacturing, 27*, 1–7.
4. Zhang, W., Tong, M., & Harrison, N. M. (2020). Scanning strategies effect on temperature, residual stress and deformation by multi-laser beam powder bed fusion manufacturing. *Additive Manufacturing, 36*, 101507.
5. Solyaev, Y., Rabinskiy, L., & Tokmakov, D. (2019). Overmelting and closing of thin horizontal channels in AlSi10Mg samples obtained by selective laser melting. *Additive Manufacturing, 30*, 100847.
6. Kowsari, K., Zhang, B., Panjwani, S., et al. (2018). Photopolymer formulation to minimize feature size, surface roughness, and stair-stepping in digital light processing-based three-dimensional printing. *Additive Manufacturing, 24*, 627–638.

Chapter 21
AM and Machining: Waste Generation Perspective

While making an object, machining creates chips and swarf that are waste. AM does not create chips. There may be a small component of machining in AM in the form of modification or post-processing that creates chips [1, 2]—an amount miniscule in comparison to that created in machining. AM is better than machining because it creates a lesser amount of chips. Since AM does not intend to do machining, therefore, if AM is free from the demerit of the machining in comparison to a process that does only machining, the comparison is not equal. The comparison would have been equal if AM were doing as much machining as a process that does only machining.

It does not imply that if a process makes a part without machining (or without creating chips similar to machining), it is not an achievement, or the contribution of machining in AM should be increased to have a fair method for comparison. But it implies that the comparison method is unfair because it hides the fact that AM, though it does not make more chips, may make more waste [3–5].

For example, if a nylon pattern is made by PBF, it will create waste by degradation of polymer powders present in a bed [6, 7]. The degradation due to PBF will be more than the waste due to the formation of chips if the pattern is instead machined from a block. Using PBF will require more energy than that required for running a CNC machine, and thus, the fabrication by PBF will waste more energy. This example does not represent all examples in AM but shows that AM is not always better than machining.

References

1. Veiga, F., Gil Del Val, A., Suárez, A., & Alonso, U. (2020). Analysis of the machining process of titanium Ti6Al-4V parts manufactured by wire arc additive manufacturing (WAAM). *Materials, 13*(3), 766.

S. Kumar, *A Concise Encyclopedia of Additive Manufacturing*,
https://doi.org/10.1007/978-3-031-98394-8_21

2. Pérez, M., García-Collado, A., Carou, D., et al. (2021). On surface quality of engineered parts manufactured by additive manufacturing and postfinishing by machining. In *Additive manufacturing* (pp. 369–394). Elsevier.
3. Jiang, J., Xu, X., & Stringer, J. (2018). A new support strategy for reducing waste in additive manufacturing. In *The 48th international conference on computers and industrial engineering (CIE 48) at: Auckland* (Vol. 48, pp. 1–7).
4. Powell, D., Rennie, A. E., Geekie, L., & Burns, N. (2020). Understanding powder degradation in metal additive manufacturing to allow the upcycling of recycled powders. *Journal of Cleaner Production, 268*, 122077.
5. Maware, C., Muvunzi, R., Machingura, T., & Daniyan, I. (2024). Examining the progress in additive manufacturing in supporting lean, green and sustainable manufacturing: A systematic review. *Applied Sciences, 14*(14), 6041.
6. Alo, O. A., Otunniyi, I. O., & Mauchline, D. (2023). Correlation of reuse extent with degradation degree of PA 12 powder during laser powder bed fusion and mechanical behavior of sintered parts. *Polymer Engineering & Science, 63*(1), 126–138.
7. Sanders, B., Cant, E., Amel, H., & Jenkins, M. (2022). The effect of physical aging and degradation on the re-use of polyamide 12 in powder bed fusion. *Polymers, 14*(13), 2682.

Chapter 22
AM is Itself Not an Advantage

If there is no difficulty in converting a design into a product (e.g., a hollow cylinder), there is no difficulty available to be removed using AM. Then AM cannot demonstrate its advantage of how it simplifies a complex design into a set of simple steps, but AM can still be used if it is better than other techniques—this is the advantage of AM over other techniques. This advantage only shows how the conversion of a design is simplified in AM. The simplification has limited value if the conversion is limited only to simple designs. This simplification has no value if the conversion is limited only to simple designs, while the conversion is sought for complex designs. Therefore, the simplification, which is the freedom to add materials layer by layer, is a characteristic rather than an advantage.

If this freedom leads to making complex products [1, 2], this is an advantage. If this freedom leads to making simple products, this is also an advantage, but this leads to disadvantages as well. For example, if AM makes a simple product, i.e., a big metallic square block that is made economically and fast by machining, then using AM is a disadvantage. Therefore, this freedom, which is AM, is itself neither an advantage nor a disadvantage.

References

1. Olsson, A., Hellsing, M. S., & Rennie, A. R. (2017). New possibilities using additive manufacturing with materials that are difficult to process and with complex structures. *Physica Scripta, 92*(5), 053002.
2. Madhavadas, V., Srivastava, D., Chadha, U., et al. (2022). A review on metal additive manufacturing for intricately shaped aerospace components. *CIRP Journal of Manufacturing Science and Technology, 39*, 18–36.

S. Kumar, *A Concise Encyclopedia of Additive Manufacturing*,
https://doi.org/10.1007/978-3-031-98394-8_22

Chapter 23
Arc

When voltage is applied between two electrodes, air molecules or atoms present between electrode tips get polarized, which causes positive and negative charges of an atom to get separated, creating positive and negative poles within the atom. If the voltage is increased further, these poles are not confined within the atomic boundary but get detached from each other. After detachment, they are not poles but charges—this conversion of non-charged entities into charges is ionization.

The air between tips consists of charges that are collectively called plasma. The charges move toward the electrodes, which gives rise to an arc that is a continuous line of charges between the electrodes. The movement faces obstacles due to the presence of other particles in the air. While overcoming the obstacles, friction occurs between moving charges and particles, which gives rise to heat. Thus, the creation of an arc can entail the creation of a heat source. Higher voltage means more ionization, stronger attraction, and a more intense arc, giving more heat. Higher current means more injection of electrons in the plasma, resulting in more ionization and a stronger arc.

Ionization not only depends on voltage and current but also on the type of gas present between the electrodes. Gas having a lower ionization potential will be ionized at a lower voltage than those having a higher ionization potential. Thus, argon gas (15.7 eV), unlike helium gas (24.5 eV), is easier to ionize, which makes argon gas suitable to be used at a lower voltage.

The type of gas not only decides the amount of ions in an arc but also the whereabouts of heat in the arc. If the gas has a high thermal conductivity, e.g., helium gas (0.151 W/mK), the heat will not be confined within the arc line but will spread out, making the influence of the arc wider, resulting in a wider welding bead. If the gas has a low thermal conductivity, such as argon gas (0.018 W/mK), the heat will not spread, which will not result in wider but deeper welding, possibly. For creating deeper welding, the arc should contain high heat, which is not possible by just having one favorable property, i.e., low thermal conductivity of the gas.

S. Kumar, *A Concise Encyclopedia of Additive Manufacturing*,
https://doi.org/10.1007/978-3-031-98394-8_23

If the gas in the arc furnishes extra heat, it will serve the purpose. Oxygen gas during ionization splits into atoms and furnishes extra heat (dissociation energy). Thus, the presence of oxygen will contribute to the arc intensity. If a small amount of oxygen is mixed with argon, the intensity will increase. Thermal conductivity of oxygen (0.027 W/mK) is slightly higher than that of argon, but it will not offset a gain in the heat intensity (by taking heat away) obtained due to the dissociation of oxygen.

Heat generated in an arc affects an electrode, which can cause it to be melted and used as a source of material for AM. Though the heat can degrade it when it is not supposed to be melted but to act only as an electrical terminal to create the arc. In gas tungsten arc welding-based AM, the tungsten electrode acts as an electrical terminal, the cooling of which is one of the ways to mitigate heat-induced degradation. Cooling happens when a gas flows by the electrode. This is the same gas that is a precursor for the arc. If the gas has a high thermal conductivity, the heat transfer will be more, and the electrode will cool more. Therefore, helium, which has a higher thermal conductivity, cools electrodes more than argon does.

When direct current is flowing and the tungsten electrode is maintained as a positive electrode, the degradation is fast because positive tungsten ions from it are pulled by the negative electrode. It gives rise to the loss of atoms from the tungsten electrode, resulting in its depletion. Using alternating current instead helps avoid this type of degradation. When it is maintained as a negative electrode, it does not have any such degradation through losing positive ions. Besides, heat is generated more at the positive than at the negative electrode. Therefore, for the same setting of current and voltage, the positive electrode becomes hotter and degrades more.

Bigger positive ions and generally smaller negative ions (electrons) are constituents of an arc. When voltage is applied across electrodes, negative ions, being smaller, move faster toward the positive electrode, while positive ions, being bigger, move more slowly toward the negative electrode. Fast electrons have a higher impact on the positive electrode, making it hotter.

Thus, a tungsten electrode that is maintained as a negative electrode is cooler and safer. Its efficiency is further improved by alloying it with thoria. When voltage is applied across electrodes, it is the negative electrode from which electrons enter into the arc; the transfer of electrons from the electrode to the arc also depends upon the crystal structure of the electrode. When tungsten is mixed with another material, which decreases its work function, the transfer of electrons becomes easier, and the efficiency of the electrode increases. Addition of 2 wt. % Thoria decreases its work function from 4.5 to 2.6 eV.

Chapter 24
Arc Welding-Based AM: GMAW-Based

Gas metal arc welding (GMAW)-based AM utilizes an arc to melt and make inexpensive parts [1]. In GMAW, out of two electrodes, one electrode is consumable and the other electrode is a workpiece or substrate (Fig. 24.1). During the creation of an arc between two electrodes, when one electrode is melted and ready to be deposited, the need for feeding materials is eliminated. The consumable electrode is a feedstock.

Since the form of a consumable electrode is usually a rod or wire, the feedstock is usually a rod or wire. It gives a more efficient transfer of heat from an arc to the wire than a process where the electrode and the wire are not the same; this results in a fast deposition.

A consumable electrode has some disadvantages. If the electrode gets consumed, the arc length will change. This change will disturb the process by changing either voltage or current, which will require further adjustment of the voltage, current, or feed rate.

The process makes a structure by transferring molten parts of an electrode to a workpiece (Fig. 24.1). The transfer depends mainly upon how molten the molten part is. If it is a little molten, mainly at a condition of low current, this will not be detached from the electrode on its own, and therefore, no deposition will occur. If it is a little more molten, mainly at a medium-current condition, this will be detached on its own, resulting in a deposition.

To be detached, the gravitational force needs to overcome surface tension. After detachment, droplets that are around the size of the electrode diameter will be deposited. If it is well molten, mainly at a high-current condition, it will not only be detached on its own but will take the shape of small droplets. Because under these conditions, there is high heat energy that decreases the surface tension, and therefore, a small gravitational force due to the droplet will be sufficient to overcome the surface tension.

With an increase in current, heat energy increases, which increases the temperature of molten material, allowing it to overcome surface tension and get fragmented

S. Kumar, *A Concise Encyclopedia of Additive Manufacturing*, https://doi.org/10.1007/978-3-031-98394-8_24

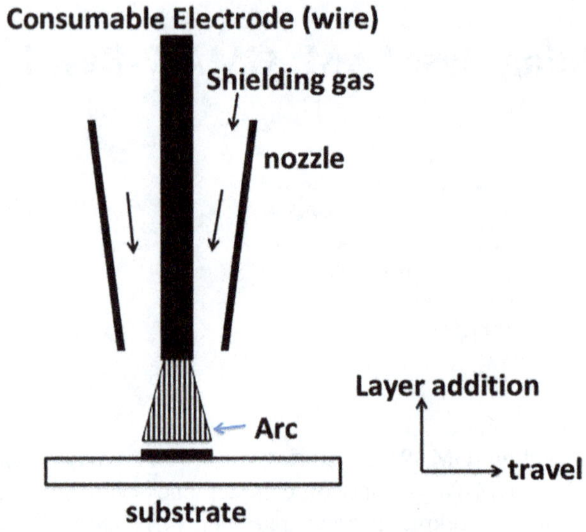

Fig. 24.1 Schematic diagram of GMAW-based deposition

into droplets. Instead of increasing the current to impart heat, it can also be imparted by other methods, such as the use of oxidizing gas. These three conditions of current show three types of respective material transfer: short-circuit transfer related to low current, globular transfer related to medium current, and spray transfer related to high current.

In a short-circuit transfer, since material is not detached on its own, it needs to be detached by initiating a short circuit between an electrode (wire) and a workpiece. When material is not detached from the wire tip, the wire is continuously fed, which continuously decreases the arc length between the wire and the workpiece until the length or gap becomes zero—this is a short circuit. A constant current then passes from the wire to the workpiece, which gives rise to resistive heating at the interface between the wire and the workpiece. This heat melts the tip.

Molten material detaches from the tip by overcoming the surface tension at the tip; this again creates an arc between the tip and the deposited material, which brings an end to the short circuit.

Thus, in this type of material transfer, a continuous cycle of creation (of the short circuit) and the end of creation gives rise to a continuous deposition that enables fabrication. In this type, uncontrolled resistive heating produces spatter, which decreases the precision of the deposition.

To avoid resistive heating and reduce spatter, a variation of this type of metal transfer, cold metal transfer, has been developed. The prefix "cold" denotes the absence of (resistive) heating. Thus, the current is set to zero and the wire is retracted the moment the semi-molten tip of the wire contacts the workpiece, and thus resistive heating is avoided. In the absence of the resistive heating, the tip does not get heat energy from the resistive heating and needs to melt without it. Thus, a low current is able to melt the tip but is not high enough to detach the molten material from the tip on its own. Thus, cold metal transfer is able to furnish 3D steel structures with high deposition rate and reproducibility [2].

References

1. Pattanayak, S., & Sahoo, S. K. (2021). Gas metal arc welding based additive manufacturing—A review. *CIRP Journal of Manufacturing Science and Technology, 33*, 398–442.
2. Ali, Y., Henckell, P., Hildebrand, J., et al. (2019). Wire arc additive manufacturing of hot work tool steel with CMT process. *Journal of Materials Processing Technology, 269*, 109–116.

Chapter 25
Arc Welding-Based AM: GTAW-Based

The process, as the name implies, uses a tungsten electrode to create an arc, while the prefix "gas" implies a shielding gas used to prevent oxidation of materials melted during welding. Shielding gas flows around the arc, which shields atmospheric gas from coming near the arc and prevents the reaction between the atmospheric gas and arc-metal interaction zones (Fig. 25.1).

Shielding gas is generally an inert gas (Ar or He) but can also be a mixture of inert and non-inert gases. Since it is the same gas from which an arc is formed, it serves the dual purpose of shielding and arc-creating. Besides, the arc type depends on the gas type. Therefore, its selection does not depend upon the sole purpose of shielding but on the arc type needed.

To create an arc, a tungsten electrode is made the cathode, and a workpiece or substrate is made the anode when direct current is used. Since 70% of heat is generated at the anode, this type of polarity helps overheat the workpiece and underheat the cathode. Since the aim is to melt the workpiece (or the material on it) rather than the tungsten electrode, this polarity helps conserve energy by channelizing the heat near the workpiece.

The material in the form of powder [1] or wire [2] is fed to be melted either directly by an arc or indirectly by a meltpool that is created by the arc. Since the stability of the arc remains largely undisturbed by its effect on melting, it gives an advantage for controlling the deposition.

S. Kumar, *A Concise Encyclopedia of Additive Manufacturing*, https://doi.org/10.1007/978-3-031-98394-8_25

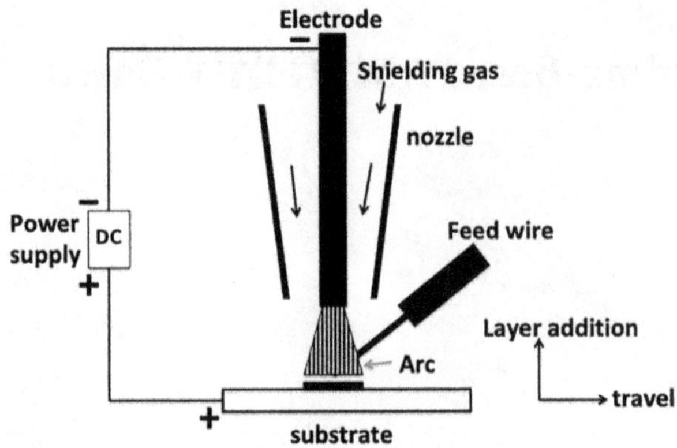

Fig. 25.1 Schematic diagram of GTAW-based deposition

References

1. Moreira, A. F., Ribeiro, K. S., Mariani, F. E., & Coelho, R. T. (2020). An initial investigation of tungsten inert gas (TIG) torch as heat source for additive manufacturing (AM) process. *Procedia Manufacturing, 48*, 671–677.
2. Yu, Y., Xiong, J., Chen, Y., & Zhao, H. (2023). Process stability control of corner structures in robotic gas tungsten arc additive manufacturing via arc sensing. *Journal of Manufacturing Processes, 101*, 156–170.

Chapter 26
Arc Welding-Based AM: PAW-Based

AM based on plasma arc welding (PAW) is widely used to make metallic parts [1–3], where the plasma arc is mainly a transferred arc [4–6], instead of a non-transferred arc.

1 How Plasma Arc Is Generated

An arc created in gas tungsten arc welding (GTAW) exists near an electrode, and thus, the operational ability of the arc is limited by its physical proximity to the electrode. If the arc were created far away from the electrode, it could have more freedom—the effect of melting due to the arc could have been planned without worrying about its negative consequence on the electrode. Besides, feeding could not have been restricted due to a constricted space between the electrode and the workpiece, and the size of the melt pool could not have been limited due to a limited variation in arc dimension. These limitations are in sharp contrast to the freedom provided by an l-beam—the effect of an l-beam spot on a workpiece does not affect a laser source.

Limitations imposed by an arc can be overcome by creating better heat sources utilizing the arc. As the arc is hot, if a gas passes through it, the gas becomes hot, which can be used as a heat source. For obtaining it, a tungsten rod is fitted inside a hollow nozzle, and while the rod usually acts as a cathode, the nozzle acts as an anode. When the gas is flown through the nozzle, the gas provides a medium to create an arc between the rod and the nozzle. Besides, the gas coming out from the nozzle is affected by the arc that is already created, i.e., the gas becomes hot and ionized, which is plasma. The plasma can reach away from the location of the cathode, and therefore, its effect can be realized at a farther distance. The reach of the plasma depends upon the gas speed, the nozzle type, and the arc temperature.

S. Kumar, *A Concise Encyclopedia of Additive Manufacturing*,
https://doi.org/10.1007/978-3-031-98394-8_26

This new heat source is due to the generated plasma and the arc struck between a nozzle and a cathode. To produce this, the location of the arc does not need to be changed. It means the arc remains maintained between the nozzle and the cathode before the flow of plasma gas (through the nozzle to the workpiece) and after its flow. If the arc remains maintained, the arc is not transferred anywhere. Therefore, this new heat source is called plasma non-transferred arc (Fig. 26.1a).

What if an arc is no longer maintained between the nozzle and the cathode, and the arc is transferred to the workpiece? Then, the arc striking and heating the workpiece is no longer the same as the plasma striking and heating the workpiece. This

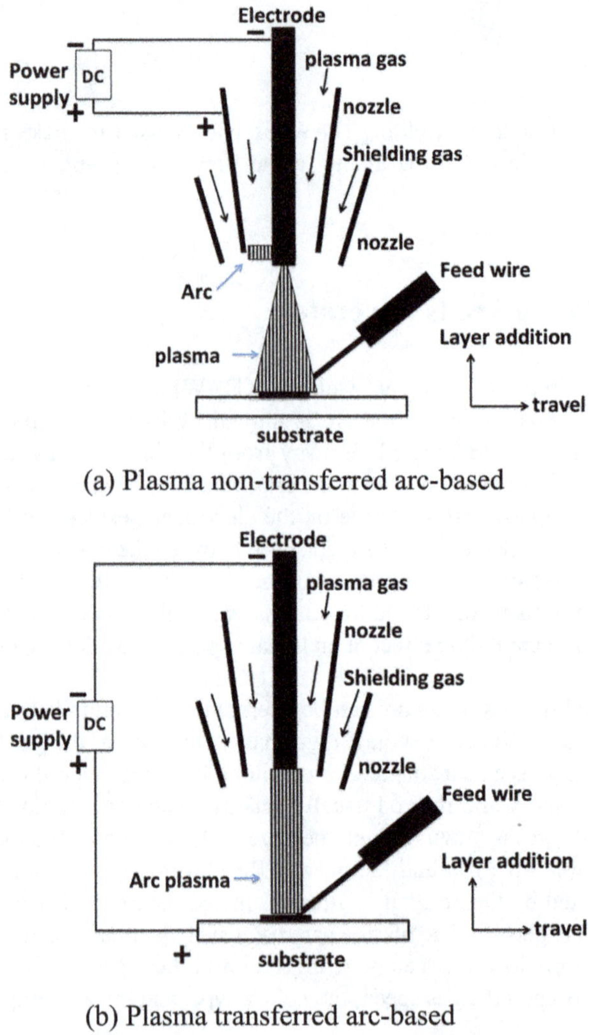

Fig. 26.1 Schematic diagrams of PAW-based deposition: (**a**) non-transferred arc, (**b**) transferred arc

method of heating the workpiece can give rise to one more type of heat source. In this method, the arc is first created between the nozzle and the cathode. Since this arc does not serve or intend to serve either directly or indirectly as a heat source, a less hot arc or low-current arc is sufficient, which is started between the nozzle and the cathode.

When the gas passes through it and the plasma strikes between the cathode and the workpiece, it is easier to create another arc between the cathode and the workpiece. Thus, when the polarity of the workpiece is changed from neutral to positive with respect to the tungsten rod, another arc is started. The initial arc created serves its purpose when this new arc is created, and the initial arc is then extinguished. The initial arc is named as the pilot arc, while the heat source due to this new arc is called the plasma transferred arc (Fig. 26.1b). The pilot arc is also started by a high-frequency generator, and in that case, the polarity of the nozzle and the tungsten electrode periodically changes.

Plasma transferred arc needs to pass through a constricted nozzle orifice and is therefore thin, resulting in the concentration of heat energy. Since plasma gas also passes through the orifice, the gas is converted into plasma due to its interaction with the arc, which results in a heat source consisting of both arc and plasma. Therefore, this heat source has a higher temperature (almost two times) than the arc in GTAW does. Besides, in plasma transferred or non-transferred arc, shielding gas is separately used from plasma gas, which provides an advantage that the striking of the arc is not affected by the shielding gas as much as it is affected in GTAW.

References

1. Artaza, T., Suárez, A., Veiga, F., et al. (2020). Wire arc additive manufacturing Ti6Al4V aeronautical parts using plasma arc welding: Analysis of heat-treatment processes in different atmospheres. *Journal of Materials Research and Technology, 9*(6), 15454–15466.
2. Kopf, T., Glück, T., Gruber, D., et al. (2024). Process modeling and control for additive manufacturing of Ti-6Al-4V using plasma arc welding-methodology and experimental validation. *Journal of Manufacturing Processes, 126*, 12–23.
3. Wu, C. S., Wang, L., Ren, W. J., & Zhang, X. Y. (2014). Plasma arc welding: Process, sensing, control and modeling. *Journal of Manufacturing Processes, 16*(1), 74–85.
4. Sadasivam, P., & Amirthalingam, M. (2022). Design and fabrication of micro-plasma transferred wire arc additive manufacturing system. *CIRP Journal of Manufacturing Science and Technology, 37*, 185–195.
5. Mercado Rojas, J. (2021). Design and system identification for a plasma transferred arc additive manufacturing technology.
6. Niedermayer, J. (2022). Parameter study for 3D printing of Al5356, Al4047 and Al4043 by directed energy deposition with transferred plasma arc. (Doctoral dissertation, Wien).

Chapter 27
Beam Deposition Process: Advantage and Disadvantage

1 Advantages

The process has advantages as it uses a beam that can.

- Make fine features and parts having high accuracy.
- Scan—otherwise, the scanning needs to be done by moving the machine table or heat source.
- Focus on a spot—a deposition with precision is obtained.
- Defocus—a fast deposition is obtained by covering the increasing area during scanning [1].
- Focus and defocus [2]—beam spot size can be increased and decreased, resulting in changes in the area on which deposition takes place.
- Access the remote site of a complex part while repairing [3].

2 Disadvantages

Beam has disadvantages: it is not absorbed by many materials, is energy-inefficient, is unsafe, etc. Since the purpose of the beam in this process is to melt, the melting gives the following disadvantages:

- **Large grain size**—melting layerwise increases the temperature of a build, providing a conducive environment for a grain to increase in size, causing a decrease in strength.
- **Generation of porosity**—high solidification rate does not let trapped gases emanate, causing the formation of circular pores [4].

S. Kumar, *A Concise Encyclopedia of Additive Manufacturing*,
https://doi.org/10.1007/978-3-031-98394-8_27

- **Cracks due to thermal expansion mismatch**—when a mixture of two materials (e.g., metal and ceramic) is melted, then, during solidification, both materials contract by different amounts depending upon their thermal expansion coefficients, causing stress and possible cracks.
- **Hot cracking**—during solidification, a high melting point material solidifies earlier, leaving a low melting point material still in a liquid state. Cavities or grain boundaries containing such liquid become the source to initiate cracks, which brings restrictions to the type of materials that can be mixed.
- **Segregation of elements**—depending upon their density and melting point, elements segregate in a melt pool, leading to their inhomogeneous distribution, which causes a change in mechanical properties.
- **Vaporization of materials**—to melt a high melting point material of a mixture of materials, if the temperature of the mixture exceeds the vaporization temperature of a low melting point material, then the low melting point material will vaporize, which will change the mixture composition.

This brings a restriction to the maximum melting point difference between two materials in a mixture. It will narrow the processing window, implying a limited variation in process parameters. If the parameter is scan speed, it will limit the maximum scan speed, slowing the fabrication speed.

- **Anisotropic properties**—a melted layer is cooled through either previously melted layers or a substrate, causing a grain growth in the build direction (columnar grain). This induces different mechanical properties—one in the grain growth direction, another in its transverse direction, etc. [5]. Although the anisotropic properties can be managed by an applied external electromagnetic field, which will initiate transition from columnar to equiaxed grains [6].
- **Oxidation**—in the presence of residual oxygen in the feedstock or the environment, the material gets oxidized. The oxidized material becomes inclusions in the melt pool, which decreases the mechanical properties [7].

References

1. Steen, W. M., & Majumder, J. (2010). *Laser material processing*. Springer-Verlag London Limited.
2. Wang, L., Zhu, G., Shi, T., et al. (2018). Laser direct metal deposition process of thin-walled parts using variable spot by inside-beam powder feeding. *Rapid Prototyping Journal, 24*(1), 18–27.
3. Chen, H., Lu, Y., Luo, D., et al. (2020). Epitaxial laser deposition of single crystal Ni-based superalloys: Repair of complex geometry. *Journal of Materials Processing Technology, 285*, 116782.
4. Zhao, T., Wang, Y., Xu, T., et al. (2021). Some factors affecting porosity in directed energy deposition of AlMgScZr-alloys. *Optics & Laser Technology, 143*, 107337.

5. Shamsaei, N., Yadollahi, A., Bian, L., & Thompson, S. M. (2015). An overview of direct laser deposition for additive manufacturing; part II: Mechanical behavior, process parameter optimization and control. *Additive Manufacturing, 8*, 12–35.
6. Zeng, C., Xue, J., Jia, Y., et al. (2024). A review of additive manufacturing of metallic materials assisted by electromagnetic field technology. *Journal of Manufacturing Processes, 131*, 920–946.
7. Tang, M., & Pistorius, P. C. (2017). Oxides, porosity and fatigue performance of AlSi10Mg parts produced by selective laser melting. *International Journal of Fatigue, 94*, 192–201.

Chapter 28
Beam Deposition Process: Powder and Wire

The difference between powder and wire as a feedstock is as follows:

1. **Cost**: Powder is more expensive than wire. Many times, powder is produced from wire, and therefore, the cost of the powder is equal to the cost of the wire plus the cost of conversion of the wire into powder. Higher cost can also be because getting a narrow powder size distribution is more difficult than getting a uniform wire diameter.

   Powder requires carrier gas to be delivered, which increases the total cost. When a big part is fabricated, the cost of feedstock increases the cost significantly, making the wire preferable due to its lower cost [1].
2. **Availability**: Materials that are used to make a wire are limited. The available wire is made mostly from metals. If the metal or alloy is malleable, it is easy to draw wire from such materials, which increases the availability. Powder has higher availability than wire because powder can be made from a higher number of techniques, such as from solution (joining atom by atom), as well as from big blocks (melting and spraying).

   If there is irregularity in the diameter of a wire, the whole wire needs to be discarded, while if some powders are irregular, these are discarded, and the remaining powders can still be used.
3. **Material efficiency**: During deposition, there is a chance of losing powders, while there is no such possibility with the wire. Thus, the powder gives lower deposition efficiency than the wire [2, 3] and a lower building rate [4]. Powders left out during deposition need to be collected, sorted, and recycled, while there is no such problem with the wire. This increases the processing cost with powders, as for creating the same size of a part, a higher amount of powder is required.

   Powder furnishes higher accuracy than the wire because powders can work with smaller melt pools. The wire needs to be thin to give a small pool. A thinner wire cannot be fed, which gives a limitation to the smaller size of the pool that

S. Kumar, *A Concise Encyclopedia of Additive Manufacturing*,
https://doi.org/10.1007/978-3-031-98394-8_28

can be obtained by decreasing the wire diameter. Besides, a small pool cannot be planned to be obtained by partially melting a thick wire, as it may cause the wire to be entangled with the build.

Wire is preferred in microgravity because the whole wire can be traced, while some powder can spill and pose safety problems [5].

4. **Processing in vacuum**: In a vacuum, wire can be fed with ease without disturbing it. Feeding the powder using carrier gas changes the vacuum pressure due to the presence of gas. Besides, the gas gets ionized, which obscures the e-beam and decreases its efficiency [6]. Vacuum allows for working with reactive materials [7].
5. **Oxidation**: For converting a material into a bead, the material needs to be melted. For the same weight, the material in the form of powder rather than wire has a higher surface area. Thus, if the powder instead of the wire is used, a higher surface area is involved, which exposes a bigger area to oxidation in an oxidative environment, making powder-based deposition more vulnerable to oxidation than wire-based deposition.
6. **Safety**: Powders have a higher surface-to-volume ratio than wire. For reactive materials, powders have a higher chance of burning. Besides, it is easy to inhale powder, which creates a health hazard. There is a limit to the maximum beam power used because a high-power beam generates x-rays that need to be shielded from coming out of the chamber.
7. **Effect on fabrication**: Powder gives flexibility to process—many types of powders can be mixed and fed through a single nozzle to make a composite [8], while the number of wires that can be fed from a wire-feeder to make a composite is limited. For example, ceramic and metal powders can be mixed and fed, while feeding ceramic and metal wires together is not equally feasible because of the difference in the stiffness of the ceramic and the metal. In the case of a bundle of ceramic wire and metal wire, both types need to be melted individually to make a composite. If the ceramic wire is not melted, it cannot contribute a ceramic material to the melt pool. Consequently, a ceramic-metal composite will not be formed. In the case of a mixture of ceramic powder and metal powder, even without melting the ceramic, a ceramic-metal composite can be formed.

   In wire-based deposition, the wire needs to be completely melted. If it is partially melted, it will attach to the bead, causing the termination of the fabrication. In powder-based deposition, when all powders are required to melt, if some do not melt, it causes part imperfections, but the fabrication is not terminated. Hence, feeding excess or insufficient powder can be a strategy for varying properties and making porous structures. Wire feeding does not have such flexibility, as the wire feed rate needs to correspond with the heat input to prevent non-melting of the wire. Therefore, the wire does not offer such property variations that can be obtained by a combination of melting and non-melting. Thus, in wire-based AM, the strength of a part is not optimized by partially melting the wire. Complete melting followed by the control of the melting is required to qualify a wire-based process [9], while complete melting is not always a requirement to make parts from powders [10].

Wire is generally fed from sideways or off-axial positions, while powder can be fed from both sideways and the coaxial position. Wire is confined to a certain point or limited area in a melt pool (Fig. 28.1), while powder can be defocused or focused to cover all areas of the melt pool.

Since wire feeding necessarily entails localization of the wire tip at a particular place in the melt pool, it gives rise to a question: which place of the melt pool is its particular place? A melt pool is created by melting the substrate (or previously melted layer) using a beam; the wire is positioned in such a way that it is to be dipped in the coming melt pool. If the wire is positioned under the beam, it will attenuate the beam. Therefore, any position that is not under the beam is the right place.

For the realization of the process, movement of a beam nozzle (relative to a substrate) or the substrate (relative to the nozzle) is used to create a melt pool. A moving pool thus created has two edges: leading and trailing. Since the leading edge is in continuous contact with the beam, it is a hot edge, while the trailing edge is relatively cold and solidifies faster than the leading edge.

Wire fed at the trailing edge has a higher chance to solidify (and become part of the solidified melt pool) before the wire acquires enough energy from the melt pool to melt itself and let unmelted feedstock wire detach from the melt pool. The detachment will not hinder the feeding nozzle from moving forward parallel to a substrate. This is the reason why the leading edge rather than the trailing edge is the preferred place where the wire is pointed (Fig. 28.1).

Direction of motion determines which edge of the melt pool is leading or trailing. If the direction reverses, the leading edge will become the trailing edge and vice versa. Therefore, if the position of the wire is not reversed with a reverse in the direction, the performance of the process in both directions will not be the same. Changing the position of the wire can be avoided if there are two wire feeders on either side of the beam, and with a change in direction, the alternate wire feeder is used.

8. **Processing an inaccessible area:** Wire is a large integrated unit in comparison to disunited powders. It requires a certain minimum stiffness to be pointed outside the wire feeder and to reach the melt pool. Without stiffness, it will not move

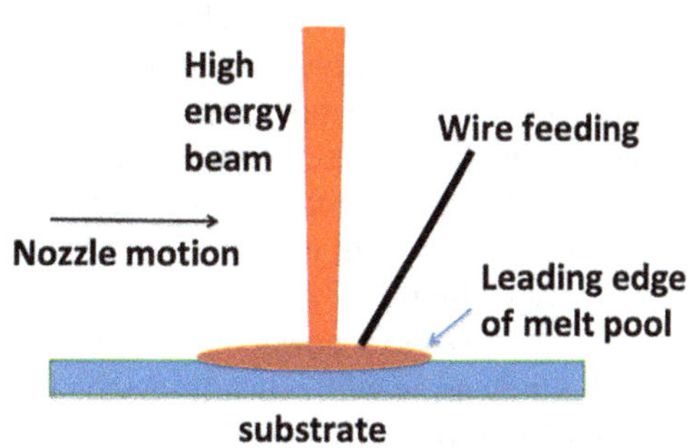

Fig. 28.1 Wire feeding at the leading edge of the moving melt pool

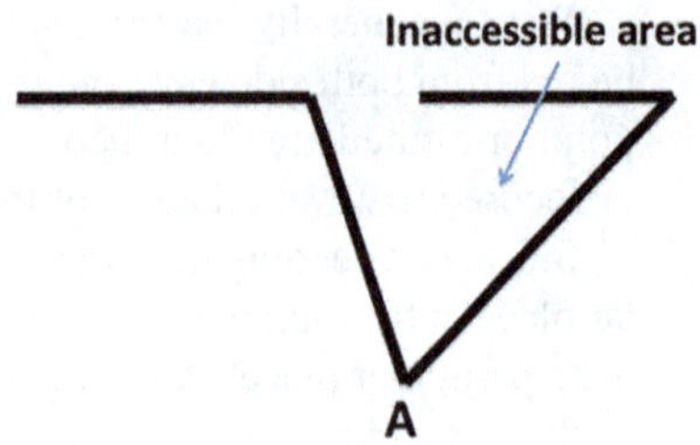

Fig. 28.2 Point A in an inaccessible area where material needs to be deposited

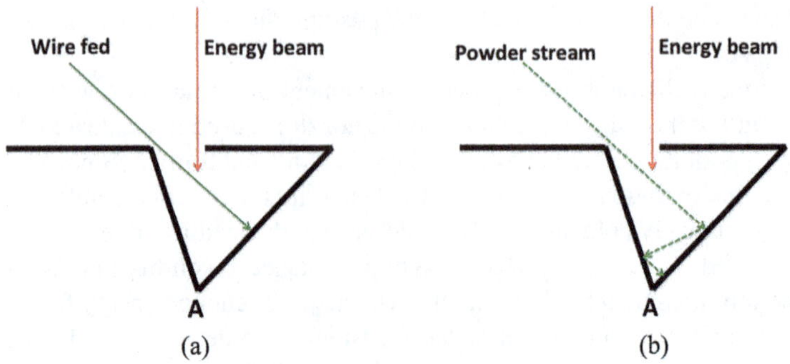

Fig. 28.3 Material deposition at point A of an inaccessible area: (**a**) wire feeding, (**b**) powder feeding

toward the target but will dangle. This stiffness, which is an essential property of the wire, is a problem when the aim is to reach an inaccessible area for processing.

Powder, unlike wire, is not an inseparable part of an integrated unit; the moment it leaves a nozzle, it no longer remains attached to the nozzle. Therefore, it has a greater possibility of reaching an inaccessible point by other means, such as rebounding and falling.

Figure 28.2 shows an inaccessible area where, at point A, material needs to be deposited. Vertical feeding is not possible due to the geometry of the structure. Because if the tip of the wire reaches point A, the l-beam, instead of entering the cavity, will be reflected away.

By feeding the wire from an off-axial position (Fig. 28.3a), the point of repair, i.e., A, is not accessible. While from the same off-axial position, powder can reach point A by rebounding from the side surface and falling (Fig. 28.3b). This allows feeding to happen at a melt pool created by the beam and a build to be completed. In the case of wire feeding, the melt pool thus created cannot be fed, and the build will not be completed. This type of powder deposition is not an ideal deposition, but it shows that what can be achieved with powder cannot be achieved with wire.

This shows that powder as a feedstock has an advantage over wire. Though deposition using wire can still be accomplished by melting the wire midway, the completion of the deposition in this way again conveys the advantage of the powder feeding.

References

1. Hassen, A. A., Noakes, M., Nandwana, P., et al. (2020). Scaling up metal additive manufacturing process to fabricate molds for composite manufacturing. *Additive Manufacturing, 32*, 101093.
2. Schmidt, M., Merklein, M., Bourell, D., et al. (2017). Laser based additive manufacturing in industry and academia. *CIRP Annals, 66*(2), 561–583.
3. Kumar, P., & Jain, N. K. (2020). Effect of material form on deposition characteristics in micro-plasma transferred arc additive manufacturing process. *CIRP Journal of Manufacturing Science and Technology, 30*, 195–205.
4. Blinn, B., Lion, P., Jordan, O., et al. (2021). Process-influenced fatigue behavior of AISI 316L manufactured by powder-and wire-based laser direct energy deposition. *Materials Science and Engineering: A, 818*, 141383.
5. Watson, J. K., Taminger, K. M. B., Hafley, R. A., & Petersen, D. D. (2002). Development of a prototype low voltage electron beam freeform fabrication system. In *SFF proceedings* (pp. 458–465).
6. Taminger, K. M. B., & Hafley, R. A. (2013). Electron beam freeform fabrication: A rapid metal deposition process. In *Proceedings of the 3rd annual automotive composites conference, Troy, MI.*
7. Guimarães, R. P. M., Rauchdobler, B., Pixner, F., et al. (2025). Ageing response of Ni-rich NiTi shape memory alloy additively manufactured by electron beam freeform fabrication. *Journal of Materials Research and Technology, 35*, 2656–2675.
8. Kumar, S. (2022). Comparison. In *Additive manufacturing solutions* (pp. 57–92). Springer.
9. Gockel, J., Beuth, J., & Taminger, K. (2014). Integrated control of solidification microstructure and melt pool dimensions in electron beam wire feed additive manufacturing of Ti-6Al-4V. *Additive Manufacturing, 1*, 119–126.
10. Sillani, F., MacDonald, E., Villela, J., et al. (2022). In-situ monitoring of powder bed fusion of polymers using laser profilometry. *Additive Manufacturing, 59*, 103074.

Chapter 29
Big Part

Large parts are usually made by conventional manufacturing (CM), i.e., forming, forging, casting, or machining. In AM, LPBF can make up to 2 m long [1], while wire arc additive manufacturing can make up to 6 m long [2]. Reasons to make big parts in AM are as follows:

1. A part for which the lead-time (time from design to fabrication) is high needs to be fabricated at the earliest.
2. There is a need to conserve materials such as tungsten [3], titanium, tantalum, and niobium that are wasted as chips in machining [4].
3. To save material [5]. For example, up to 500 kg per part can be saved [6].
4. A secret part is to be made, e.g., aircraft frames for defense. AM requires one or a few pieces of equipment, which makes it a preferred choice.
5. Fabrication requires several subprocesses in CM.
6. Fabrication by CM is difficult [3].
7. Fabrication is less expensive in AM, e.g., fabrication of a single part, for which CM is expensive due to tooling or machining cost [4, 7], or when a low-cost feedstock is used in AM [8].
8. If it is made from recycled materials, it will utilize a large amount of recycled materials, which will decrease cost and make AM sustainable [9, 10]. The fabrication works with various recycled materials [11–13].

References

1. Yang, Y., Jiang, R., Han, C., et al. (2024). Frontiers in laser additive manufacturing technology. *Additive Manufacturing Frontiers, 3*(4), 200160.
2. Yi, H., Jia, L., Ding, J., & Li, H. (2024). Achieving material diversity in wire arc additive manufacturing: Leaping from alloys to composites via wire innovation. *International Journal of Machine Tools and Manufacture, 194*, 104103.

S. Kumar, *A Concise Encyclopedia of Additive Manufacturing*,
https://doi.org/10.1007/978-3-031-98394-8_29

3. Marinelli, G., Martina, F., Ganguly, S., & Williams, S. (2019). Development of wire + arc additive manufacture for the production of large-scale unalloyed tungsten components. *International Journal of Refractory Metals and Hard Materials, 82*, 329–335.
4. Dominguez, L. A., Xu, F., Shokrani, A., et al. (2020). Guidelines when considering pre & post processing of large metal additive manufactured parts. *Procedia Manufacturing, 51*, 684–691.
5. Kokare, S., Oliveira, J. P., & Godina, R. (2024). Exploring the environmental and economic benefits of wire arc additive manufacturing compared to subtractive manufacturing. *Procedia CIRP, 130*, 3–8.
6. Chen, Z., & Soh, G. H. (2023). Wire arc additive manufacturing: Systems, microstructure, defects, quality control, and modeling. In K. Zhou (Ed.), *Additive manufacturing* (pp. 205–244). Springer.
7. Roschli, A., Post, B., & Heineman, J. (2024). *Utility of big area additive manufacturing for part production for low-head hydropower* (Vol. No. ORNL/TM-2024/3376, p. NFE-21-08560). Oak Ridge National Laboratory (ORNL).
8. Hassen, A. A., Noakes, M., Nandwana, P., et al. (2020). Scaling up metal additive manufacturing process to fabricate molds for composite manufacturing. *Additive Manufacturing, 32*, 101093.
9. Goh, G. D., Wong, K. K., Tan, N., et al. (2024). Large-format additive manufacturing of polymers: A review of fabrication processes, materials, and design. *Virtual and Physical Prototyping, 19*(1), e2336160.
10. Romani, A., & Levi, M. (2024). Large-format material extrusion additive manufacturing for circular economy practices: A focus on product applications with materials from recycled plastics and biomass waste. *Sustainability, 16*(18), 7966.
11. Walker, R., Korey, M., Hubbard, A. M., et al. (2024). Recycling of CF-ABS machining waste for large format additive manufacturing. *Composites Part B: Engineering, 275*, 111291.
12. Burgos Pintos, P., Marzo Gago, P., Fernández Delgado, N., et al. (2024). Sustainable product design by large format additive manufacturing of cork composites. *Virtual and Physical Prototyping, 19*(1), e2386106.
13. Korey, M., Rencheck, M. L., Tekinalp, H., et al. (2023). Recycling polymer composite granulate/regrind using big area additive manufacturing. *Composites Part B: Engineering, 256*, 110652.

Chapter 30
Binder Jet AM and Selective Inhibition Sintering

Binder jet AM (BJAM) [1], also known as binder jet three-dimensional printing (BJ3DP), is a powder bed non-fusion process in which binder from an inkjet printhead is used to join powders (Fig. 30.1). The process is similar to powder bed fusion (PBF) except that the job of the beam of PBF is done by a binder jet. It is also called 3D printing, but this name can be confused with general 3D printing, a synonym for AM [2], and therefore, BJ3DP or BJAM is used instead.

Binder does not melt powders, i.e., powders are joined without going through any type of melting. Hence, different types of powders, irrespective of their melting points, can be processed. Since the binder is usually not a constituent of a final part, it needs to be removed from the part in such a way that the act of removing does not inadvertently change the part's shape and dimensions. This requires post-processing with an aim to fulfill the following: 1) removal of the binder, so a part will not continue to remain contaminated with it, 2) solid state sintering, so the structure of the part does not change, and 3) infiltration, so density and strength are enhanced. The requirement of post-processing means BJ3DP is inferior to those AM processes that give final products without post-processing.

1 Role of Binder

Binder is jetted on a powder bed; jetting implies that the binder is either liquid or air. There is no air-based binder. Thus, all jetted materials are liquid. Binder comes from a nozzle of an ink jet printhead—whatever is jetted from the nozzle is supposed to act as a binder (to join) or to act as a partial binder (to cause to join). The binder implies that it is sufficient to bind powders without seeking an active contribution from the powder itself, e.g., joining of iron or copper powders by an organic binder.

The binder joins iron powders as efficiently as it joins copper powders. If there is a difference in the joining strength, it is due to the difference in the size or surface

S. Kumar, *A Concise Encyclopedia of Additive Manufacturing*,
https://doi.org/10.1007/978-3-031-98394-8_30

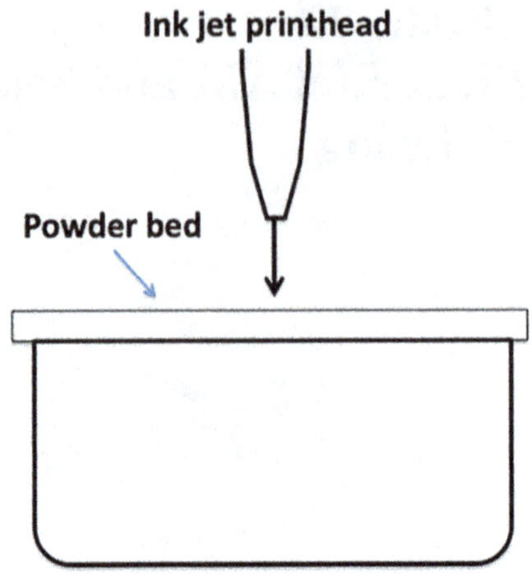

Fig. 30.1 Schematic diagram of BJ3DP

texture of the powder, not due to the different role of iron as a material than copper as a material. Thus, the powder is neutral and does not contribute to increasing the capability of binders. Hence, the situation will remain the same if iron or copper is replaced with aluminum or tungsten.

What will happen if the powder is not neutral? Then, its constituents might be different. Possibly, it is made of binders, or is itself a binder, or has a few percentages of binder in it, or plays different roles, and one of its roles is to act as a binder in the presence of certain materials. In these cases, whatever is jetted from a nozzle may not be a complete but a partial binder, because if it does not trigger a reaction and cause binding to happen on PB, the nozzle will not be able to make a pattern and help create a 3D shape. An example of a partial binder is using water as a binder on cement powders. Water reacts with cement and binds cement powders; the same water may not be used as a binder for iron powders.

A partial binder is used if a binder.

1. Solely jetted from a nozzle is not able to go through the thickness of a layer; some other binder present in the layer is required to make up for the binding process.
2. Clogs a nozzle.
3. Does not improve the rheology of a liquid jet or help form better drops.
4. Leaves residue after post-processing.
5. Is not required, e.g., water as a partial binder on cement or plaster.
6. Is not intended to be removed during post-processing but is required as a constituent of a final part.

The jetted material aims to bind those areas of PB where it is jetted. But if it does just the opposite, i.e., it does not bind only those areas, then an object that is the negative 3D image of a 3D object is formed. This process is called selective inhibition sintering (SIS), implying the jetted material inhibits binding [3]. This is possible if the jetted area has a lower ability to bind while the other area is full of binder, so during post-processing heat treatment, the other area, unlike the jetted area, binds, as the jetted area is immune to the high temperature.

The main difference between BJ3DP and SIS:

1. If a big part is to be made, then in BJ3DP, it requires a lot of binder to be jetted, while in SIS, it requires it to be jetted only on boundaries. Thus, more binder is saved using SIS.
2. If a complex part is to be made, then in BJ3DP, the part bound with binder (a green part) is taken to a furnace for post-processing, while in SIS, the whole powder container containing the jetted area and surrounding powder is taken. In most geometries, surrounding powder only on one side of the jetted line or area is intended to become a part, while that of the other side becomes waste. Thus, in SIS, the powder that does not contribute to the final part is also subjected to furnace treatment and becomes unusable. Thus, more powder is saved using BJ3DP.

Figure 30.2 shows the difference between BJ3DP and SIS. Figure 30.2a shows binder deposition in BJ3DP; the binder is deposited in a yellow area, and the part forms (Fig. 30.2b) as per the deposited area. Figure 30.2c shows inhibitor deposition in the yellow area; the part forms (Fig. 30.2d) from the area that surrounds the powder (red in color). The part consists of two components: a rectangular area outside the yellow area and a square inside the yellow area. The rectangular component may not be used and becomes waste.

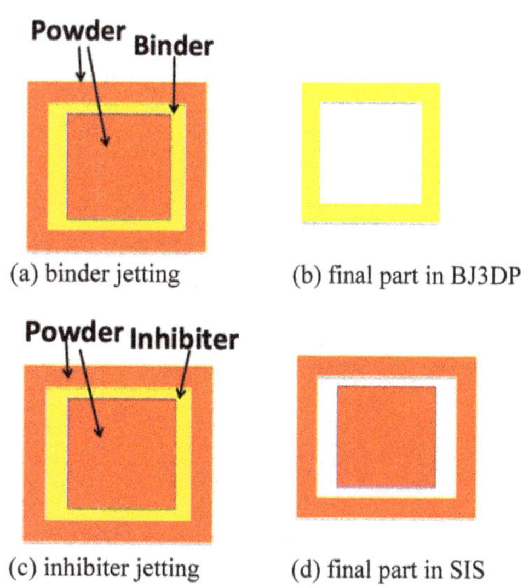

Fig. 30.2 Role of binder and inhibitor. (**a**) binder jetting. (**b**) final part in BJ3DP. (**c**) inhibiter jetting. (**d**) final part in SIS

References

1. Yang, M., Keshavarz, M. K., Vlasea, M., et al. (2023). Supersolidus liquid phase sintering of water-atomized low-alloy steel in binder jetting additive manufacturing. *Heliyon, 9*(3), e13882.
2. Kumar, S. (2022). Synonym. In *Additive manufacturing solutions* (pp. 1–6). Springer.
3. Khoshnevis, B., Zhang, J., Fateri, M., & Xiao, Z. (2014). Ceramics 3D printing by selective inhibition sintering. In *2014 international solid freeform fabrication symposium* (pp. 163–169).

Chapter 31
Binder Jetting

Binder jetting (BJT) is a type of AM technique in which a binder is jetted on a powder bed to consolidate selectively as per the 2D design of the 3D model of a product [1]. BJT is also known as binder jet AM [2], binder jetting AM [3], binder jet 3D printing [4], or binder jetting 3D printing [5].

References

1. Sarila, V., Koneru, H. P., Pyatla, S., et al. (2024). An overview on 3D printing of ceramics using binder jetting process. *Engineering Proceedings, 61*(1), 44.
2. Gilmer, D., Kim, S., Goldsby, D. J., et al. (2024). Predictive binder jet additive manufacturing enabled by clean burn-off binder design. *Additive Manufacturing, 80*, 103955.
3. Du, W., Hayes, J., Myers, K., et al. (2025). Development of a high-temperature Inconel 625 heat exchanger by model design and binder jetting additive manufacturing. *Materials & Design, 251*, 113333.
4. Wagner, J. J., & Higgs, C. F., III. (2024). Coupled CFD-DEM simulation of interfacial fluid–particle interaction during binder jet 3D printing. *Computer Methods in Applied Mechanics and Engineering, 421*, 116747.
5. Wang, Y., Genina, N., Müllertz, A., & Rantanen, J. (2024). Binder jetting 3D printing in fabricating pharmaceutical solid products for precision medicine. *Basic & Clinical Pharmacology & Toxicology, 134*(3), 325–332.

S. Kumar, *A Concise Encyclopedia of Additive Manufacturing*,
https://doi.org/10.1007/978-3-031-98394-8_31

Chapter 32
Binding Mechanism

1 Liquid Phase Sintering

Liquid phase sintering (LPS) is a well-known binding mechanism in powder metallurgy [1]. Taking the inspiration from powder metallurgy, this mechanism is used to explain the consolidation that happens in various powder-based AM techniques, such as powder bed fusion [2], binder jet AM [3], and laser deposition process [4].

It is frequently used in laser-based powder bed techniques [5], where some powders are converted into liquid during laser treatment. The liquid flows and fills up the pores made by adjacent powders, causing the powders to be joined. Therefore, melting some rather than all powders is sufficient for the consolidation to occur.

LPS is also referred to as partial melting because, for joining, part of a single powder or part of a bunch of powders is melted. Partial melting is considered a case of full melting in which laser energy is either insufficient or intentionally set insufficient (by adjusting parameters) to melt powders completely.

Powders can be engineered in the following way to facilitate consolidation:

- **Coating of a powder**: powders are coated with a low melting point material so the coating preferentially melts. The molten liquid then acts as a binder to bind unmelted cores. Examples of coated powders are steel coated with polymer, diamond coated with copper [6], copper coated with polymer [7], sand coated with phenol, carbide coated with nickel [8], etc.
- **Powder mixture**: Low-melting-point material is mixed with a high-melting-point material so the low-melting-point material melts and joins the remaining material. In this type, the former material acts as a binder, while the latter acts as a structural material. Examples are WC + Co, SiC + PA, and stainless steel + polymer. In these examples, WC, SiC, and stainless steel are structural materials, while Co, PA, and polymer are binder materials [9].
- In another type of mixing, the same material having two sizes (small and big) can be mixed. Small-sized powder preferentially melts because of its high

S. Kumar, *A Concise Encyclopedia of Additive Manufacturing*,
https://doi.org/10.1007/978-3-031-98394-8_32

surface-to-volume ratio and acts as a binder,, while large-sized powder acts as a structural material.
- **Composite powder**: Different powders of a powder mixture are segregated during powder bed making, giving rise to the nonhomogeneous melting of powders, responsible for nonuniformity in properties. Making a composite powder by mechanically alloying different powders (instead of just mixing them) helps avoid the problem [10].

LPS economizes laser energy required for consolidation, which is also accomplished, without engineering powders, by adjusting process parameters so that only that amount of melting takes place that is sufficient for consolidation. Low melting gives rise to a higher dimensional accuracy because high melting causes an increased melt flow, resulting in the loss of a dimensional accuracy.

This mechanism gives a message that the complete melting can be avoided by only partially melting a feedstock if its complete melting gives a problem. This can be applied in another type of feedstock, i.e., wire, in another type of AM. For example, if a titanium alloy is completely melted, it decreases the ductility of a product made. Therefore, if only the surface of the wire of the alloy is melted, it will increase the ductility [11].

2 Full Melting

In full melting [12], as the name implies, the feedstock is completely melted to consolidate separate feedstocks. Complete melting gives rise to a melt pool that, on solidification, forms a unit that adds on with progress in scanning, giving a desired geometry. During the melt pool formation, powders melt and disappear, leaving no powders in the melt pool. Therefore, in this mechanism, unlike LPS, the problem of binding powders does not exist. The problem that exists is how to avoid binding with those powders present adjacent to melt pools, or underneath the melt pool in case of overhangs. The binding of the melt pool with other powders depends upon the size, viscosity, surface tension, and dwell time of the melt pool. Besides, it depends on the capillary that is determined by packing, size, and size distribution of powders. At the boundary of the geometry, the interaction of the melt pool with powders causes an increase in surface roughness (of the side surface), resulting in a high difference in the surface roughness of the side surface and the top surface. An increase in such binding decreases the geometrical accuracy. Therefore, this binding mechanism demands to have control to prevent such ill effects of the binding. In the case of the melt pool formed on overhangs facing powders underneath, or in the case of the melt pool formed on the first layer facing powders underneath, minimization of the binding of the powders with the melt pool furnishes a smooth surface.

This mechanism provides high density, high strength, variable strengths, a range of ductility, reduced porosity, better microstructure than that of wrought or cast materials, and novel microstructures not possible through conventional manufacturing.

References

1. German, R. M., Suri, P., & Park, S. J. (2009). Liquid phase sintering. *Journal of Materials Science, 44*, 1–39.
2. Kruth, J. P., Mercelis, P., Van Vaerenbergh, J., et al. (2005). Binding mechanisms in selective laser sintering and selective laser melting. *Rapid Prototyping Journal, 11*(1), 26–36.
3. Liu, J., Kannan, R., Zhang, D., et al. (2022). Multi-scale characterization of supersolidus liquid phase sintered H13 tool steel manufactured via binder jet additive manufacturing. *Additive Manufacturing, 56*, 102834.
4. Zhou, S., Liang, Y. J., Zhu, Y., et al. (2021). Ultrashort-time liquid phase sintering of high-performance fine-grain tungsten heavy alloys by laser additive manufacturing. *Journal of Materials Science and Technology, 90*, 30–36.
5. Navarrete-Segado, P., Frances, C., Tourbin, M., et al. (2022). Powder bed selective laser process (sintering/melting) applied to tailored calcium phosphate-based powders. *Additive Manufacturing, 50*, 102542.
6. Zhang, L., Li, Y., Li, S., Gong, P., et al. (2022). Fabrication of titanium and copper-coated diamond/copper composites via selective laser melting. *Micromachines, 13*(5), 724.
7. Fu, Y., Yan, C., Yang, X., et al. (2021). Preparation and selective laser sintering of nylon-12 coated copper powders. *Rapid Prototyping Journal, 27*(7), 1355–1362.
8. Ergüder, T. O., Güler, O., & Yıldız, F. (2024). Determination of selective laser melting process parameters of tungsten carbide powders coated with nickel via electroless plating for improved interface properties. *International Journal of Refractory Metals and Hard Materials, 122*, 106735.
9. Jucan, O. D., Gădălean, R. V., Chicinaş, H. F., et al. (2021). Study on the indirect selective laser sintering (SLS) of WC-co/PA12 powders for the manufacturing of cemented carbide parts. *International Journal of Refractory Metals and Hard Materials, 96*, 105498.
10. Colucci, G., Lupone, F., Bondioli, F., & Messori, M. (2024). 3D printing of PBAT-based composites filled with agro-wastes via selective laser sintering. *European Polymer Journal, 215*, 113197.
11. Lv, J., Li, B., Li, Z., et al. (2025). Resistive thermal fusion interface: A novel additive manufacturing process of titanium alloy. *Journal of Materials Processing Technology*, 118773.
12. Kruth, J. P., Levy, G., Klocke, F., & Childs, T. H. (2007). Consolidation phenomena in laser and powder-bed based layered manufacturing. *CIRP Annals, 56*(2), 730–759.

Chapter 33
Binding Mechanism: Misnomer

There are various examples where complete melting of feedstock has not taken place [1–10], whether they belong to liquid phase sintering (LPS) or full melting.

Checking the processed material just after the processing and before consolidation will clarify. The processed material looks like it is a mixture of liquid and unmelted solid. It gives a false impression that the presence of liquid and solid automatically implies that LPS has taken place. LPS holds good only when the liquid is responsible for giving rise to a structure made up of solid particles by holding the solid particles. This is possible only when the majority of the structure is made up of solid and the liquid is distributed well enough. This is not possible when the majority of the structure is made up of liquid and the solid is in the minority. Because when the liquid is in the majority, there are not enough solid particles for it to hold and make the structure. The structure is then made not because liquid binds solid particles, but because the solidification of the liquid takes place. The solidification of the liquid does not necessarily consist of solid particles. Where the solidification consists of solid particles, the solidification does not necessarily bind two solid particles but traps each single solid particle.

When the solidification takes place without necessarily binding two solid particles because the processed material does not have enough solid particles, this case is not different from the mechanism of full melting. The difference between LPS and full melting is what holds the structure together. In LPS, it is the binding by liquid because liquid is required to hold two solid particles. In full melting, it is the absence of binding by liquid because liquid is not required to hold two solid particles. When melting has taken place and there is an absence of LPS, it implies that there is a presence of full melting as a binding mechanism. It means that when partial melting has taken place and the partial melting does not lead to LPS, there is a presence of full melting as a binding mechanism. It means there is a difference between full melting of the processed material (a mechanism that is full melting) and partial melting of the processed material (a mechanism that is full melting as well).

S. Kumar, *A Concise Encyclopedia of Additive Manufacturing*,
https://doi.org/10.1007/978-3-031-98394-8_33

References

1. Li, K., Hu, Z., Yang, W., et al. (2024). Selective laser melting and mechanical behavior of Mo-coated diamond particle reinforced metal matrix composites. *Diamond and Related Materials, 144*, 110952.
2. Zhang, S., Tang, H., Tang, D., et al. (2024). Effect of fabrication process on the microstructure and mechanical performance of carbon fiber reinforced PEEK composites via selective laser sintering. *Composites Science and Technology, 246*, 110396.
3. Han, S., Li, S., Song, X., et al. (2024). Carbon nanotubes/α-ZrP sheets for high mechanical performance and flame-retarding polyamides using selective laser sintering. *Virtual and Physical Prototyping, 19*(1), e2368644.
4. Azam, M. U., Schiffer, A., & Kumar, S. (2024). Mechanical and piezoresistive properties of GNP/UHMWPE composites and their cellular structures manufactured via selective laser sintering. *Journal of Materials Research and Technology, 28*, 1359–1369.
5. Mingione, E., Salvi, D., Almonti, D., & Ponticelli, G. S. (2025). Improvement of thermal, electrical, and tribological performances of GnPs composites produced by selective laser sintering. *Polymer Composites, 46*(9), 7924–7938.
6. Hu, J., Lin, X., & Hu, Y. (2024). High wear resistance and strength of Hastelloy X reinforced with TiC fabricated by laser powder bed fusion additive manufacturing. *Applied Surface Science, 648*, 159004.
7. Jiang, M., Li, B., Chen, X., et al. (2024). Enhanced surface finish and reduced porosity of TiC nanoparticles reinforced 2219 aluminum alloy deposit fabricated via oscillating laser-arc hybrid additive manufacturing. *Journal of Manufacturing Processes, 120*, 414–425.
8. Tan, J., Chen, X., Peng, K., et al. (2024). Effect of Y2O3 nanoparticles on powder plasma arc additive manufacturing of CoCrFeNiMn high-entropy alloys: Microstructure evolution and mechanical properties. *Journal of Alloys and Compounds, 997*, 174933.
9. Li, Z., Gain, A. K., Cui, Y., & Zhang, L. (2025). Wear mechanism and subsurface microstructure of high entropy alloy reinforced with TiC nano-particles fabricated by additive manufacturing. In *Wear* (p. 205934).
10. Wei, W., Li, Y., Wei, B., et al. (2025). The influence of TiC nanoparticles addition on the microstructure and mechanical properties of IN738LC alloy prepared by EB-PBF. *Materials & Design, 251*, 113723.

Chapter 34
Classification

The ISO/ASTM 52900 standard classifies AM into the following seven categories:

1. Binder jetting (BJT).
2. Directed energy deposition (DED).
3. Material extrusion (MEX).
4. Material jetting (MJT).
5. Powder bed fusion (PBF).
6. Sheet lamination (SHL).
7. Vat photopolymerization (VPP).

AM can be classified into two categories [1, 2]:

1. Bed process.
2. Deposition process.

These two processes can be subclassified: bed process (Fig. 34.1) and deposition process (Figs. 34.2, 34.3 and 34.4).

In Fig. 34.1, SLM is selective laser melting, EBM is electron beam melting, SLS is selective laser sintering, HSS is high speed sintering, BJ3DP is binder jet three-dimensional printing, and CLF is ceramic laser fusion.

In Fig. 34.4, PJ is polymer jetting, IJP is ink jet printing, and RFP is rapid freeze prototyping.

S. Kumar, *A Concise Encyclopedia of Additive Manufacturing*,
https://doi.org/10.1007/978-3-031-98394-8_34

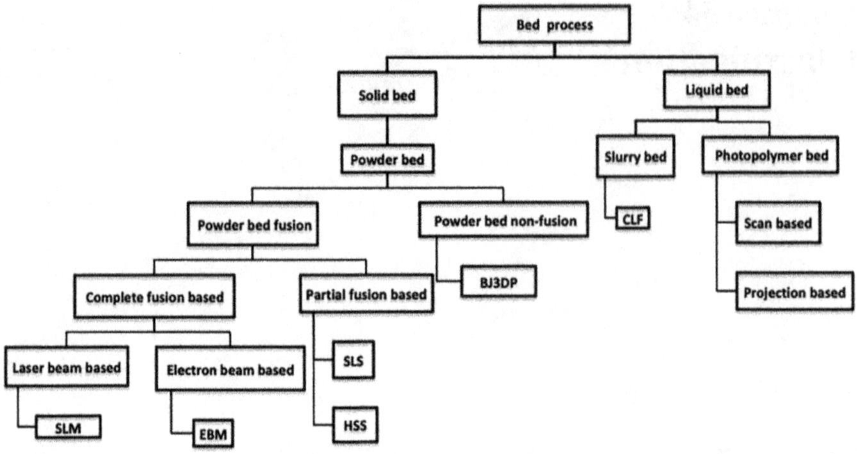

Fig. 34.1 Classification of bed process

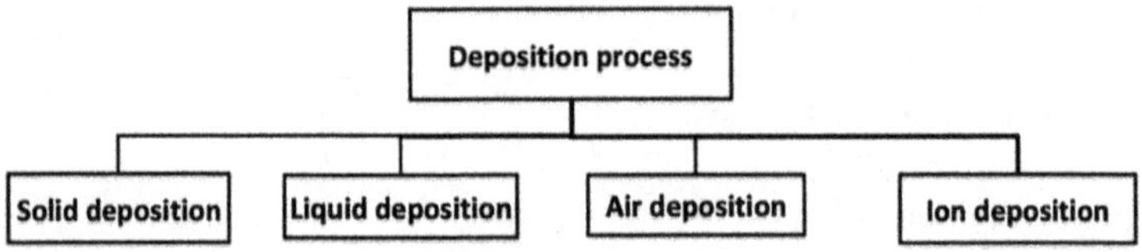

Fig. 34.2 Classification of deposition process

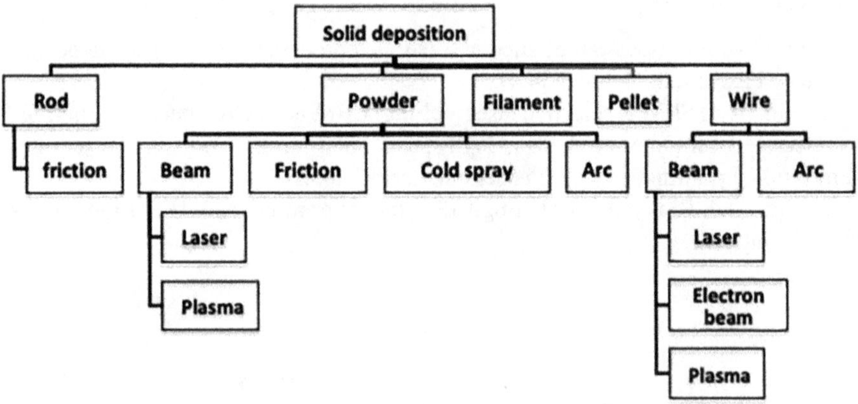

Fig. 34.3 Classification of solid deposition process

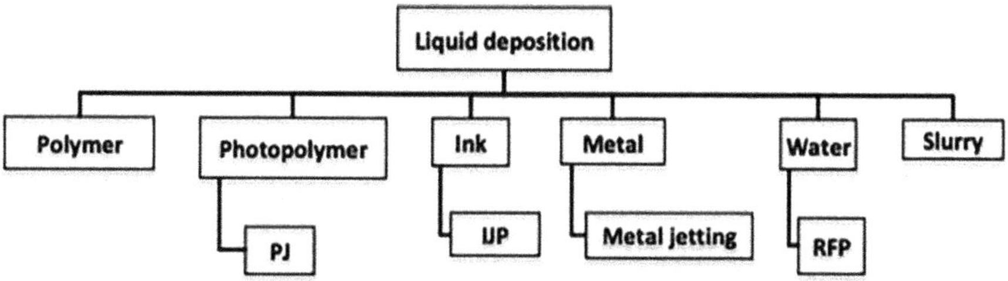

Fig. 34.4 Classification of liquid deposition process

References

1. Kumar, S. (2020). Classification. In *Additive manufacturing processes* (pp. 21–40). Springer.
2. Kumar, S. (2022). Process and classification. In *Additive manufacturing classification* (pp. 1–42). Springer.

Chapter 35
Classification: Air and Ion Deposition

The classification of air and ion deposition processes is given in Fig. 35.1 [1].

In Fig. 35.1, ECAM is electrochemical additive manufacturing.

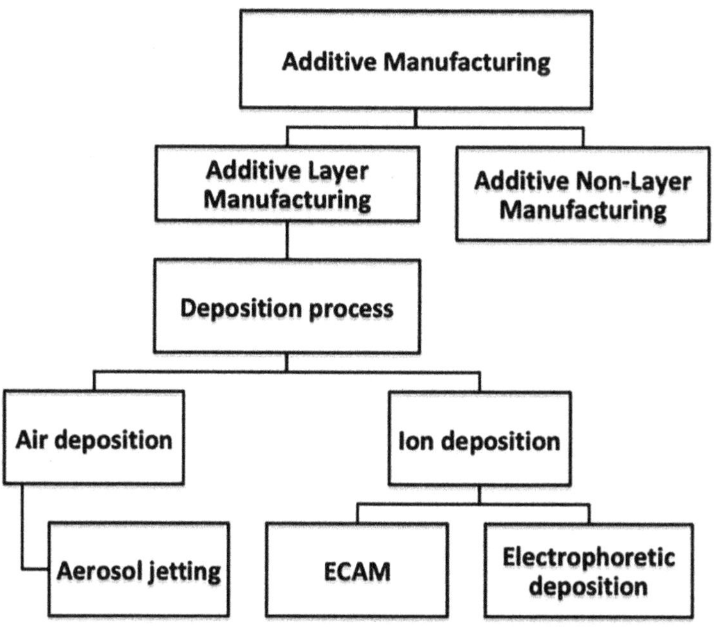

Fig. 35.1 Air and ion deposition

S. Kumar, *A Concise Encyclopedia of Additive Manufacturing*,
https://doi.org/10.1007/978-3-031-98394-8_35

Reference

1. Kumar, S. (2020). Air and ion deposition processes. In *Additive manufacturing processes* (pp. 147–157). Springer.

Chapter 36
Classification: Arc-Based Deposition

Classification of arc-based deposition is given in Fig. 36.1. The arc created between two electrodes is used as a heat source in metal welding, which is applied in AM. The processes are based on gas tungsten arc welding (GTAW), plasma welding (plasma non-transferred arc, plasma transferred arc), and gas metal arc welding (GMAW). The plasma is less intense and collimated than other beams, such as electron beam or laser. But the plasma beam can still be used to distinguish it from other non-plasma beam arc-based heat sources [1].

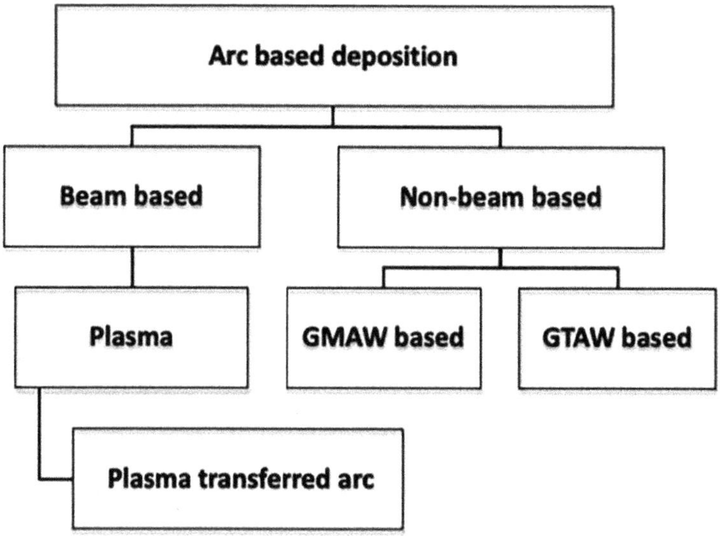

Fig. 36.1 Classification of arc-based deposition

S. Kumar, *A Concise Encyclopedia of Additive Manufacturing*,
https://doi.org/10.1007/978-3-031-98394-8_36

Reference

1. Kumar, S. (2022). Deposition process. In *Additive manufacturing classification* (Synthesis lectures on engineering, science, and technology) (pp. 95–150). Springer.

Chapter 37
Classification: Non-beam Solid Deposition

The classification of non-beam solid deposition is given in Fig. 37.1.

Deposition processes based on arc or heater come as an alternative to beam-based deposition, as they do not use expensive beams and can melt higher melting point materials. But they have to deal with melting-induced problems: porosity, residual stress, cracks, anisotropy, partial evaporation, etc. Processes based on cold spray or friction do not depend on melting and are, therefore, free from such problems. Besides, there are processes based on filament extrusion, which melt polymers and, therefore, need a heater and not an arc or beam.

Friction-based deposition, which makes parts without raising the temperature more than 90% of the melting point of the material, is additive friction stir deposition (AFSD) and friction surfacing-based AM (FSBAM). AFSD is derived from friction stir processing (FSP), while FSBAM is akin to FSP minus stirring.

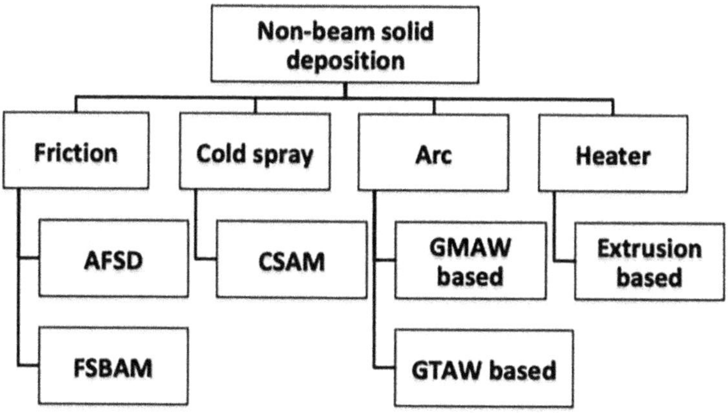

Fig. 37.1 Classification of non-beam solid deposition

S. Kumar, *A Concise Encyclopedia of Additive Manufacturing*,
https://doi.org/10.1007/978-3-031-98394-8_37

Chapter 38
Classification: Powder Bed Process

Bed process is of three types: powder, photopolymer, and slurry (Fig. 38.1), which can be both layerwise and layerless [1].

Powder bed process (PBP) is a generic name for AM processes in which a powder bed (PB) is created and selectively joined to make a part. PB means a thin layer of powder spread on a platform (or horizontal surface). Variation in PB or joining brings variation in PBP. For example, if PB is compacted by applying pressure, it is no longer similar to PB created by spreading powders. The compaction creates another process. In other examples of variation, if powders are joined by two types: 1) completely melting using a high-energy beam and 2) interlocking using a binder, then these two types are not the same process.

PBP is found to be of two major types: 1) powder bed fusion (PBF) and 2) powder bed non-fusion (Fig. 38.2). As per the Oxford dictionary, fusion is the process of causing a material or object to melt with intense heat so as to join with another.

In PBF, powders are partially or fully melted to join them. Selective laser sintering (SLS) [2] and selective laser melting (SLM) [3] are laser powder bed fusion [4], while electron beam melting (EBM) [5] is electron beam powder bed fusion [6].

In PB non-fusion, powders are not melted; binders are used to join them, e.g., binder jet 3D printing (BJ3DP) [7] or binder jet AM [8].

S. Kumar, *A Concise Encyclopedia of Additive Manufacturing*,
https://doi.org/10.1007/978-3-031-98394-8_38

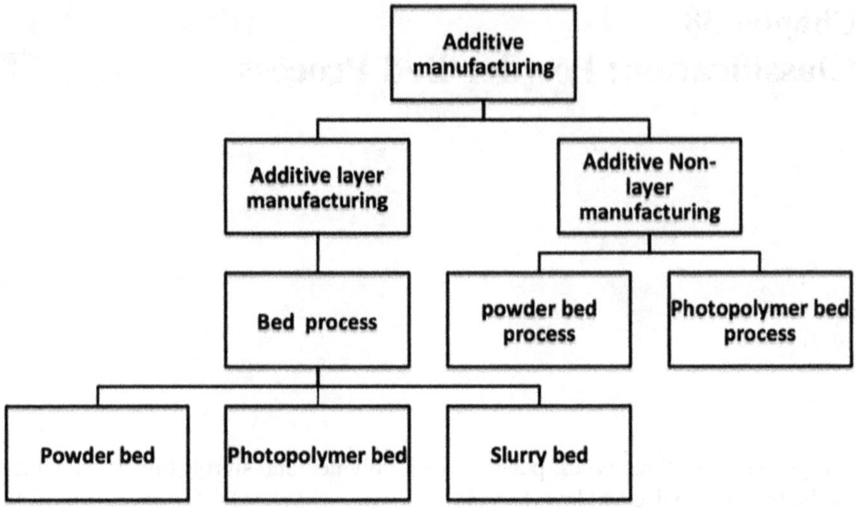

Fig. 38.1 Classification of bed process

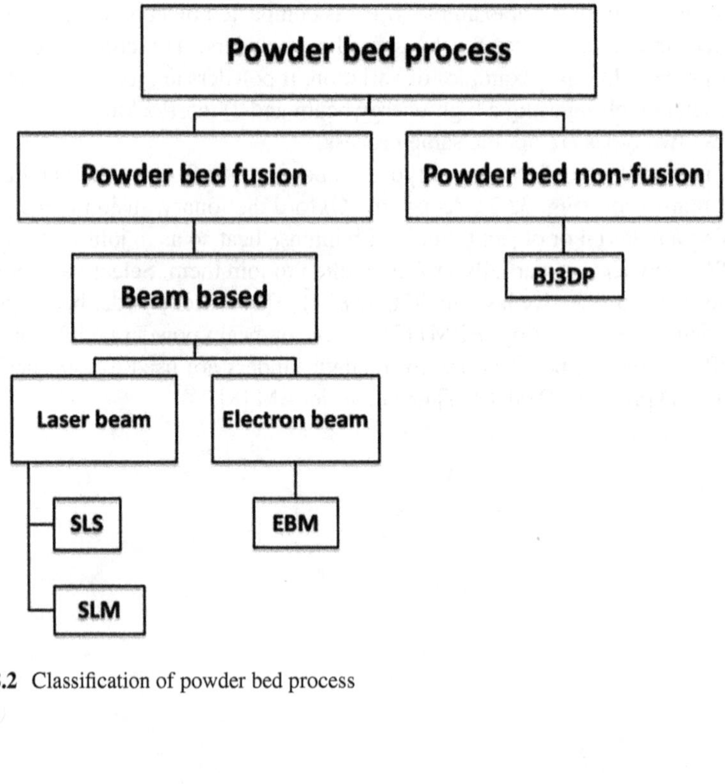

Fig. 38.2 Classification of powder bed process

References

1. Kumar, S. (2022). Bed process. In *Additive manufacturing classification* (Synthesis lectures on engineering, science, and technology) (pp. 43–94). Springer.
2. Yehia, H. M., Hamada, A., Sebaey, T. A., & Abd-Elaziem, W. (2024). Selective laser sintering of polymers: Process parameters, machine learning approaches, and future directions. *Journal Of Manufacturing And Materials Processing, 8*(5), 197.
3. Aufa, A. N., Hassan, M. Z., Ismail, Z., et al. (2024). Current trends in additive manufacturing of selective laser melting for biomedical implant applications. *Journal of Materials Research and Technology, 31*, 213–243.
4. De Leon, E., Riensche, A., Bevans, B. D., et al. (2025). A review of modeling, simulation, and process qualification of additively manufactured metal components via the laser powder bed fusion method. *Journal Of Manufacturing And Materials Processing, 9*(1), 22.
5. Kolamroudi, M. K., Asmael, M., Ilkan, M., & Kordani, N. (2021). Developments on electron beam melting (EBM) of Ti–6Al–4V: A review. *Transactions of the Indian Institute of Metals, 74*, 783–790.
6. Bhatt, B., Martucci, A., Biamino, S., et al. (2025). Current trends in electron beam and laser powder bed fusion additive manufacturing of copper alloys: Composition, defects, properties, and challenges. In *Materials & design* (p. 113742).
7. Chen, Z., Wan, B., Liu, J., et al. (2024). Sintering densification mechanism of binder jet 3D printing 316L stainless steel parts via dimensional compensation technology. *Journal of Materials Research and Technology, 33*, 3296–3307.
8. Paudel, B. J., & To, A. C. (2025). Spatial green density variation and its effect on distortion prediction in binder jet additive manufacturing. *Additive Manufacturing, 98*, 104640.

Chapter 39
Classification: Solid Deposition

Solid deposition process is a family of AM processes in which solid materials are used as a feedstock and brought as a solid to a point on or near a platform before they are converted into a desired shape. Feedstocks in the form of powder, wire, and rod are used. To transform the feedstock, energy is required, which comes from a beam (laser, electron, plasma), arc, heater, friction, or motion.

Examples are

- Laser beam (l-beam) based: laser engineered net shaping (LENS) [1], laser powder deposition, laser wire deposition, laser-based AM.
- Electron beam (e-beam)-based: wire e-beam AM (WEAM).
- Arc-based: wire arc AM (WAAM).
- Plasma-based: plasma welding AM.
- Friction-based: additive friction stir deposition (AFSD).
- Heater-based: fused deposition modeling (FDM), fused pellet modeling (FPM).
- Motion-based: cold spray AM (CSAM).

The process is realized through both layerwise and layerless. Layerless FDM is layerless, while all other examples of the process are layerwise (Fig. 39.1).

The process can be divided into two broad categories based on beam and non-beam. The former can be divided into three categories based on l-beam, e-beam, and plasma beam, while the latter can be divided into four categories based on friction, cold spray, arc welding, and heater.

S. Kumar, *A Concise Encyclopedia of Additive Manufacturing*,
https://doi.org/10.1007/978-3-031-98394-8_39

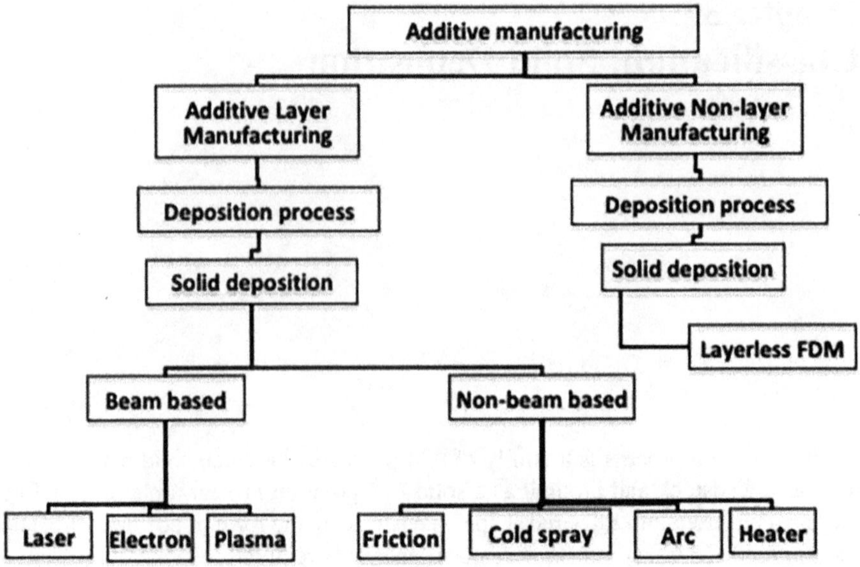

Fig. 39.1 Classification of solid deposition

Reference

1. Shi, W., Ren, Z., He, W., et al. (2021). A technique combining laser spot thermography and neural network for surface crack detection in laser engineered net shaping. *Optics and Lasers in Engineering, 138*, 106431.

Chapter 40
Classification of Deposition Process and Mechanism

On the basis of states of matter, deposition process can be divided into four types: solid, liquid, air, and ion (Fig. 40.1a). This division guarantees that no process will be left out from a classification, as every process deals with a feedstock that has a state of matter.

What if a process deals with more than one state of matter? For example, extrusion-based deposition process works with polymer that has a solid state. Besides, the process works with ink and liquid metal that have liquid states. The process, therefore, will come under both solid and liquid deposition, resulting in it not having a unique place in the classification (Fig. 40.2a). Thus, using the state of the matter first and then using the process type does not lead to a unique place. Why not classify the other way round—process type first and then the state of matter? How to know a process type? To know it, it needs to know what the deposition mechanism for a deposition process is.

There are six types of deposition mechanisms (melting, extrusion, deformation, jetting, spraying, and field force), which classify deposition process into the following types: 1) melting-based, e.g., LENS [1], WAAM [2], 2) extrusion-based, e.g., FDM [3], 3) deformation-based, e.g., AFSD [4], 4) jetting-based, e.g., ink jet printing (IJP) [5], 5) spraying-based, e.g., CSAM [6], and 6) field-based, e.g., electrolytic or colloidal solution-based AM [7] (Fig. 40.1b).

What if a process is classified based on a deposition mechanism first and then based on a state of matter? Then there will be no problems related to the classification of extrusion-based deposition, but another problem will start. A known process may be working with more than one mechanism, and the process does not have different names due to different mechanisms. For example, IJP works with two mechanisms: jetting and extrusion. For both mechanisms, the name of the process is IJP. Therefore, IJP does not have a unique place but two places in the classification (Fig. 40.2b).

Thus, there is a problem in the classification (Fig. 40.2b), but this problem is not because a process works with two mechanisms when the name of the process is

S. Kumar, *A Concise Encyclopedia of Additive Manufacturing*,
https://doi.org/10.1007/978-3-031-98394-8_40

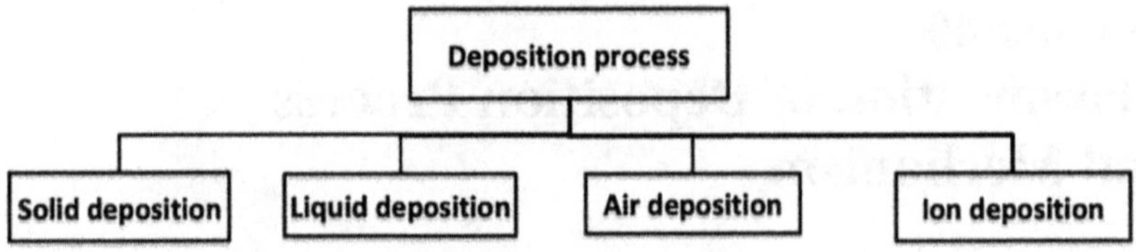

(a) Classification due to states of matter

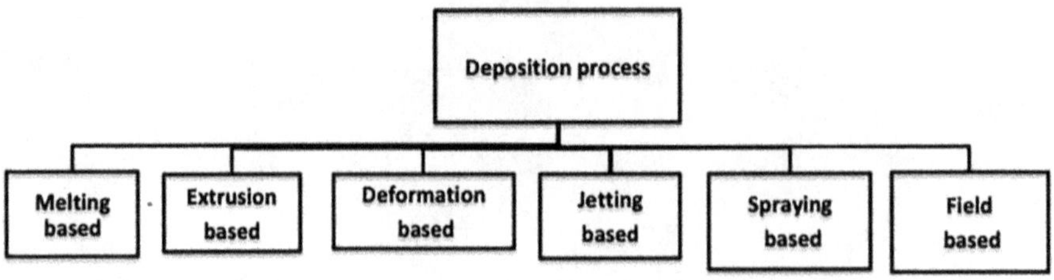

(b) Classification due to deposition mechanisms

Fig. 40.1 Classification on the basis of: (**a**) state of matter, (**b**) deposition mechanism

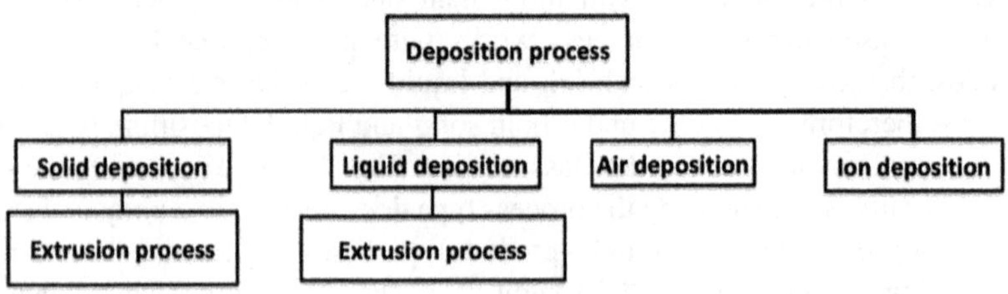

(a) Same process under different states of matter

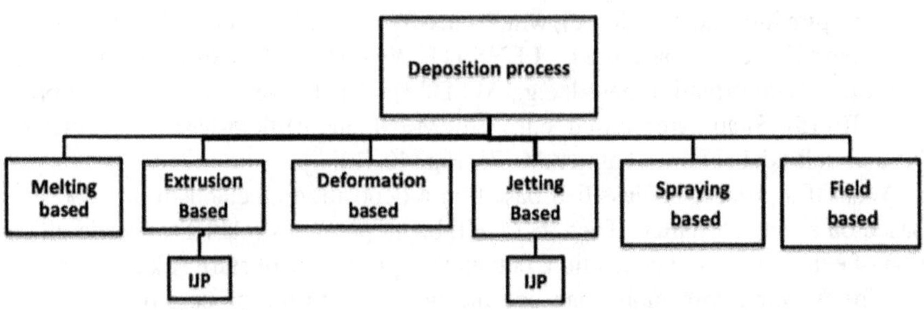

(b) Same process under different mechanisms

Fig. 40.2 Problems in classification made based on: (**a**) state of matter, (**b**) deposition mechanism

changed. But because two processes are working with different mechanisms, the names of the two different processes are not different. There would not be any problem if IJP working due to extrusion had a name, IJP-extrusion, and IJP working due to jetting had a name, IJP-jetting.

Therefore, the problem in the classification (Fig. 40.2b) is not because the classification is not able to provide a unique place for a process, but because the process used in the classification does not have a unique name. Consequently, if the classification based on mechanism (Fig. 40.1b) does not provide a unique place for a process, it is because AM is not free from arbitrary process names.

Though there is again a problem in the classification (Fig. 40.1b), it is based on the mechanism. What if the number of mechanisms increases in the future? Then, the classification needs to be modified. It shows that the classification (Fig. 40.1b) is only for the present time, while the classification (Fig. 40.1a), though having problems, is not restricted only to the present time, as the number of states of matter will not increase in the future.

Mechanisms are explained below:

1 Melting

In this type, the feedstock is melted. The melting is a precondition for deposition. It comprises all processes that are beam-based, such as melting due to a laser beam [8], an electron beam [9], or melting due to an arc [10]. The feedstock can be powder [8], wire [8–10], rod, or solid of any other shape.

Since melting is possible with only one state of matter, i.e., solid, the classification can work in either way. If a deposition process is first classified based on states of matter and then the mechanism, this will give a unique place for the process. Similarly, if the process is first classified based on the mechanism and then the states of matter, it will also work.

2 Extrusion

As per the Oxford dictionary, the meaning of extrusion is "the act of forcing or pushing something out of something." A material will be forced or pushed out of something (die, nozzle, or orifice). If the material is unable to come out on its own, it needs to be plastically deformed to flow out.

Thus, anything (solid metal, solid polymer, viscous liquid) that can be plastically deformed in a nozzle to get a shape will fulfill the condition of this mechanism.

In the case of polymer, it is heated so it acquires deformability and then forced by a piston so it flows out of a nozzle. If the material is not pure polymer but a mixture of polymer and some other material, the mixture must have deformability to be extruded. The mixture will get deformability if it has enough polymers. This is the polymer that gives extrudability to the polymer-based material. In AM, extrusion is known mainly for its relation with solid polymer because of the widespread use of fused deposition modeling (FDM) systems [11].

Extrusion can be achieved in a metallic alloy by heating it at a higher temperature and then forcing it through a die [12], while it can be achieved in ceramic by mixing with water and then extruding [13]. Thus, the extrusion mechanism can be applied for both states of matter, i.e., solid and liquid.

3 Deformation

In this type, e.g., Additive friction stir deposition (AFSD), the feedstock needs to be plastically deformed [4, 14]. If feedstock (wire, rod, powder) is placed on a substrate and is plastically deformed so that it will attach to the substrate, the process will come under this mechanism.

Since deformation works with only one state of matter, i.e., solid, the classification can work in either way—first process, then state of matter, or vice versa.

4 Comparison between Deformation and Extrusion

Plastic deformation of a feedstock and then its deposition come under extrusion. But the deformation takes place within a nozzle, giving rise to a shape to the deformed material.

Deformation-based deposition does not rely on deformation before the material reaches a substrate. Thus, there is no effort to deform the feedstock and give it a shape before it reaches the substrate. While in extrusion-based deposition, there is no effort to deform the extrudate after it is deposited.

In extrusion-based deposition, the plastic deformation is meant to get a shape of the extrudate that, in turn, gives a 3D structure. If there is no optimum extrusion, it will affect the accuracy of the structure. Getting a well-defined extrudate is the primary aim of this mechanism. The property of the structure is important, but the property is not achieved at the cost of its accuracy. For example, if there is a possibility to obtain an extrudate that will not be able to give a well-defined structure but will give good properties, this is not preferred in this mechanism.

In deformation-based deposition, the plastic deformation (or extrusion on a substrate) is meant to achieve a mechanical property. This is the primary aim of this mechanism. The deformation is increased or decreased depending upon its effect on the dynamic recrystallization and the grain size, but the aim of the deformation is not to get a particular shape of an extrudate with an aim to fulfill the accuracy of a certain 3D structure (as happened in extrusion-based deposition). This mechanism cannot ignore mechanical properties to get a well-defined structure.

AFSD cannot expect to make a delicate structure by deforming, as the act of deforming requires a certain minimum strength from a previous structure. If the previous structure is delicate, it will yield before further deformation on it will be complete. This requirement means it cannot compete with FDM that does not rely on deformation to carry out the building of a structure. FDM relies on laying out extrudate on previous structures to carry forward the ongoing build. This act of laying does not demand as much strength from the previous structure as the act of deforming demands.

Though placing an extrudate in FDM may not be free from deformation, the act of placing may cause deformation on a previous structure and on the extrudate itself. It will depend upon the temperature during placement, the cooling ability of the previous structure, the time of deposition, and the material. And this deformation needs to be taken into account for attaining the accuracy. The extent of the deformation may depend upon a case-by-case basis, but the deformation is not required as much as in AFSD. If there is not even this deformation, there will not be a source of any inaccuracy in FDM due to the deformation. Comparing FDM with AFSD, as far as the source of inaccuracy due to the deformation is concerned, FDM is better than AFSD, as AFSD cannot avoid deformation.

Though deformation during laying an extrudate in FDM can be a source of inaccuracy, the deformation during the same step in AFSD is a reason for building a structure. It is not an aim in AFSD to build a structure whose accuracy is guaranteed only when no deformation is taking place—the same cannot be an aim in FDM.

Though there can be a process that relies on both mechanisms (deformation and extrusion), i.e., deformation within a nozzle to get an extrudate, and deformation of this extrudate on the substrate to get bonding [15].

5 Jetting and Spraying

As per the Oxford dictionary, jetting means quick expulsion of a stream of liquid or gas from an opening, while spraying means sending very small drops of liquid through air.

Spraying requires the help of a gas to send droplets. The gas has no other purpose than to transport drops from one point to another. It is the droplet that has utility at the site. It is different from jetting, where a liquid drop can go from one point to another point without the help of a carrier.

Thus, ink jetting-based AM implies that whatever ink is jetted from a nozzle or printhead becomes part of the ongoing build [16], while spraying-based AM implies that whatever comes out from the nozzle will not necessarily become part of the ongoing build [17]. Thus, cold spray-based AM means that only particles and not the carrier gas will become constituents of the build [18].

Spraying differs from jetting, as the former implies spreading out liquid drops or particles—this spreading will not help get deposition in AM. If the spraying-based process is used not for spreading or diverging but for converging the spray, then spraying in this case will not be different from jetting when the two are compared in terms of the spreading of particles or droplets. Thus, spraying, which sends a stream of droplets, is not doing anything different from jetting, which also sends a stream of droplets, except that the former does it with the help of a carrier gas while the latter does it without it. This is the reason why aerosol jetting or aerosol jet printing, which is an aerosol spraying technique, uses "jetting" instead of "spraying."

Since cold spray-based AM relies on spraying but not on spreading, it will not be wrong if, taking the inspiration from the name of aerosol jetting, the process can instead be "cold jet-based AM." Similarly, aerosol jetting relies on spraying but not on spreading; it will not be wrong if, taking the inspiration from cold spray-based AM, the process can instead be 'aerosol spraying.'

Irrespective of the name, there are two mechanisms: one mechanism that does not rely on carrier gas to send droplets, while the other relies on the carrier gas. Consequently, based on these two mechanisms, there will be two deposition-based AM.

Jetting-based deposition, such as inkjet printing (IJP), depends on the quick expulsion of droplets. There are methods to accomplish the quick expulsion—it can be by creating a sudden pressure or temperature in a printhead so the droplets will not be able to hold themselves at an orifice and will leave as per impulse on it.

The quick expulsion is necessary as it provides a droplet the ability to strike a target and not drift away from the intended direction. This ensures repeatability in action by droplets as they will have sufficient momentum to furnish repeatable motion. Thus, quick expulsion will provide precision along with a high fabrication speed.

What if the expulsion is not quick because quick expulsion is not required? Is the precision of deposition not required? Consider a liquid in a syringe, which is expelled dropwise. Pressure on the liquid is required so that the liquid will leave the syringe opening drop by drop. This pressure is responsible for creating a stream of drops. But this drop, which is part of such a high-speed stream, is not required, and therefore, such pressure is not required. But a drop is required. A drop will not leave the syringe on its own, and therefore, a minimum pressure is necessary so the drop will detach from the syringe opening. This type of minimum pressure does not create a condition that comes under jetting.

Thus, without jetting, the liquid drop reaches the desired site. This movement of drops also does not constitute an action that is called spraying. Thus, there is a nozzle that drops a liquid, and a product is formed—this process will not come under any deposition mechanism, as the mentioned mechanisms are not sufficient to account for it.

Therefore, this type of process is loosely categorized as extrusion deposition. However, it is not illogical if this type of process can be classified as an extrusion deposition process that has a special condition that there is no extrusion taking place. Similarly, a syringe-based process where a drop is coming from the spraying without applying much pressure can be classified as a jetting deposition that has the special condition that there is no jetting taking place.

6 Field Force

Some processes do not require a nozzle or channel to send material for deposition. Material can move from one point to another if there is a force acting on it, e.g., an electrochemical force [19]. To take into account the ionic or charge movement, a mechanism is required, which is the application of a field force. This force can again be increased, e.g., by applying magnetic fields [20].

References

1. Gzik-Zroska, B., Joszko, K., Piątek, A., et al. (2025). The influence of hot isostatic pressing on the mechanical properties of Ti-6Al-4V samples printed using the LENS method. *Materials, 18*(3), 612.
2. Mishra, S. K., Sreeraj, P. R, , Sajin Jose Silva., & Upadhyay, R. K. (2025). Wire arc additive manufacturing: Materials, processes and its constraints. Materials and manufacturing processes, 1–18.
3. Raza, U., Ahmed, A., Waheed, S., et al. (2025). Recent advancements in fused deposition modeling. *Polymers for Advanced Technologies, 36*(1), e70028.
4. Huang, Y., Liu, Q., Zhang, K., et al. (2025). Investigation of three-dimensional forces during additive friction stir deposition—How could force signals reveal the deposition quality? *International Journal of Machine Tools and Manufacture, 204*, 104234.
5. Chang, W., Oh, K. H., Kang, S. M., et al. (2025). Nanostructural engineering of porous perovskite cathodes for high-performance protonic ceramic fuel-cells using piezoelectric inkjet-printing. *Additive Manufacturing, 97*, 104606.
6. Sirvent, P., Lozano, A., Garrido-Maneiro, M. A., et al. (2025). Surface topography analysis in cold spray additive manufacturing. *Precision Engineering, 92*, 207–218.
7. Rad, S. M. M., Hajialimohammadi, A., & Abedini, V. (2025). Manufacturing of copper-chromium composite using electrochemical additive manufacturing method: A screening study. *Progress in Additive Manufacturing, 10*(2), 1435–1446.
8. Peng, H., Xie, S., Niu, P., et al. (2021). Additive manufacturing of Al0. 3CoCrFeNi high-entropy alloy by powder feeding laser melting deposition. *Journal of Alloys and Compounds, 862*, 158286.
9. Pixner, F., Warchomicka, F., Peter, P., et al. (2020). Wire-based additive manufacturing of Ti-6Al-4V using electron beam technique. *Materials, 13*(15), 3310.
10. Zhang, G., Xu, Z., Jiang, F., et al. (2025). Metal transfer and forming behavior of bypass-coupled variable polarity plasma arc additive manufacturing. *Journal of Manufacturing Processes, 133*, 510–523.
11. Rajan, K., Samykano, M., Kadirgama, K., et al. (2022). Fused deposition modeling: Process, materials, parameters, properties, and applications. *The International Journal of Advanced Manufacturing Technology, 120*(3), 1531–1570.
12. Grong, O., Sandnes, L., & Berto, F. (2019). A status report on the hybrid metal extrusion & bonding (HYB) process and its applications. *Material Design & Processing Communications, 1*(2).
13. Blackburn, S., & Szymiczek, M. (2021). Extrusion. In M. Pomeroy (Ed.), *Encyclopedia of materials: Technical ceramics and glasses* (pp. 162–178). Elsevier.
14. Yu, H. Z., & Mishra, R. S. (2021). Additive friction stir deposition: A deformation processing route to metal additive manufacturing. *Materials Research Letters, 9*(2), 71–83.
15. Blindheim, J., Grong, O., Welo, T., & Steinert, M. (2020). On the mechanical integrity of AA6082 3D structures deposited by hybrid metal extrusion & bonding additive manufacturing. *Journal of Materials Processing Technology, 282*, 116684.

16. Yi, H., Guo, X., Chang, F., et al. (2025). Ink-jetting-based conformal additive manufacturing: Advantages, opportunities, and challenges. *International Journal of Extreme Manufacturing, 7*, 032002.
17. Falco, R., & Bagherifard, S. (2025). Cold spray additive manufacturing: A review of shape control challenges and solutions. *Journal of Thermal Spray Technology*, 1–19.
18. Tregenza, O., Hutasoit, N., Palanisamy, S., & Hulston, C. (2025). Air-based cold spray: An advanced additive manufacturing technique for functional and structural applications. *The International Journal of Advanced Manufacturing Technology, 136*, 1–38.
19. Hengsteler, J., Kanes, K. A., Khasanova, L., & Momotenko, D. (2023). Beginner's guide to micro-and nanoscale electrochemical additive manufacturing. *Annual Review of Analytical Chemistry, 16*(1), 71–91.
20. Li, L., Ma, B., Cao, J., et al. (2024). Magnetic field-assisted electrochemical additive manufacturing of nickel structure: Growth mechanism and microstructural evolution. *Materials Today Communications, 40*, 110030.

Chapter 41
CNC Accumulation

In this process, a substrate is immersed in photopolymer, and a part is fabricated inside the liquid (Fig. 41.1). To start curing the liquid from the substrate, a light beam needs to reach the substrate without solidifying the liquid on the way. It is possible if the beam is guided through an optical fiber. The beam will then solidify the liquid between the fiber tip and the substrate, which gets attached to the substrate and the fiber tip. Teflon film on the tip is used to decrease the adhesion between the solidified material and the tip, causing the material to attach solely to the substrate. An increment in the attached material happens as per the requirement rather than as per the pre-set layerwise values, thus making the process additive layerless manufacturing.

The attached material grows in the direction of the tip. When the tip moves, the solidified material will follow it, i.e., controlling the movement will create a desired structure. For the tip to be able to move, it needs to be stiff. The tip is supported inside a rigid plastic body and is called a tool. Movement of the tool creates a structure and can repair parts. The movement of this tool is different from the movement of a tool in powder bed fusion (PBF).

While creating a structure, the movement of the tool is not confined to a layer or a plane, which can furnish many build directions, and, therefore, the process is relatively free from staircase effects. This movement creates structures on many sides of a substrate, at least three sides of a rectangular substrate, which gives the process the ability to repair a part having damages on many sides without changing the orientation. While that movement creates structures only on the upper side of a substrate. This ability gives the process an edge over PBF. Though this ability does not give an edge over the deposition process, as these processes can facilitate multi-orientation deposition [1].

The movement of the tool is facilitated by attaching it to a CNC machine, which is meant to add materials. A tool comprising a large beam diameter and high laser power can add more material and speed up the fabrication, while a tool comprising

S. Kumar, *A Concise Encyclopedia of Additive Manufacturing*, https://doi.org/10.1007/978-3-031-98394-8_41

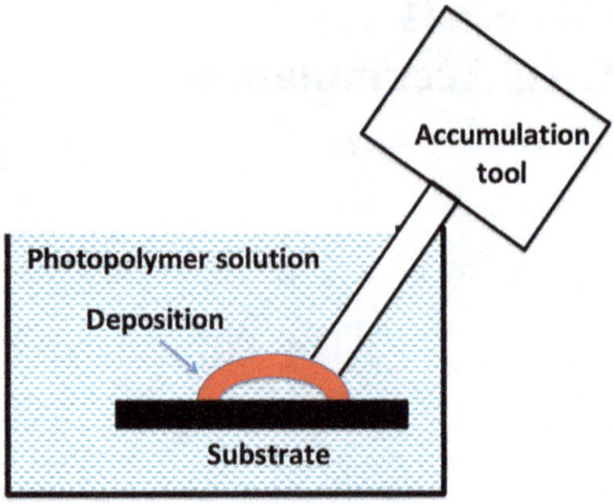

Fig. 41.1 Schematic diagram of CNC accumulation

a small beam diameter will add a small amount, causing an increase in resolution [2]. A combination of tools can thus enable the fabrication of various geometries.

References

1. Yang, Y., Fuh, J. Y., Loh, H. T., & Wong, Y. S. (2003). Multi-orientational deposition to minimize support in the layered manufacturing process. *Journal of Manufacturing Systems, 22*(2), 116–129.
2. Pan, Y., Zhou, C., Chen, Y., & Partanen, J. (2014). Multitool and multi-axis computer numerically controlled accumulation for fabricating conformal features on curved surfaces. *Journal of Manufacturing Science and Engineering, 136*(031007), 1–14.

Chapter 42
Cold Spray

If a high-speed gas is available and some powder is injected into it, this has the potential to be cold spray. High-speed gas coming from a pressurized cylinder or centralized pressurized gas source fulfills one of the basic conditions for cold spray generation. The maximum pressure of the gas thus obtained is limited, which limits the highest speed obtained therefrom. The speed obtained may not be sufficient. If gas is heated on its way, the pressure and the speed obtained can be further increased. The speed obtained again may not be sufficient. This brings up the question of whether there is a method to increase the speed of the gas without using increasingly powerful cylinders and hotter heaters.

If gas and/or gas plus accelerated powders exit through a cylindrical pipe to the substrate, the speed will remain the same because the pipe will not increase the speed. If the diameter of the pipe at the end of the cylinder is constricted, the speed will increase if the gas is not compressible. Since there is no gas generated inside a pipe, the amount of gas entering the pipe per unit time must be equal to that leaving per unit time.

The amount of gas that can be accommodated in a pipe depends upon its diameter—if the diameter of the pipe is decreased, it can accommodate less gas. If the flow rate does not increase, the amount of gas entering per unit time will be more than the amount of gas leaving per unit time, and therefore the rate will increase. Therefore, by taking a cylindrical pipe having a constricted diameter at one end or taking a conical nozzle, the gas speed can be increased without resorting to extra cylinders or heaters. But the speed thus achieved will still be limited. Moreover, the gas will diverge, which will increase the diameter of the deposited tracks, which will not be suitable for making small features.

If gas is compressible, the amount of gas that can be accommodated in a pipe will depend not only on the diameter of the pipe but also on the compressibility of the gas. Thus, for a compressible gas, the same advantage with a conical nozzle is not obtained. Therefore, an increase in the speed of the compressible gas may not be as high as an increase in the speed of an incompressible gas.

S. Kumar, *A Concise Encyclopedia of Additive Manufacturing*,
https://doi.org/10.1007/978-3-031-98394-8_42

However, if an increase in speed is sufficient enough to increase the speed of a compressible gas to more than or equal to the speed of sound, this high-speed gas (supersonic gas) increases the compressibility of the gas. Since compressibility increases, the gas no longer follows the principle of an incompressible gas. The principle is that when an incompressible gas passes through a convergent nozzle, its speed increases, and when the same gas passes through a divergent nozzle, its speed decreases.

But the gas follows the principle of a very compressible gas. The principle is that when a compressible gas passes through a convergent nozzle, its speed decreases, and when the same gas passes through a divergent nozzle, its speed increases. At such supersonic speed, resulting in high compressibility, the speed of the compressible gas further increases when it enters a divergent nozzle or when it exits from a convergent section to a divergent section of a convergent-divergent nozzle.

Thus, a gas coming from a compressed cylinder gets its speed increased in the convergent section of a convergent-divergent nozzle or de Laval nozzle because the gas is slightly compressible (behaving like an incompressible gas), and the speed gets further increased in the divergent section of the de Laval nozzle because the gas is no longer just slightly compressible but has become highly compressible because of a change in the behavior of the gas at such high speed [1].

Thus, using the right type of nozzle is another method to increase the speed of a gas. When a particle is injected into this gas, the particle is accelerated at high speed and can be used for cold spray AM.

Reference

1. Papyrin, A., Kosarev, V., Klinkov, S., et al. (2006). *Cold spray technology*. Elsevier.

Chapter 43
Cold Spray AM

Cold spray AM (CSAM) is a process in which powder particles are accelerated at high speed to deposit on a substrate to make a structure (Fig. 43.1) [1]. Cold spray means powder particles are not melted, and thus, there is no provision for utilizing high-temperature sources to cause melting. On the contrary, in thermal spray, partially or fully molten drops are used, and thus, high-temperature sources such as arc, direct current plasma, radio frequency plasma, and flame are required to melt wire, powder, or rod to make molten drops.

The American Welding Society classifies cold spray as part of thermal spray due to functional similarities. In the cold spray, the temperature is increased to 1000 °C, but its purpose is to increase the speed of the gas. It also does not mean the cold spray has nothing to do with the melting—partial melting of powder may happen when the speed of the powder is high and the substrate is at a higher temperature—partial melting will provide metallurgical bonding between powders and the substrate or a deposited layer. Thus, the cold spray deposition is not so cold. It, like a thermal spray process, is not hot enough to be able to melt powders before their deposition. Partial melting or liquid-phase sintering also happens when post-processing heat treatment is used to improve properties [2].

Cold spray relies on the speed of particles to create a bond between the particles and a substrate. Kinetic energy of a particle provides binding energy, while in thermal spray, kinetic plus thermal energy provides binding energy. If the speed is high, the particle will deform, which will break its oxidized layer, exposing its surface to the substrate for a metallurgical bond to occur. The particle can also get interlocked in the surface roughness or nonuniformity, giving rise to a mechanical bond, and can be trapped inside a microservice of the substrate surface, resulting in a bond.

If the speed is high, but neither the particle nor the substrate is deformable, the particle will rebound, and there will be no bonding. Thus, a combination of high speed and ductility of the material is required. If the substrate is not ductile, by localized heating, its non-ductility can be decreased to some extent. If the particle is not ductile, by combining it with another ductile material, the ductility of the

S. Kumar, *A Concise Encyclopedia of Additive Manufacturing*, https://doi.org/10.1007/978-3-031-98394-8_43

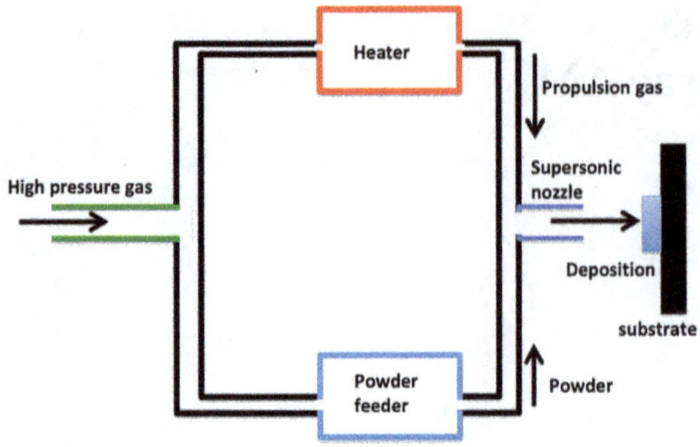

Fig. 43.1 Schematic diagram of CSAM

resulting composite (ductile plus non-ductile material) can be made workable. Thus, a composite powder made from the ductile metal and the non-ductile ceramic can be deposited where the metal will give rise to bonding while the ceramic will impart strength to deposited layers.

A powder that gets deposited without being melted gives certain advantages [3] over a powder that is partially or fully molten during deposition. These advantages are the absence of.

1. Phase transition of materials.
2. Residual stress caused by solidification.
3. Need for heat management.
4. Grain growth, thus the nanostructure of the material can be retained.
5. The possibility of deleterious intermetallic compound formation.
6. Splat.
7. Oxidation in the presence of an oxidizing gas.
8. Need for a processing chamber, therefore, no limitation in part size.
9. Thermal expansion mismatch during multi-material deposition.

Hence, CSAM has certain advantages over powder fusion-based deposition. However, it has disadvantages such as low dimensional accuracy and spatial resolution [4, 5].

References

1. Yin, S., & Lupoi, R. (2021). *Cold spray additive manufacturing* (Springer tracts in additive manufacturing). Springer.
2. Prasad, K., Khalik, M. A., Rashid, R. R., et al. (2025). Repeatability and reproducibility of liquid-phase sintered Al-bronze 3D-printed parts using cold spray additive manufacturing. *Next Materials, 6*, 100312.

3. Gärtner, F., Stoltenhoff, T., Schmidt, T., & Kreye, H. (2006). The cold spray process and its potential for industrial applications. *Journal of Thermal Spray Technology, 15*, 223–232.
4. Wu, H., Xie, X., Liu, M., Verdy, C., et al. (2020). Stable layer-building strategy to enhance cold-spray-based additive manufacturing. *Additive Manufacturing, 35*, 101356.
5. Falco, R., & Bagherifard, S. (2025). Cold spray additive manufacturing: A review of shape control challenges and solutions. *Journal of Thermal Spray Technology, 34*, 1023–1041.

Chapter 44
Complexity

Complexity is generally given by the complexity of a geometry [1, 2]. But this definition of complexity does not work in AM [3]. Therefore, complexity must be defined as equal to fabrication problems. If there is no problem with fabricating a complex part, the part is not complex. A geometry is simple or complex—it depends upon whether fabricating it is easy or difficult. If a part is complex in other techniques, it does not imply that the part is also complex in AM when there is no problem in fabricating it in AM. Similarly, if a part is easy to make in AM, it does not imply that the part will not be complex in other techniques.

References

1. Touzé, S., Rauch, M., & Hascoët, J. Y. (2024). Methodology for complexity and cost comparison between subtractive and additive manufacturing processes. *Journal of Intelligent Manufacturing, 35*(2), 555–574.
2. Amor, S. B., Tahan, A., & Louhichi, B. (2019, July). Proposition of a geometric complexity model for additive manufacturing process based on CAD. In *2019 23rd international conference information visualisation (IV)* (pp. 442–448). IEEE.
3. Kumar, S. (2023). Advantage and Disadvantage. In *Additive manufacturing advantage, synthesis lectures on engineering, science, and technology* (pp. 1–60). Springer.

S. Kumar, *A Concise Encyclopedia of Additive Manufacturing*,
https://doi.org/10.1007/978-3-031-98394-8_44

Chapter 45
Complexity in Products

A simple product has a simple design, e.g., a hollow cylinder, while a complex product has a complex design, e.g., a turbine blade.

A hollow cylinder has a simple design; therefore, it can be fabricated easily. Vice versa, if the cylinder can be fabricated easily, it is a confirmation that its design is simple for AM. To verify its fabrication, it can be checked how an extrusion deposition system will fabricate it. The system's nozzle will deposit a material by moving in a circle and then repeating the step by moving up—this is simple, requiring simple, predictable data. The data will consist of the diameter of the circle and the height of each increment—both values remain the same for each layer. Therefore, there is no need to make a design; the operator can visualize it without the help of software. Again, there is no need for software to convert the design into machine-readable instructions; the operator can provide the instructions by inputting the data.

What if the operator is asked to make a turbine blade? The operator cannot visualize the design in detail because the design has many channels and curves that change in dimension with each layer. These will create enormous data that is difficult to operate. Thus, when design becomes complex, the difficulty for the operator increases. This is the difference between a simple and a complex product. The difficulty faced by an operator is a measure of the complexity of the product. If the operator faces more difficulty, the product is more complex.

The difficulty can also increase when the material is not compatible, giving inadequate bonding, or the nozzle does not move accurately, giving inaccurate dimensions. But when the difficulties due to machine or material are overcome, there is still difficulty if the turbine blade, instead of a cylinder, is made. This shows that the difficulty of an operator can be related to an increase in the design complexity. This difficulty is not related to the skill or experience of an operator but is equal to the number of tasks that need to be performed. If the design is simple, tasks are few, i.e., machine-readable instructions are few. When the design becomes complex, the instructions increase.

S. Kumar, *A Concise Encyclopedia of Additive Manufacturing*, https://doi.org/10.1007/978-3-031-98394-8_45

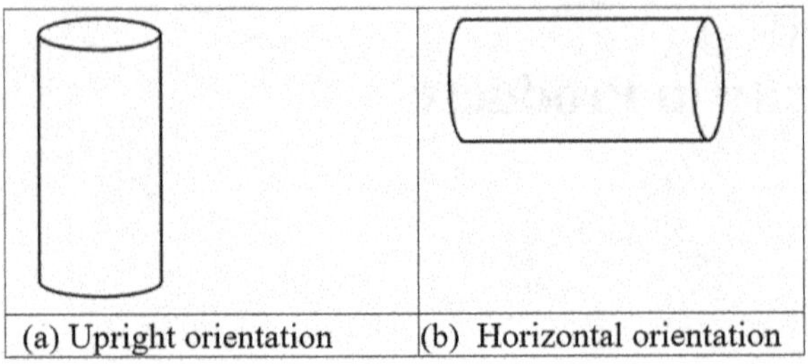

Fig. 45.1 Cylinder orientations. (**a**) Upright orientation. (**b**) Horizontal orientation

What if the same cylinder, which is in an upright orientation, is made in a horizontal orientation (Fig. 45.1)? Then, instructions will be more because the dimensions of the layers will change with a change in height. More instructions will not lead to a complex product, but the same product. This demonstrates that complexity in fabrication does not lead to complexity in design. This demonstration seems to contradict that the complexity of a product must be related to the number of instructions. But this demonstration clarifies that the number of instructions changes with a change in process steps, material, orientation, etc.; therefore, equating the number of instructions to complexity in design will be misleading unless the fabrication is optimized. When the fabrication for a cylinder is optimized, the cylinder will only be selected in an upright orientation to compare with the turbine blade.

Consider a situation where making an upright cylinder is more difficult than making a turbine blade in an AM technique. Then, which one is simple? Since it is the difficulty in fabrication that decides which design is simple or complex, the turbine blade will be a simple product, while the cylinder will be a complex product. Thus, depending on the type of AM techniques, a design or product can be simple in one AM technique while complex in another AM technique. But the reason why it is simple or complex will be the same in all AM techniques.

In the beginning, it is mentioned that a cylinder is simple while a turbine blade is complex. What is its basis? It is based on how a design is seen through conventional manufacturing. A cylinder is simpler because it is made with less difficulty than a turbine blade using conventional manufacturing. Details about the design and perspective are given elsewhere [1].

Reference

1. Kumar, S. (2023). Advantage and Disadvantage. In *Additive manufacturing advantage, synthesis lectures on engineering, science, and technology* (pp. 1–60). Springer.

Chapter 46
Composite

A composite material is made from two material types to achieve properties not achievable by either of the two. A composite works because the load gets transferred from one material to another. For example, the transfer of load from ceramic to the metal in a ceramic composite (metal is in the minority) saves the composite from getting fractured. In a metal matrix composite, the load transferred from the metal to the ceramic saves the composite from being elongated.

A composite gives high strength if there is a sufficient load transfer due to the presence of one material to form a matrix and another to disperse uniformly as an additive. If the amount of any material is low, the combination of the two is still composite but does not give the required strength. Thus, there are two types of composites: one is a material (metal, ceramic) matrix composite, while the other is without 'matrix.'

Materials are added to increase not only strength but other properties as well, such as thermal conductivity, electrical resistivity, heat resistance, etc. In this case, the material combination is composite because of non-mechanical properties.

1 Extension of Concept

In a metal matrix composite, the transfer of load happens within a few or a hundred microns. What if the transfer is desired within a few millimeters or centimeters? If the scale becomes big, the components of a composite need to become big.

Since particles or fibers are not available as components of big sizes, big components need to be made. The composite is made to transfer the load, which is relevant when one component transfers while another receives. It implies one component is weaker than the other. Thus, making big components is not sufficient; they should have different strengths.

S. Kumar, *A Concise Encyclopedia of Additive Manufacturing*,
https://doi.org/10.1007/978-3-031-98394-8_46

The sizes of big, medium, and small are relative to each other. When components are big and a medium-sized product is made (Fig. 46.1a), the shape of the components can change the accuracy of a product shape. Thus, a big component will not work to make all product shapes. New component shapes need to be searched. When components are small, their shapes do not affect the product shape (Fig. 46.1b); the selection of a component shape is then done on the basis of how it affects properties (tensile strength, load sharing, fracture toughness, hardness, etc.). But when

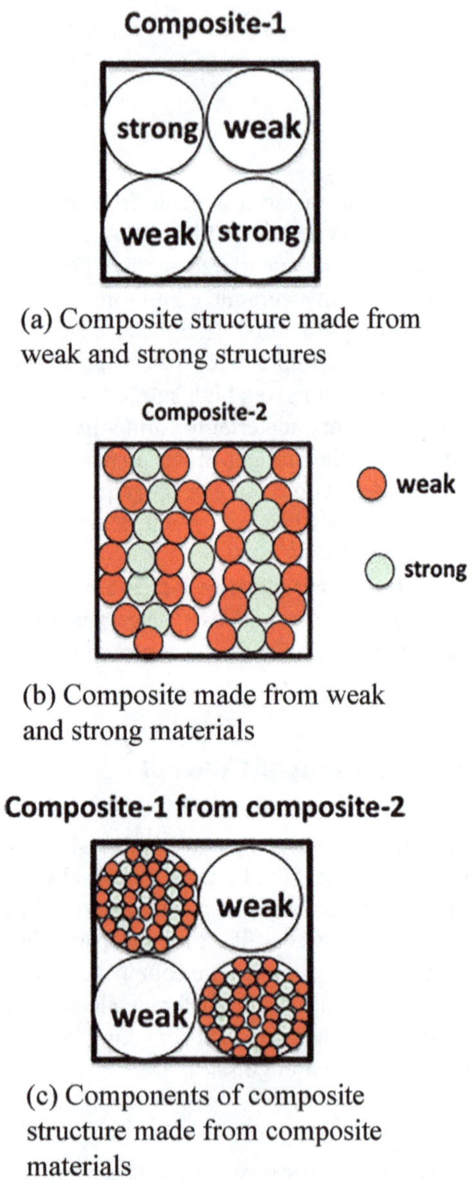

(a) Composite structure made from weak and strong structures

(b) Composite made from weak and strong materials

(c) Components of composite structure made from composite materials

Fig. 46.1 Schematic diagrams of composite-1 and composite-2. (**a**) Composite structure made from weak and strong structures. (**b**) Composite made from weak and strong materials. (**c**) Components of composite structure made from composite materials

the component is big, this cannot be the only basis for the selection; the selection is done on the basis of how it affects both product shapes and properties.

When components are big, the load is transferred at a big scale, and the resulting combination makes a composite having the property of a composite, e.g., lighter and stronger, not achievable by either of the components. When a component is big, it can be made by many small components (Fig. 46.1c). If small components are of two types (weak, strong), they combine to give a composite, which is due to the load transfer on a small scale.

2 Composite-1 and Composite-2

There are thus composites due to the load transfer at a big scale (composite-1) and at a small scale (composite-2). Composite-1 can be made of many materials, including composite-2. Composite-2 leads to many products but can also lead to various shapes that will lead to composite-1 as well (Fig. 46.1).

Composite-1 is made by changing the structure at a big scale to make a composite-1 part, while composite-2 is either procured or developed in an AM system to make a composite-2 part. Examples of composite-1 parts are metallic sandwich structures [1] and continuous fiber reinforced sandwich structures [2], while an example of composite-2 part is metal matrix composite using TiB_2 reinforced Al powder [3].

3 Composite Structure

AM can make shapes of various strengths. For example, if a shape is made up of lattice, the shape can have strengths depending upon the characteristics of the unit cell of the lattice, such as its type, size, density, and thickness. Thus, a structure of defined weakness can be made. Making a dense plate will furnish a strong structure. Alternately changing these weak and strong structures will make a composite structure (Fig. 46.2), e.g., a sandwich structure made of core and panel using either metal [1] or polymer [4] (Fig. 46.3).

Fig. 46.2 Composite structure, property changes in z-direction

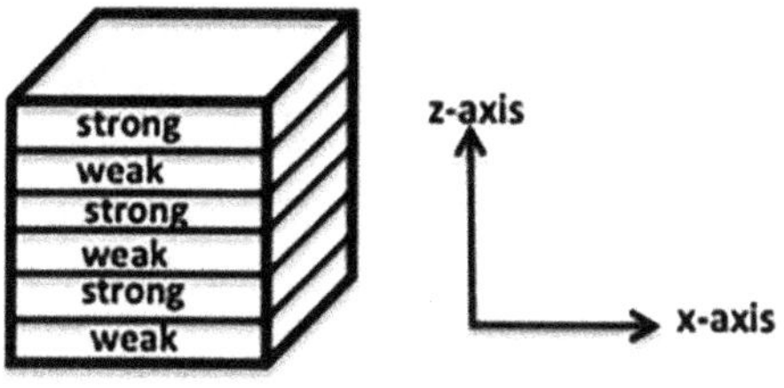

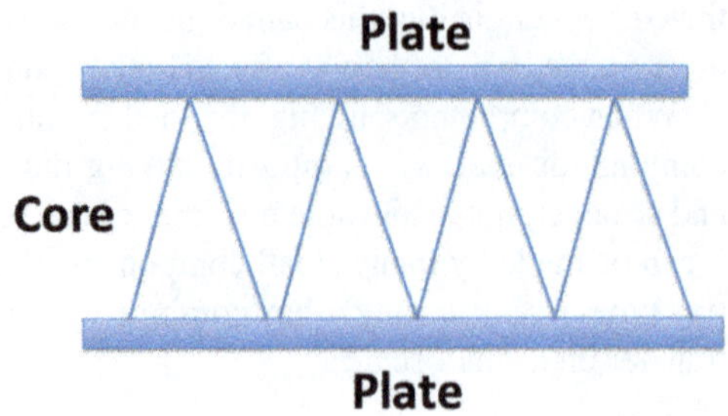

Fig. 46.3 Schematic diagram of a sandwich structure

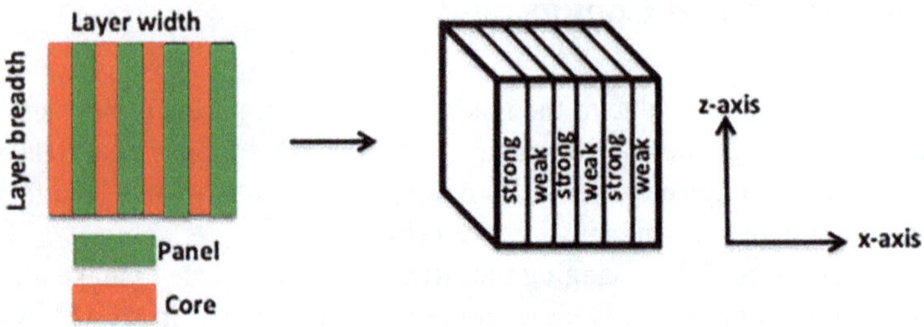

Fig. 46.4 Arrangement of two different structures alternately in the same layer

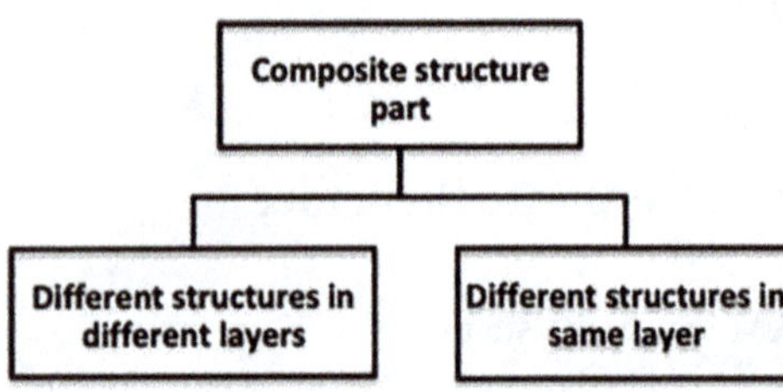

Fig. 46.5 Methods to fabricate a composite structure part

If one or several layers are core and another or several layers are panel, pursuing layerwise fabrication leads to an alternate arrangement of weak-strong structures. Due to the similarity between panel and layer, AM will provide less complexity in process planning.

If the design (Fig. 46.2) is turned by 90° so the direction of the panel progression changes from the z-to the x-direction, a single layer will not be meant for either core or panel. But each layer will comprise some part of each panel and core (Fig. 46.4). This is another method for fabricating a composite structure in which properties change in the x-direction (Fig. 46.5).

References

1. Bühring, J., Nuño, M., & Schröder, K. U. (2021). Additive manufactured sandwich structures: Mechanical characterization and usage potential in small aircraft. *Aerospace Science and Technology, 111*, 106548.
2. Hou, Z., Tian, X., Zhang, J., & Li, D. (2018). 3D printed continuous fiber reinforced composite corrugated structure. *Composite Structures, 184*, 1005–1010.
3. Xie, X., Ma, Y., Chen, C., et al. (2020). Cold spray additive manufacturing of metal matrix composites (MMCs) using a novel nano-TiB2-reinforced 7075Al powder. *Journal of Alloys and Compounds, 819*, 152962.
4. Marchese, S. S., Epasto, G., Crupi, V., & Garbatov, Y. (2025). Feasibility study on additive-manufactured honeycomb sandwich structural solutions for a fast patrol vessel. *Composite Structures, 351*, 118607.

Chapter 47
Composite Fabrication in Deposition Process

1 Fiber-Reinforced Composite

A composite filament having continuous fiber can make continuous fiber-reinforced composite [1]. But the composite filament restricts the fiber-polymer ratio, which limits the properties achieved. Using separate fiber and polymer overcomes the limitation, which can occur in two ways: (1) a single nozzle to deposit separate fiber and polymer filament simultaneously (Fig. 47.1a), and (2) separate nozzles to deposit separate fiber and filament (Fig. 47.1b).

When a single nozzle is used, the deposited polymer is controlled by changing the temperature, feed rate, nozzle speed, and diameter. By using different fibers and polymers, various composites are formed, e.g., a wire-reinforced composite to be used as sensors [2]. A single nozzle compared to two nozzles is simple to plan for depositing two material types (polymer and fiber) but has limitations in changing the fiber-polymer ratio. Besides, the fiber orientation in a layer is similar to the orientation of the polymer bead, as the fiber remains at the core of the bead.

Using two nozzles allows materials to be independently varied and polymer beads to be deposited at any orientation with respect to the deposited fiber. Thus, the effect of various orientations on the property can be realized. Two nozzles can control property by depriving one layer of fiber while the other of polymer. Depositing only fiber on a solidified layer does not work, as the fiber does not bond to the solid layer. Thus, a pure fiber contains glue to facilitate the bonding. Commercial fibers, though named as pure fibers, contain polymers, e.g., epoxy. Thus, a fiber-reinforced composite made by two nozzles is a mixture of polymer due to one nozzle and glue plus fiber due to the other.

S. Kumar, *A Concise Encyclopedia of Additive Manufacturing*,
https://doi.org/10.1007/978-3-031-98394-8_47

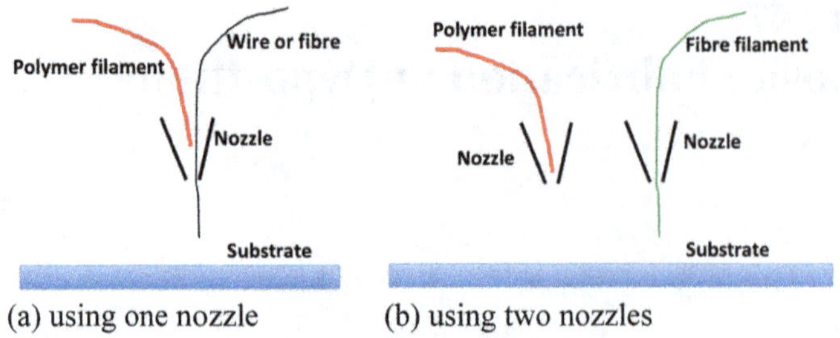

Fig. 47.1 Composites due to non-composite feedstock. (**a**) using one nozzle. (**b**) using two nozzles

Fig. 47.2 Fabrication using composite feedstocks: (**a**) powder or drop, and (**b**) wire or filament. (Heat source not shown)

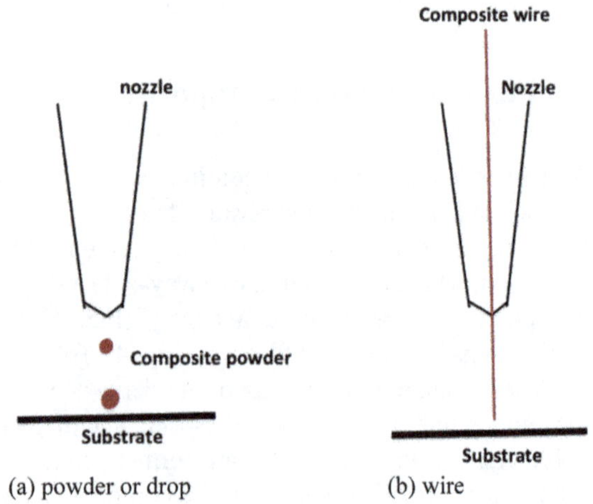

2 Non-fiber Composite

Non-fiber composites are formed using the following:

- Composite feedstock, e.g., composite powder (Fig. 47.2a) and filament [3] or wire (Fig. 47.2b). If the composite wire consists of a reactive metal, the wire will need less external heat to continuously melt the wire, as the exothermic reaction due to the reactive metal will provide heat [4].
- Composite ink [5, 6].
- Two types of non-composite wires (Fig. 47.3a) [7, 8]—Different wire feed rates will give different contributions in a resulting composite.
- Two types of non-composite powders (Fig. 47.3b) [9, 10].
- Non-composites: wire and powder (Fig. 47.3c) [11, 12].
- Deposition on a modified substrate [13].
- Post-processing, such as infiltration and sintering [14].
- Wire filled up with powder [15].

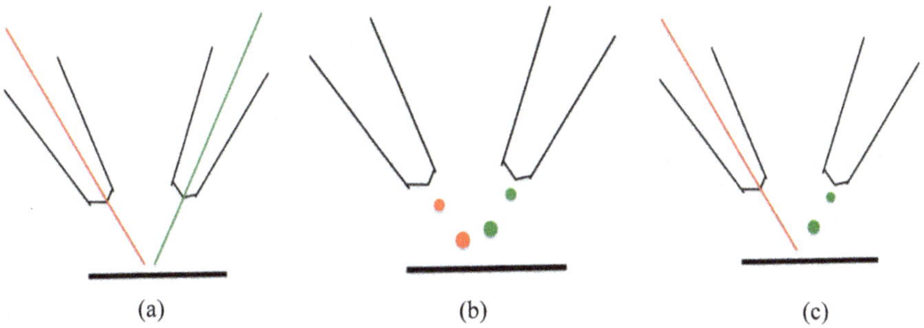

Fig. 47.3 Fabrication using non-composite feedstocks: (**a**) wires or filaments, (**b**) powders or drops, and (**c**) wire and powder

References

1. Deng, K., Khan, M. H. U., Park, S., et al. (2025). Additive manufacturing of continuous carbon fiber/epoxy composites with structured core-shell towpreg: Methods, characterization, and mechanics. *Composites Part B: Engineering, 291*, 112001.
2. Ibrahim, Y., Melenka, G. W., & Kempers, R. (2018). Additive manufacturing of continuous wire polymer composites. *Manufacturing Letters, 16*, 49–51.
3. Silva, L. R., Bertolim, L. V., Stefano, J. S., et al. (2025). New route for the production of lab-made composite filaments based on soybean oil, polylactic acid and carbon black nanoparticles, and its application in the additive manufacturing of electrochemical sensors. *Electrochimica Acta, 513*, 145566.
4. Liu, R., Gao, C., Hou, A., & Wang, S. (2023). Ni/Al foil-based reactive additive manufacturing with fast rate and high energy-efficiency. *Journal of Materials Processing Technology, 321*, 118167.
5. Das, R., Katiyar, N. K., Khan, A., et al. (2025). Enhanced mechanical, tribological, and thermal properties of copper-graphene composites using additive manufacturing. *Journal of Manufacturing Processes, 136*, 380–390.
6. Ameri, B., Taheri-Behrooz, F., & Ghahari, M. (2025). Enhancing scaffold strength through rheological profiling of hydroxyapatite/nano-silica composite ink for advanced additive manufacturing ofbone-mimicking ceramic scaffolds. *Rapid Prototyping Journal, 31*(2), 447–462.
7. Wang, J., Pan, Z., Wang, Y., et al. (2020). Evolution of crystallographic orientation, precipitation, phase transformation and mechanical properties realized by enhancing deposition current for dual-wire arc additive manufactured Ni-rich NiTi alloy. *Additive Manufacturing, 34*, 101240.
8. Cai, X., Dong, B., Yin, X., et al. (2020). Wire arc additive manufacturing of titanium aluminide alloys using two-wire TOP-TIG welding: Processing, microstructures, and mechanical properties. *Additive Manufacturing, 35*, 101344.
9. Syed, W. U. H., Pinkerton, A. J., Liu, Z., & Li, L. (2007). Single-step laser deposition of functionally graded coating by dual 'wire–powder' or 'powder–powder' feeding—A comparative study. *Applied Surface Science, 253*(19), 7926–7931.
10. Bandyopadhyay, A., & Heer, B. (2018). Additive manufacturing of multi-material structures. *Materials Science and Engineering: R: Reports, 129*, 1–16.
11. Teli, M., Klocke, F., Arntz, K., et al. (2018). Study for combined wire + powder laser metal deposition of H11 and niobium. *Procedia Manufacturing, 25*, 426–434.
12. Guo, X., Ma, M., Zhang, S., & Wei, Z. (2025). Wear-resistant Ti/TiCp composite coatings via synchronous powder supplied wire arc additive manufacturing. *Journal of Materials Research and Technology, 34*, 761–775.

13. Zhang, Z., Wan, L., Xie, Y., et al. (2025). Steel grid-reinforced aluminum matrix composite prepared via wire-based friction stir additive manufacturing. *Virtual and Physical Prototyping, 20*(1), e2451114.
14. Hamza, A., Azam, M. U., Raza, A., et al. (2025). Additive manufacturing of ceramic composite cellular structures by spontaneous infiltration of copper oxide in alumina. *Journal of Materials Research and Technology, 34*, 1539–1548.
15. Yi, H., Jia, L., Ding, J., & Li, H. (2024). Achieving material diversity in wire arc additive manufacturing: Leaping from alloys to composites via wire innovation. *International Journal of Machine Tools and Manufacture, 194*, 104103.

Chapter 48
Composite Feedstock

Composite feedstock consists of one material or form type (Fig. 48.1), whose examples are as follows:

- **Fiber-reinforced powder**: Powder contains another component in the form of a fiber. Fiber-based composite forms when the powder melts, leaving the fiber unmelted. The use of the fiber inside the powder allows it to carry the fiber wherever it moves during processing. This allows the fiber to be distributed in a layer as uniform as the powder, which is not possible when a mixture of separate fiber and powder is used, as they may separate during layer formation. Fibers are useful to develop feedstocks both for powder- and filament-based AM [1].
- **Composite powder**: It is made from a composite material. The material is fabricated by mechanical alloying, melting, atomization, and spraying. Pulverizing a composite part and converting the pieces into the right size also helps make the material. A polymer composite powder may require phase separation [2]. The advantage of a composite powder is that, during processing, its shaping is the only concern and not whether the processing is able to make a composite out of the powder.
- For example, a composite powder made of Ti and TiC forms a Ti-TiC composite part [3, 4]. A single composite powder may consist of either separate powders (Ti and TiC) or a single Ti-TiC powder. If, instead of a single powder, separate powders are taken, they give freedom to make composites from materials unavailable as a composite feedstock. But the spreading of a layer and their consolidation may disturb their uniform distribution in the layer, which can change the component ratio in a composite.
- **Coated powder**: It has more than one material (coating plus bulk), necessary to form a composite. Since the coating remains thinner than the powder diameter, the coating provides a lower amount of material, which limits the material type used as a coat. The coating can have another function, e.g., to provide a binder.

S. Kumar, *A Concise Encyclopedia of Additive Manufacturing*,
https://doi.org/10.1007/978-3-031-98394-8_48

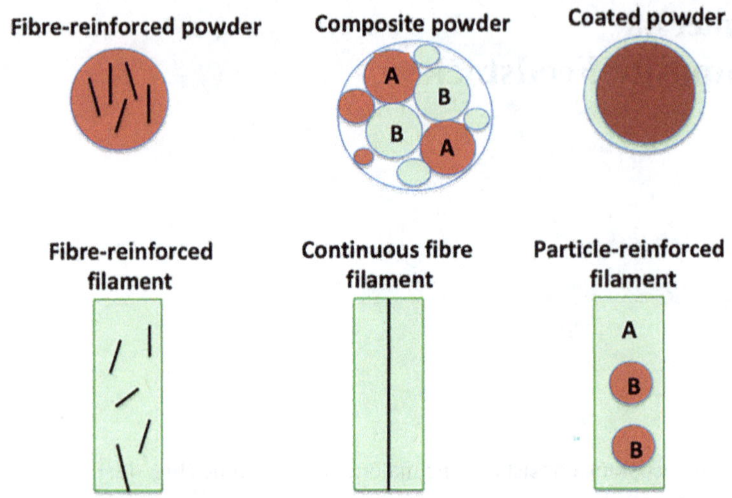

Fig. 48.1 Types of composite feedstocks

Grinding polymer powder and metal flakes can give polymer coating on metal flakes [5].

- **Reinforced filament**: In extrusion AM, a polymer filament can consist of small fibers (whiskers), platelets [6], or particles. Fiber is made of carbon, thermotropic liquid crystalline polymer [7], glass [8], cellulose [9], aramid or Kevlar [10], etc. Most polymers are polyamide, polylactic acid, and ABS, while most reinforcements are glass fiber, carbon fiber, and ceramics [11].

Using small fibers helps make a composite 30 to 40% stronger [12]. But the fiber volume fraction in a filament remains small, which restricts the maximum strength achieved. Replacing small fibers with a continuous fiber helps increase the fiber fraction, which increases the strength [13]. Besides, a continuous fiber, unlike a whisker, gives strong isotropic properties.

The job of a filament ends when a layer is made. The job again starts when a new layer over the previous layer is made. It is a discontinuous job. To maintain the discontinuity, at the end of a layer deposition, there should not remain any connection between the layer and the filament. If there is any connection, it needs to be broken for depositing the next layer.

When a filament reinforced with a continuous fiber is used, the fiber does not melt. Therefore, the continuity will not be broken unless the fiber breaks. The fiber needs to be weak, and the deposited layer needs to quickly solidify, so when the formation of the layer is over, the upward movement of the nozzle should be enough to break the fiber attached to the layer. This limits the diameter (around 10 μm) and the number (around 1000) of fibers [14] used in a filament [10]. Post-processing removes any remnant of fibers.

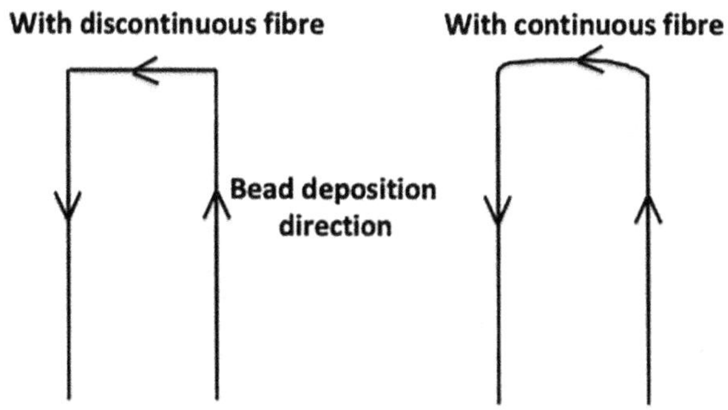

Fig. 48.2 Deposition of extruded material reinforced with discontinuous and continuous fiber

Due to the stiffness of the continuous fiber, the circular movement of a nozzle is restricted, which restricts the fineness of a feature made. The nozzle needs to make a big circle when it turns to scan. Thus, it will not be able to take a sharp turn [7] (Fig. 48.2).

- **Composite pellet:** A composite pellet makes a polymer composite part. It also makes those metal or ceramic composites that require polymer to facilitate processing. For example, a metallic alloy-based composite magnet is shaped because the polymer of the composite facilitates shaping [15]. The role of the polymer is more to shape than to contribute a desired mechanical property to the magnet. Other examples are a polymer composite made up of polymers and a composite (coffee grounds) for pellet extrusion [16].

References

1. Hu, Y., Lin, Y., Yang, L., et al. (2024). Additive manufacturing of carbon fiber-reinforced composites: A review. *Applied Composite Materials, 31*(2), 353–398.
2. Hung, Y. M. X. H., Talou, M. H., & Camerucci, M. A. (2024). Additive manufacturing of porous ceramic structures by indirect powder bed fusion with laser beam using a novel polyamide/alumina-based feedstock. *Ceramics International, 50*(18), 32039–32050.
3. Gu, D., Hagedorn, Y. C., Meiners, W., et al. (2011). Nanocrystalline TiC reinforced Ti matrix bulk-form nanocomposites by selective laser melting (SLM): Densification, growth mechanism and wear behavior. *Composites Science and Technology, 71*(13), 1612–1620.
4. Zhang, L. C., & Attar, H. (2016). Selective laser melting of titanium alloys and titanium matrix composites for biomedical applications: A review. *Advanced Engineering Materials, 18*, 463–475.
5. Schwenger, M. S., Kaminskyj, M. S., Haas, F. M., & Stanzione, J. F., III. (2024). Mixed-material feedstocks for cold spray additive manufacturing of metal–polymer composites. *Journal of Thermal Spray Technology, 33*(2), 619–628.
6. Caminero, M. Á., Chacón, J. M., García-Plaza, E., et al. (2019). Additive manufacturing of PLA-based composites using fused filament fabrication: Effect of graphene Nanoplatelet reinforcement on mechanical properties, dimensional accuracy and texture. *Polymers, 11*(5), 799.

7. Ansari, M. Q., Redmann, A., Osswald, T. A., et al. (2019). Application of thermotropic liquid crystalline polymer reinforced acrylonitrile butadiene styrene in fused filament fabrication. *Additive Manufacturing, 29*, 100813.
8. Carneiro, O. S., Silva, A. F., & Gomes, R. (2015). Fused deposition modeling with polypropylene. *Materials and Design, 83*, 768–776.
9. Wang, Z., Xu, J., Lu, Y., et al. (2017). Preparation of 3D printable micro/nanocellulose-polylactic acid (MNC/PLA) composite wire rods with high MNC constitution. *Industrial Crops and Products, 109*, 889–896.
10. Dickson, A. N., Barry, J. N., McDonnell, K. A., & Dowling, D. P. (2017). Fabrication of continuous carbon, glass and Kevlar fiber reinforced polymer composites using additive manufacturing. *Additive Manufacturing, 16*, 146–152.
11. Rodrigues, A. D. S. L., Pires, A. C. B., Barbosa, P. A., & Silveira, Z. D. C. (2024). Polymeric composites in extrusion-based additive manufacturing: A systematic review. *Polymer Composites, 45*(8), 6741–6770.
12. Parandoush, P., & Lin, D. (2017). A review on additive manufacturing of polymer-fiber composites. *Composite Structures, 182*, 36–53.
13. Werken, N. V. D., Koirala, P., Ghorbani, J., et al. (2021). Investigating the hot isostatic pressing of an additively manufactured continuous carbon fiber reinforced PEEK composite. *Additive Manufacturing, 37*, 101634.
14. Tian, X., Liu, T., Yang, C., et al. (2016). Interface and performance of 3D printed continuous carbon fiber reinforced PLA composites. *Composites Part A Applied Science and Manufacturing, 88*, 198–205.
15. Li, L., Tirado, A., Nlebedim, I., et al. (2016). Big area additive manufacturing of high performance bonded NdFeB magnets. *Scientific Reports, 6*, 36212.
16. Romani, A., Paramatti, M., Gallo, L., & Levi, M. (2024). Large-format material extrusion additive manufacturing of PLA, LDPE, and HDPE compound feedstock with spent coffee grounds. *The International Journal of Advanced Manufacturing Technology, 134*(3), 1845–1861.

Chapter 49
Conformal Cooling Channel

A conformal cooling channel made in a part allows it to cool uniformly, which increases its efficiency and reduces energy consumption [1]. For example, if the part is a mold, its production cycle and quality will increase [2]. The channel lies equidistant from a cavity contour, which tries uniform cavity cooling due to the proximity of the channel to the contour. A channel is at different distances from planes A, B, and C of the cavity—a non-conformal cooling channel (Fig. 49.1a) [3]. Another channel bent twice to follow the contour of the cavity, so all three planes (A, B, and C) have the same distance from the channel—a conformal cooling channel (Fig. 49.1b).

1 Drilling

Drilling makes a straight channel. Therefore, if a curved channel is required, an approximate curved channel is fabricated by drilling a block from various orientations and then filling up extra gaps [4].

To make a cooling channel (white color) in a metallic block (black color), two steps are required (Fig. 49.2a). The first step is to drill the block, which is shown in Fig. 49.2b by four arrow marks at four locations of the block. The second step is to fill up the extra drilled channel with filler (Fig. 49.2c, red color).

If a channel is conformal and curved, drilling holes and filling up extra holes will not work; many tools along with their planning and setting fixtures are required. If a block has some other features, drilling from all orientations without damaging the features is not possible. If the channel diameter is changing, even using tools of various diameters will pose difficulties, and drilling will be possible only by breaking the block into two. Thus, drilling has limitations to make a curved channel.

S. Kumar, *A Concise Encyclopedia of Additive Manufacturing*,
https://doi.org/10.1007/978-3-031-98394-8_49

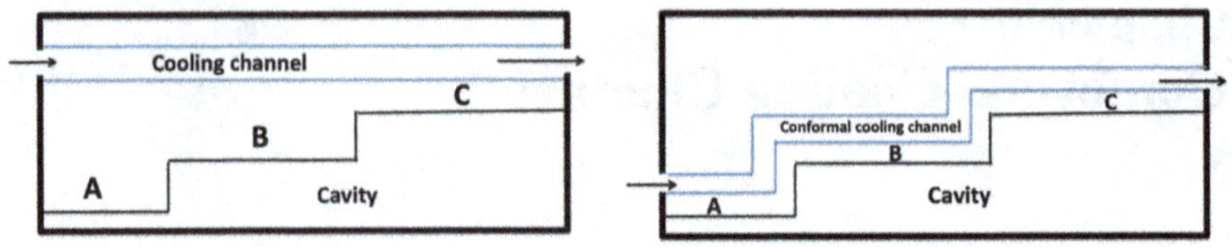

(a) non-conformal (b) conformal

Fig. 49.1 Schematic diagrams of cooling channels. (**a**) non-conformal. (**b**) conformal

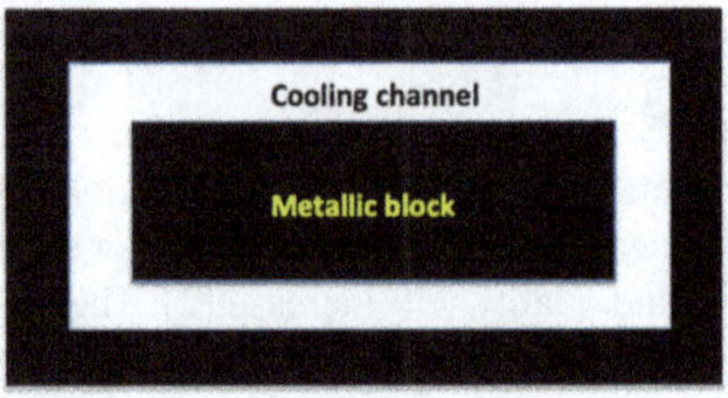

(**a**) Design of a cooling channel (the connection to an external water source not shown)

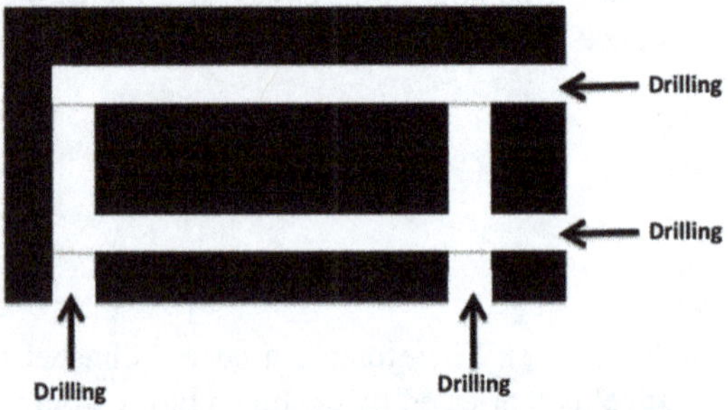

(**b**) Drilling at four places from two sides of the block

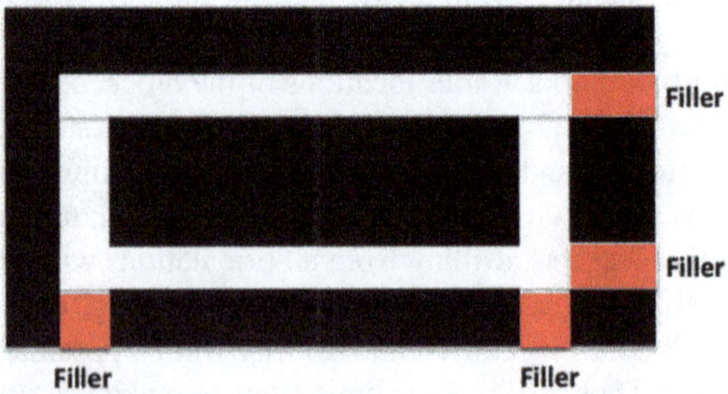

(**c**) Filling gaps

Fig. 49.2 Fabricating by drilling. (**a**) Design of a cooling channel (the connection to an external water source not shown). (**b**) Drilling at four places from two sides of the block. (**c**) Filling gaps

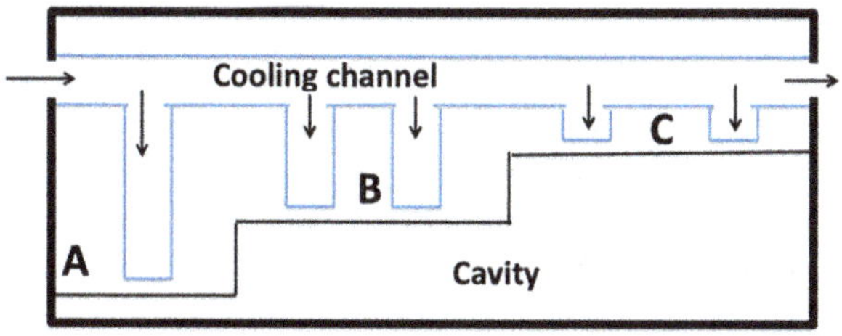

Fig. 49.3 Vertical plus horizontal cooling channels

A vertical channel along with the horizontal cooling channel can help reach the contour of a cavity [5]. Figure 49.3 shows a channel consisting of vertical plus horizontal sections. In the absence of free flow, there will be limited benefit from this channel.

2 Casting

Casting [6], preferably investment casting (IC), makes internal channels. IC requires many steps to make a part. First, a duplicate part from a low melting point material (e.g., wax) is made—this is the pattern. The aim is to convert the pattern into a part using required material, which requires a negative impression from the pattern.

To make an impression, liquid ceramic is poured surrounding the pattern. When liquid ceramic is solidified. A shell is formed, partially covering the pattern. The shell plus pattern is then heated so the pattern drains out. This shell is a mold, which is filled with the required material. When the material dries, the shell is broken to let the dried material or part out.

To make a complex part, a complex pattern needs to be fabricated. If the part is more complex, simpler patterns need to be fabricated and arranged to make the complex one. Such an arrangement compensates for the need to make a complex pattern, but the patterns must not shift during pouring. If many parts of the same design need to be made, a permanent pattern needs to be fabricated to replicate many wax patterns. The process thus takes many steps, where each step is prone to inaccuracy.

IC has limitations—pattern wax needs to be drained out from a shell; the process may leave residue that may react with casted liquid to form defects. Moreover, the surface finish of a part depends upon a smooth ceramic shell. Setting time and selection of binder for ceramic slurry are required to fabricate a shell having optimized properties. Casting and thereafter its removal by breaking the shell limits the fabrication of the fine features.

AM is free from such limitations. If a complex channel is not formed, it is not because AM has to pass through many dependable steps, as IC has. Processing in IC depends on materials (pattern, shell) that do not end up in a product but get wasted, which does not contribute to sustainability. Besides, AM parts have better mechanical properties than cast parts have [7].

3 Laser Powder Bed Fusion

Laser powder bed fusion is used to make complex channels [8, 9]. It makes a channel by melting an area outside the channel contour. Melting gives rise to the diameter of the channel, while the unprocessed area inside the contour gives the hollowness of the channel. A big or narrow channel diameter is made by leaving, respectively, the big or narrow area unprocessed.

To fabricate a narrow channel, a beam of defined spot size is required to make a small melt pool. The effect of the spot size on a layer thickness at a certain scan speed is required to make an unmelted line. If the line is narrow, the surrounding area will have low surface roughness. Otherwise, the channel will become narrower, which will not facilitate powder being drained out [10]. This channel furnishes a lower finish and accuracy when compared with that made by drilling.

Thus, AM can facilitate making a channel not possible by drilling. Its position, size, and shape do not affect its fabrication in AM in the same way as they do in drilling. But if a complex channel is not required, drilling or casting is better as they provide better quality.

References

1. Pusateri, V., & Olsen, S. I. (2024). LCA and LCC of wire arc additively manufactured and repaired parts compared to conventional fabrication techniques. *Procedia CIRP, 122*, 491–496.
2. Andronov, V., Pitrmuc, Z., Zajíc, J., et al. (2025). Conformal cooling as a support tool for eliminating local defects in high-pressure die casting series production. *Progress in Additive Manufacturing, 10*(2), 1511–1528.
3. Shinde, M. S., & Ashtankar, K. M. (2017). Additive manufacturing-assisted conformal cooling channels in mold manufacturing processes. *Advances in Mechanical Engineering, 9*(5), 168781401769976.
4. Diegel, O., Schutte, J., Ferreira, A., & Chan, Y. L. (2020). Design for additive manufacturing process for a lightweight hydraulic manifold. *Additive Manufacturing, 36*, 101446.
5. Feng, S., Kamat, A. M., & Pei, Y. (2021). Design and fabrication of conformal cooling channels in molds: Review and progress updates. *International Journal of Heat and Mass Transfer, 171*, 121082.
6. Kuo, C. C., Jiang, Z. F., & Lee, J. H. (2019). Effects of cooling time of molded parts on rapid injection molds with different layouts and surface roughness of conformal cooling channels. *The International Journal of Advanced Manufacturing Technology, 103*, 2169–2182.
7. Scudino, S., Unterdörfer, C., Prashanth, K. G., et al. (2015). Additive manufacturing of cu–10Sn bronze. *Materials Letters, 156*, 202–204.
8. Çalışkan, C. İ., Khan, H. M., Özer, G., et al. (2022). The effect of conformal cooling channels on welding process in parts produced by additive manufacturing, laser powder bed fusion. *Journal of Manufacturing Processes, 83*, 705–716.
9. Kanbur, B. B., Zhou, Y., Shen, S., et al. (2022). Metal additive manufacturing of conformal cooling channels in plastic injection molds with high number of design variables. *Materials Today: Proceedings, 70*, 541–547.
10. Sai, M. M. K., Mandal, A., Bishwakarma, H., et al. (2024). Laser powder-bed fusion printability of micro-holes in SS316L (B) and post-processing through electrochemical micromachining. *Progress in Additive Manufacturing*, 1–18.

Chapter 50
Continuous Liquid Interface Production

Continuous liquid interface production (CLIP) is a layerless AM. In inverse stereolithography, where a part is made upside down, a film is attached on the inside of an exposure window so the solidified layer can be detached from the window before resin flows for the next layer to be formed. In CLIP [1, 2], an oxygen-permeable window instead of a film is used (Fig. 50.1). Oxygen gas entering through the window reacts with free radicals and neutralizes them so that they will not be free for further polymerization on or near the window.

In the absence of polymerization, there will not be any solidified material on the window that required to be detached from. This makes the process free from the following problems: detachment, repositioning the substrate after detachment, breakage of features during detachment, and not making a large area because of the problem of detachment. Utilization of oxygen molecules to inhibit photopolymerization makes the process different from stereolithography, where inhibition by such a method is not used.

CLIP is a continuous process—there is no recurring phenomenon of layer formation, and there is no recurring detachment. Therefore, to make a part continuously without being interrupted layerwise, photopolymer through a window is continuously exposed to beams while a substrate is continuously pulled up.

Continuous pulling up the substrate ensures that the photopolymer continuously flows in between the window and the substrate, which is continuously cured by beams. Fast pulling up will not let complete curing, which means there will be non-uniformity or porosity in a part. Slow pulling up will cause overcuring, making the part brittle, and decrease the fabrication speed. If the photopolymer is additionally injected using a needle, the printing speed will further increase [3].

Since the process does not proceed layerwise, the process saves time that could have been spent during inter-layer processing, i.e., the process saves processing time after completion of one layer and before starting the next layer, which increases the fabrication speed. Since there is no interlayer joining, there are no seams between layers that could influence mechanical properties or change isotropicity.

S. Kumar, *A Concise Encyclopedia of Additive Manufacturing*,
https://doi.org/10.1007/978-3-031-98394-8_50

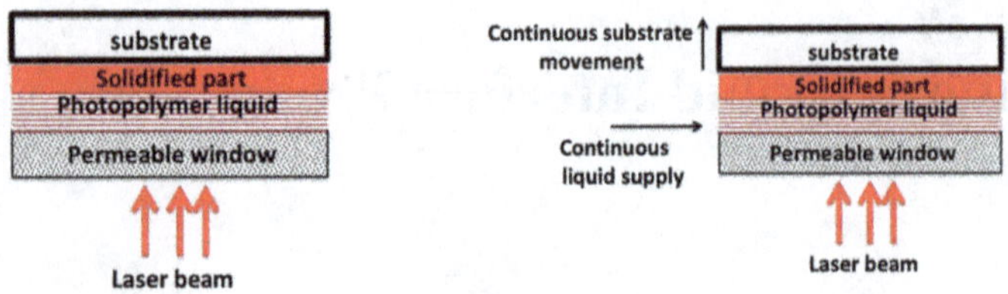

(a) Solidification without solidifying on window (b) Continuous solidification without layerwise

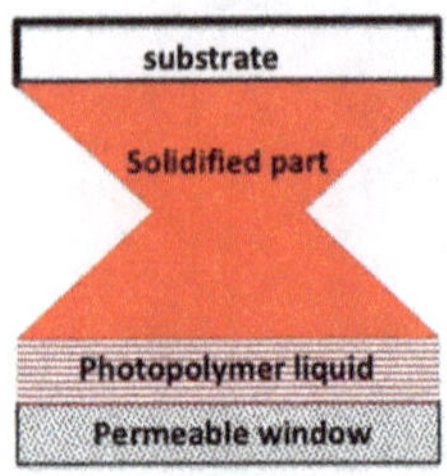

(c) Part fabrication

Fig. 50.1 Schematic diagram of CLIP. (**a**) Solidification without solidifying on window. (**b**) Continuous solidification without layerwise. (**c**) Part fabrication

However, the process is not free from a layerwise concept. The exposure on photopolymer or on a window by beam makes a 2D contour of a part in the same way as happens in a layerwise process. For example, if a solid cone is to be made, the exposure on the window will be of circular shape in either process (CLIP or layerwise). After some duration of the exposure, the exposure by beam will no longer be the same size as the circle; it will change to fit the contour of the cone. When it changes, it will again be the same to some nth layer of the layerwise process. In layerwise, the geometry of the exposure by beam changes after a progress of a minimum of one layer thickness, while the size of layer thickness is determined by the machine.

In CLIP, the geometry of exposure by beam can change early or fast because it is not related to an actual layer thickness a machine can allow and is not limited by the machine. Since build direction does not change in both processes, there should be a staircase effect in both processes. If the staircase effect is not visible in CLIP, it is due to minimization of equivalent layer thickness in CLIP. This type of minimum layer thickness cannot be set in an actual layerwise machine because of the limitations or tolerance of a machine. CLIP thus demonstrates how the limitation of an AM machine can be overcome. This also shows that the staircase effect is a limitation of a process, but this limitation is pronounced not due to the limitation of the process itself but due to that of the machine.

In layerwise, if a 3D solid cone is fabricated, the first layer will be made by scanning or exposing the photopolymer by a beam in a circular area. The effect of the scanning is the formation of a solid cylinder having a height equal to the layer thickness; this cylinder instead of a cone is the natural outcome of a layerwise process.

Though the aim is to make a cone, the cone is not made; it is actually approximated by making a number of cylinders, where each cylinder is due to one layer.

What if cone upon cone is made to make a cone? The layerwise process does not make a cone because the process does not see a cone during processing; the process only sees a circle and is not allowed to see the extension of a circle in 3D. CLIP is not different; it does not try to look beyond the plane. Though it is a continuous process, the beam can look beyond the plane; it can penetrate through the plane and solidify as per Beer–Lambert exponential law of absorption [4]. But CLIP does not use the beam to go along the contour of a cone or to penetrate obliquely to make the cone directly rather than indirectly by approximation. If CLIP uses beam penetration as per the contour of a part extended in 3D, the process will remove the staircase effect. CLIP does not use the beam differently than how it is used in other layerwise processes. It is no exaggeration if CLIP is considered a very fast layerwise process having no time to make a layer.

The process is used to make microneedles of the size of 400 μm tall with an aspect ratio of 3. These needles are used to transfer drugs to skin without causing any pain to the skin [5].

References

1. Janusziewicz, R., Tumbleston, J. R., Quintanilla, A. L., et al. (2016). Layerless fabrication with continuous liquid interface production. *PNAS, 11*(42), 11703–11708.
2. Tumbleston, J. R., Shirvanyants, D., Ermoshkin, N., et al. (2015). Continuous liquid interface production of 3D objects. *Science, 347*(6228), 1349–1352.
3. Lipkowitz, G., Samuelsen, T., Hsiao, K., et al. (2022). Injection continuous liquid interface production of 3D objects. *Science Advances, 8*(39), eabq3917.
4. Jacobs, F. P. (1992). *Rapid prototyping and manufacturing: Fundamentals of stereolithography*. Society of Manufacturing Engineers.
5. Johnson, A. R., Caudill, C. L., Tumbleston, J. R., et al. (2016). Single-step fabrication of computationally designed microneedles by continuous liquid interface production. *PLoS One, 11*(9), e0162518.

Chapter 51
Cost and Sustainability

If AM wastes material while parts are cost-effective [1], there will be no incentive to discontinue manufacturing to prevent wastage. If an AM machine is expensive, machine time will be expensive. Hence, if the material is not significantly expensive, the cost of a part will be governed by the machine time [2]. The cost of the material lost during manufacturing, therefore, will not be high enough to increase the part cost. This is the case with metal AM machines, where the cost of machine time is high compared to the cost of feedstock (e.g., steel) [3]. Hence, the loss of the steel does not increase the cost of a part too high. If an expensive material replaces the steel and there is a material loss, the loss will be expensive enough to discourage manufacturing.

What if the cost of a metal AM machine becomes low, so the material cost is not insignificant compared to machine time? Then, the cost of the part will be low, which will fluctuate with the material cost. If there is a material loss, the part cost will increase. Since the cost of the machine time is low, the material loss during manufacturing will not make the part expensive. And the material loss will not discourage manufacturing.

The cost to manufacture can be an overriding factor in deciding whether to manufacture, but this cannot always be. Consider an example where only AM can manufacture. This can be due to the part being complex, consisting of many materials, or having varied properties distributed within it. What if their fabrications do not save material? Even then, they can be made because they cannot be made by other techniques.

Consider a part that both AM and conventional manufacturing (CM) can manufacture, but when it becomes complex, only AM can. Therefore, it is not possible to know for a complex part how much its fabrication in AM could have saved material because there is no complex part in CM to compare with. In the absence of such knowledge, even if there is an overconsumption of material, it is difficult to term the overconsumption as wasting.

S. Kumar, *A Concise Encyclopedia of Additive Manufacturing*,
https://doi.org/10.1007/978-3-031-98394-8_51

References

1. Bassoli, E., Defanti, S., Tognoli, E., et al. (2021). Design for additive manufacturing and for machining in the automotive field. *Applied Sciences, 11*(16), 7559.
2. Schröder, M., Falk, B., & Schmitt, R. (2015). Evaluation of cost structures of additive manufacturing processes using a new business model. *Procedia CIRP, 30*, 311–316.
3. Piili, H., Happonen, A., Väistö, T., et al. (2015). Cost estimation of laser additive manufacturing of stainless steel. *Physics Procedia, 78*, 388–396.

Chapter 52
Design-Specific Tools: Role

In injection molding (IM), a non-AM additive process, a mold is required to make a part of one design. A number of parts can be made, but all must have the same design. If a part of another design is required, a mold of another design is required. Thus, for making parts of different designs, different molds are required.

In laser metal deposition (LMD) [1], an AM process, a part is made by depositing material along a certain track as per design. If a part of another design is required, the material is deposited along another track as per the new design—it is the pattern of deposition that determines the types of designs to be made. Thus, material is deposited differently, and a new part having a new design is formed. Therefore, a number of different parts can be made without having the need for different design-specific molds.

This is how LMD differs from IM, or how the role of a design-specific tool is different in AM and non-AM additive processes.

Some AM processes require tools. For example, additive friction stir deposition (AFSD) [2], an AM process, requires tools to accomplish the process. Then, how is the requirement of tools in AM different from that in a non-AM additive process, e.g., IM?

In AFSD, if the dimension of a design is big, so one tool will wear out before it makes a complete part, more tools will be required to complete the task. If some features of the design are small, the tool of the same size will be oversized and be unable to make the part without compromising its accuracy. Hence, a small tool is required for making a small feature.

In IM, the design of a part is the same as the design of the cavity of a mold, while in AFSD, the design of a part does not need to match the shape of a tool. Thus, in AFSD, tools are required so fabrication as per any design will be completed with the help of either one or more tools, while in IM, molds are required so the fabrication as per a particular design will be completed with the help of a particular mold.

In AFSD, a tool is a rod or pin used to convert powders to non-powder solid material using friction energy. The conversion is the primary aim of the tool. The

S. Kumar, *A Concise Encyclopedia of Additive Manufacturing*,
https://doi.org/10.1007/978-3-031-98394-8_52

second aim is to create a well-defined track along with the conversion. The conversion is primary because if there is no conversion, no part will be formed, and if the conversion is partial, a part having incomplete property will be formed. But the conversion itself is not adequate; it must lead to a well-defined track that, on adding up, will give a well-defined part. Otherwise, the part will require post-processing to overcome the demerit of an inadequate conversion, as is done in AM.

For making a small feature, e.g., a thin wall, a small tool that does not make a wider track than the width of the wall is required, while for a big feature, a big tool is required, as the same small tool will either decrease the fabrication speed or break due to overstrain.

Thus, if a design does not consist of a small or big feature, one tool is sufficient for converting the design into a part; otherwise, two tools are required. If the design becomes more complex and consists of many features of sizes far different from each other, more tools are required. This situation reminds of milling, where tools of various sizes are needed to remove materials to convert a complex design into a part.

Thus, AFSD does not have an edge over milling for fulfilling the goal of toolless manufacturing. Though the purpose of tools in milling is in sharp contrast to that in AFSD. If milling uses tools for subtraction, AFSD uses them for addition.

Considering another AM where tools are used. Selective laser melting (SLM) uses two lasers for processing: one of big diameter for making big features or speeding up production, and the other is of small diameter for making small features. Or, one for heating and another for melting [3]. In these examples, SLM can be considered to be using two tools for processing.

Then, why should SLM be considered toolless manufacturing while machining should not? This is because the tool of SLM, unlike a machining tool, does not wear out while interacting with material. Consequently, the performance of SLM is not restricted because of the loss of its tool caused by tool–material interaction, while machining performance is restricted by the wear of its tool due to the interaction. Thus, tool–material interactions have the ability to degrade the performance of machining but not of SLM, making SLM relatively toolless manufacturing when comparing it with machining. Though the tool of SLM, which is a laser beam, degrades. But it degrades not because it interacts with a hard material but because it has a specific life-time irrespective of the types of materials it interacts with. In machining, the degradation of a tool depends upon the type of materials it interacts with, making its tool, unlike an SLM tool, not independent of the materials.

There can be a question: why is SLM relatively toolless manufacturing when comparing it with machining, and why not instead make it absolutely toolless manufacturing? It is because, though SLM is free from conventional tools used in manufacturing, it is not free from the concept of tools.

What if a tool in a particular AM starts to wear in the same way a tool wears in conventional manufacturing? Then, this particular AM is not toolless manufacturing, e.g., AFSD. But AFSD, being not toolless manufacturing, does not mean it does not have an advantage over machining that is also not toolless manufacturing. It implies this advantage does not make AFSD different from machining when compared on the basis of not being toolless manufacturing.

References

1. Wang, T., Wang, Z. Y., Wang, R., et al. (2025). Effect of solution aging treatment on corrosion resistance and erosion resistance of laser metal deposition 17-4PHss. *Engineering Failure Analysis, 169*, 109167.
2. Wang, H., Li, Y., Yang, B., et al. (2025). Understanding the processing, microstructure, and deformation behavior of AZ31B Mg alloy fabricated by additive friction stir deposition. *Journal of Materials Processing Technology, 338*, 118781.
3. Heeling, T., & Wegener, K. (2018). The effect of multi-beam strategies on selective laser melting of stainless steel 316L. *Additive Manufacturing, 22*, 334–342.

Chapter 53
Direct Digital Manufacturing

Earlier, rapid prototyping or AM used to be called direct digital manufacturing (DDM) [1]. The meaning of DDM was a type of manufacturing that converts a digital file directly into a product. The meaning was not different from the meaning of AM. AM does the same; thus, the term DDM was redundant in the same way as the terms rapid manufacturing, rapid tooling, rapid prototyping, and 3D printing were redundant. The use of DDM then was more to appreciate and emphasize the digital aspect of AM. It was not conveying a new approach other than what AM was already conveying. Its use was more to convey the promise of AM.

The meaning of DDM evolved with the advancement of AM. Henceforth, DDM meant a method that is a type of AM that allows many products to be formed. Since AM can design many products, DDM meant a method for mass customization. But many AM processes were not able to fulfill this criterion; DDM inadvertently represented a few AM processes. The meaning of DDM in this case again was not different from the meaning of AM. This meaning was more to advertise the production capability of AM rather than initiating a new manufacturing paradigm. Thus, DDM was again redundant.

At that time, DDM had nothing to do with the automation. If AM as DDM was expected to do mass customization, it meant only that an AM system would make a number of customized products. The meaning of DDM did not include how materials will be automatically transferred to an AM machine or how products made will be automatically collected—these were not the expectations from AM. If an AM system is making a number of products, it is already fulfilling the criteria of DDM. The automation only implied how an isolated AM system functioned well without the need for manual input or how various components of the isolated system were automated without having any connections with peripheral units or cyber-physical systems. This automation was the minimal automation. Not emphasizing automation beyond an isolated AM system was also for a practical reason—an accepted product needs to be formed first before an extended automation is sought.

S. Kumar, *A Concise Encyclopedia of Additive Manufacturing*,
https://doi.org/10.1007/978-3-031-98394-8_53

As AM was growing, DDM had to grow as well. DDM started to acquire a new meaning. Henceforth, if an AM system can be connected to the internet and a product can be formed by a distant customer, this can be DDM [2]. But connection with the Internet and many other systems is what digital manufacturing is. DDM was no longer equal to the direct conversion of a digital file to a product but was equal to the direct conversion of a file plus digital manufacturing.

The direct conversion to product is not the privilege of only AM; other conventional manufacturing (CM) processes such as machining do it as well. The direct conversion means AM does not need to make a mold for making a product, while mold-based CM (such as injection molding, die casting, sintering, etc.) needs to make a mold for making a product. Thus, the direct conversion takes place in AM, while direct conversion does not take place in mold-based CM. Consequently, AM is better than mold-based CM in those situations where direct conversion is a virtue.

But AM does not have any advantage over mold-free CM (such as machining, drilling, milling, etc.) that is also capable of doing direct conversion and does not require any mold to carry out conversion. If AM has an advantage over mold-free CM, it is because AM can make some complex products that mold-free CM cannot make. This advantage is not because mold-free CM lacks the quality of direct conversion but because mold-free CM is different. Being different from AM does not make CM less efficient in all cases.

It can be argued that though mold-free CM does not require molds, it still requires design-specific tooling, while AM does not require design-specific tooling. Therefore, mold-free CM is unable to perform direct conversion of the file, while AM does direct conversion. Consequently, AM is better than mold-free CM. This inference is based on the fact of what will happen when a complex product needs to be formed. When a complex product is required, mold-free CM requires design-specific tooling, while AM does not require design-specific tooling; therefore, mold-free CM lacks the quality of direct conversion, while AM does not lack it.

Thus, this inference takes help from complexity to prove why mold-free CM lacks the quality. This inference will not hold good if the complexity does not arrive. The complexity arrives at the next stage. Before the next stage, mold-free CM can make a number of products without requiring more than one design-specific tool. If a mold-free CM is using only one design-specific tool, this process is free from the obstacle of having more design-specific tools to enable direct conversion of the file. The quality of AM is free from the obstacle of having design-specific tools. When a mold-free CM is using only one design-specific tool, it possesses the same quality that AM possesses. But a mold-free CM lacks the quality when complexity arrives. What happens at the next stage cannot deny what happened before the next stage. Similarly, what will happen at the advanced stage cannot undo what happened at the earlier stage. Both processes, mold-free CM and AM, are basically the same, which cannot be negated because some complex examples give the impression otherwise.

It can be argued that what happens at the earlier stage is not important, but what happens at the advanced stage is more important because the aim is to make complex products; the aim is not to make a simple product, and that is why what happens for the fabrication of a simple product is not important. This argument is meant to evade the basic issue in the name of complexity. Though the complexity is important, and therefore a process has to prove its efficiency by making complex products. But the problem starts when AM cannot always make complex products, while CM can many times.

As direct conversion is concerned, both AM and mold-free CM stand on equal footing. Some advantages of AM over mold-free CM and vice versa do not change the fact that both AM and mold-free CM equally possess the basic characteristics.

Hence, DDM should be equal to digital manufacturing plus only AM was not a logical conclusion. If it was digital manufacturing plus any manufacturing process, it would have been more logical. DDM then became digital manufacturing plus any manufacturing process, albeit with some conditions [3]. It did not mean the evolution of DDM had been leading to any consensus on the use of DDM. As a result, it was again used as its oldest version, i.e., AM [4–6].

References

1. Chiu, W. K., & Yu, K. M. (2008). Direct digital manufacturing of three-dimensional functionally graded material objects. *Computer-Aided Design, 40*(12), 1080–1093.
2. Chen, D., Heyer, S., Ibbotson, S., et al. (2015). Direct digital manufacturing: Definition, evolution, and sustainability implications. *Journal of Cleaner Production, 107*, 615–625.
3. Holmström, J., Liotta, G., & Chaudhuri, A. (2017). Sustainability outcomes through direct digital manufacturing-based operational practices: A design theory approach. *Journal of Cleaner Production, 167*, 951–961.
4. Rojas-Nastrucci, E. A., Ramirez, R. A., & Weller, T. M. (2018). Direct digital manufacturing of mm-wave vertical interconnects. In *IEEE 19th WAMICON* (pp. 1–3).
5. Kacar, M., Wang, J., Mumcu, G., et al. (2019). Phased Array antenna element with embedded cavity and MMIC using direct digital manufacturing. In *2019 IEEE international symposium on antennas and propagation and USNC-URSI radio science meeting* (pp. 81–82).
6. Kumar, S. (2022). Future. In *Additive manufacturing solutions* (pp. 169–184). Springer.

Chapter 54
Directed Energy Deposition

It is the type of AM in which material deposition occurs where external energy is directed. For example, in laser powder deposition, material deposition with the help of powder occurs at a particular place on a substrate where a laser beam as an external energy is directed [1–3]. In additive friction stir deposition, material deposition with the help of a rod occurs where friction energy as an external energy is directed [4, 5]. Other examples are laser wire deposition [6, 7], electron beam wire deposition [8] or electron beam wire-fed deposition [9], wire arc directed energy deposition [10] or wire arc additive manufacturing [11], etc. Directed energy deposition implies directed energy material deposition. Since deposition can happen when the deposition of only a material happens, "deposition" in "directed energy deposition" is sufficient to connote the material deposition. Therefore, "directed energy deposition" has "material" hidden between "energy" and "deposition."

Laser powder bed fusion (LPBF) [12, 13] can come under directed energy deposition. Laser connotes an external energy, which, during scanning, is directed where material deposition needs to be facilitated. Therefore, LPBF relies only on the directed energy for the deposition to occur. There is no other way LPBF can happen. Therefore, LPBF is technically a directed energy deposition technique. But LPBF is not understood as a directed energy technique because LPBF facilitates directed energy deposition through the creation of a powder bed. This is the powder bed that brings a difference between the family of directed energy deposition techniques and that of powder bed fusion techniques. Therefore, knowing directed energy deposition techniques requires the prior knowledge of those techniques that are technically directed energy deposition but do not come under the directed energy deposition techniques. Due to this dependency, directed energy deposition has no independent meanings. It will have an independent meaning when it is said that "deposition" of "directed energy deposition" does not mean the consolidation of a material where the external energy is directed but means only the consolidation of those materials that are transferred when the external energy is directed.

S. Kumar, *A Concise Encyclopedia of Additive Manufacturing*, https://doi.org/10.1007/978-3-031-98394-8_54

References

1. Wang, A., Wang, H., Wu, Y., & Wang, H. (2021). 3D printing of aluminum alloys using laser powder deposition: A review. *The International Journal of Advanced Manufacturing Technology, 116*(1), 1–37.
2. Jiang, Y. L., Fang, J. X., Ma, G. Z., et al. (2021). Microstructure and properties of an as-deposited and post treated high strength carbide-free bainite steel fabricated via laser powder deposition. *Materials Science and Engineering A, 824*, 141791.
3. Gradl, P., Cervone, A., & Colonna, P. (2025). Enhancement of friction factors for microchannels fabricated using laser powder directed energy deposition. *Materials & Design, 251*, 113673.
4. Dong, X., Zhou, M., Han, Y., et al. (2025). Additive friction stir deposition of WE43: The influence of rotation speed on microstructure, mechanical and corrosion properties. *Journal of Materials Research and Technology, 36*, 1585–1596.
5. Broadhead, J., Miles, M., Hovanski, Y., et al. (2025). Simulation of the additive friction stir deposition process using a directed energy deposition approach. In *TMS annual meeting & exhibition* (pp. 181–188). Cham.
6. Wang, X., Zhang, L. J., Ning, J., & Na, S. J. (2022). Effect of cu-induced eutectoid transformation on microstructure and mechanical properties of Ti–6Al–4V alloy by laser wire deposition. *Materials Science and Engineering A, 833*, 142316.
7. Zhang, Q., Sikan, F., Chekir, N., & Brochu, M. (2024). Thermo-microstructural-mechanical modeling on effect of travel speeds on thin Ti–6Al–4V deposits developed by laser wire deposition. *Journal of Materials Research and Technology, 33*, 7961–7976.
8. Gong, S., Liu, J., Yang, G., & Xu, H. (2022). *Electron beam wire deposition technology and its application*. Springer.
9. Sikan, F., Wanjara, P., Atabay, S. E., et al. (2023). Evaluation of electron beam wire-fed deposition technology for titanium compressor blade repair. *Materials Today Communications, 35*, 105701.
10. Li, K., Chen, W., Gong, N., et al. (2023). A critical review on wire-arc directed energy deposition of high-performance steels. *Journal of Materials Research and Technology, 24*, 9369–9412.
11. Treutler, K., & Wesling, V. (2021). The current state of research of wire arc additive manufacturing (WAAM): A review. *Applied Sciences, 11*(18), 8619.
12. Nabavi, S. F., Farshidianfar, A., & Dalir, H. (2025). Comprehensive review: Advancements in modeling geometrical and mechanical characteristics of laser powder bed fusion process. *Optics & Laser Technology, 180*, 111480.
13. Kosiba, K., King, W., Han, J., et al. (2025). Fabrication of filigree parts via laser powder bed fusion: From melt spots to stents. *Optics & Laser Technology, 180*, 111485.

Chapter 55
Disadvantage

The following are four disadvantages of AM:

1. Energy-intensive process—it takes more energy to make [1].
2. Lack of processable materials—not many materials [2].
3. Lack of repeatability—difficulty in making two products alike [3].
4. Slow process—it does not compete with CM to make products fast.

References

1. Jung, S., Kara, L. B., Nie, Z., et al. (2023). Is additive manufacturing an environmentally and economically preferred alternative for mass production? *Environmental Science & Technology, 57*(16), 6373–6386.
2. Aversa, A., & Fino, P. (2020). Special issue on materials development by additive manufacturing techniques. *Applied Sciences, 10*(15), 5119.
3. Venturi, F., & Taylor, R. (2023). Additive manufacturing in the context of repeatability and reliability. *Journal of Materials Engineering and Performance, 32*(15), 6589–6609.

S. Kumar, *A Concise Encyclopedia of Additive Manufacturing*,
https://doi.org/10.1007/978-3-031-98394-8_55

Chapter 56
Electrolytic Solution-Based Deposition

Electrolytic solution-based AM, or electrochemical AM, is a technique that makes a part atom by atom, promising to make precision parts [1]. The quality of the parts is found to be similar to that made traditionally [2].

When metal ions move and get deposited on a cathode, they form a layer on it. If the cathode is a big plate and the anode is a small rod facing a small area of the cathode, the formation of the layer will be limited to that area (Fig. 56.1a). The exact size of the layer will depend upon variables: gap between two electrodes, throwing power of the electrode, applied voltage, form of the voltage, insulation on the side surface of the anode, concentration of electrolytes, etc. But changing the size of the anode provides a way to localize the formation of layers or deposition of materials at some selected area [3, 4].

If the anode moves parallel to the cathode, the deposited layer moves and creates a layer in the form of a line having a width corresponding to the width of the anode (Fig. 56.1b). If the anode moves away from the cathode, the deposited material starts growing away from the cathode. If there is more growing, a pillar is made on the cathode (Fig. 56.1c). After the formation of the pillar, if the anode again moves parallel to the cathode, another pillar will form on the earlier pillar at a right angle to it (Fig. 56.1d). With a change in the direction of the movement of the anode, many such pillars can be formed. This will result in a structure similar to a pillar having many arms—these are overhang structures that, in other AM processes, require either support structures or a change in the orientation of the geometry. But in this process, these are formed without such a requirement [5].

The addition of materials happens ion by ion, and when these ions are getting added on the side surface of a pillar to make an arm, they have no such possibility to succumb to gravitational force and collapse as happens in a drop-by-drop AM process. It can collapse if the addition does not furnish enough strength to the structure, but the structure will not collapse for the reason that the ions could not be placed on the side surface.

S. Kumar, *A Concise Encyclopedia of Additive Manufacturing*,
https://doi.org/10.1007/978-3-031-98394-8_56

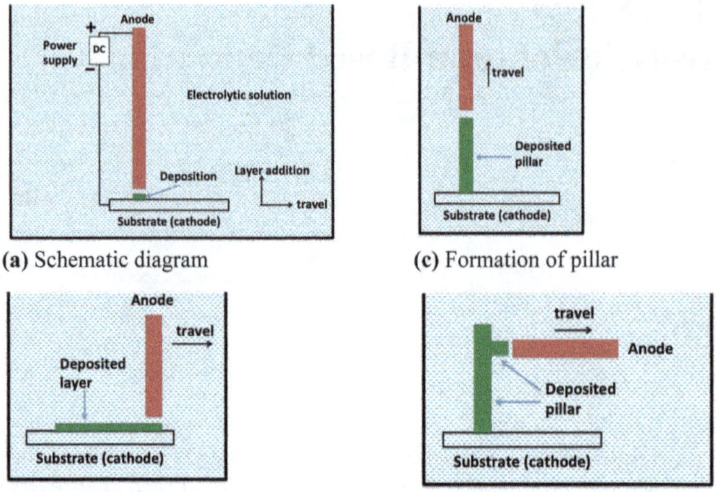

(a) Schematic diagram

(c) Formation of pillar

(b) Formation of layer

(d) Formation of another pillar

Fig. 56.1 Electrolytic solution-based deposition: (**a**) diagram, (**b**) layer, (**c**) pillar, and (**d**) another pillar

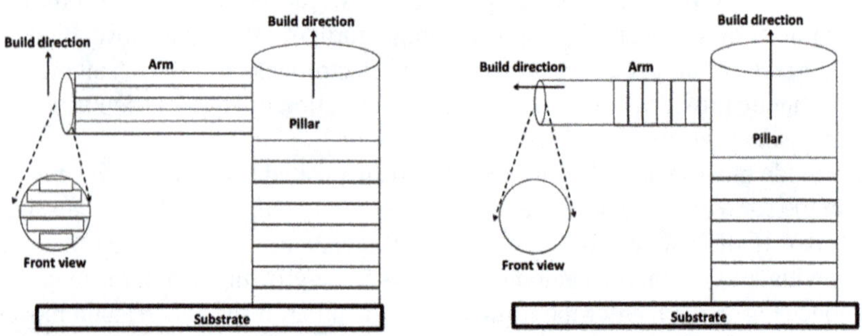

(a) Staircase effect due to same build direction of arm and cylinder

(b) No staircase effect due to different build directions

Fig. 56.2 Effect of build directions in layerwise process. (**a**) Staircase effect due to same build direction of arm and cylinder. (**b**) No staircase effect due to different build directions

The process gives the opportunity to add materials wherever it is required, leading to a method to make a complex part [6]. Hence, materials can be added anywhere on a pillar to make arms at various angles. If layerwise fabrication is followed, the pillar can be fabricated well, but the arm will show gaps between two layers (Fig. 56.2a). Thus, during the fabrication of the arm, the disadvantage of the layerwise fabrication will be visible.

If, after the fabrication of the pillar, the tool does not follow layerwise fabrication in that build direction, there will not be any disadvantage due to the layerwise fabrication. After the layerwise fabrication of the pillar—the pillar acts as if it is a new cathodic plate, the fabrication of the arm becomes the fabrication of the pillar, and the tool still follows the layerwise method—the resulting structure (arm) will not show any gap because the layerwise fabrication changes the build direction by 90° (Fig. 56.2b).

The present example of a pillar and an arm shows how the disadvantage of the layerwise method can be overcome by changing the build direction. But it will require software that will slice a CAD model both in horizontal and vertical directions [7].

Commercial AM systems have one build direction [8]. Though changing the build direction during a build is doable—(1) after fabrication of a pillar, if the position of the cathodic plate is changed by 90° and the build is continued, (2) after fabrication of a pillar, a new CAD model is used for fabricating the arm, and the model is sliced vertically instead of horizontally. A new tool path is created, which gives a build direction that is at 90° to the old build direction that was used to create the pillar. The CAD model is positioned at the position of the arm of the old CAD model, and the build is continued. Both options show it is possible to manufacture one pillar and arm.

But what if the part is not so simple, the part does not consist of just one pillar and arm but has hundreds of pillars and arms at different angles, the above options will still work but will require process planning.

The problem associated with changing a build direction increases when layerwise fabrication needs to be maintained while changing the direction. One of the reasons for the increase is that a layer is inflexible. If the size of a part is big or the size of its cross-section is big, the size or perimeter of the layer will be big. The perimeter of a layer is inflexible and constant; it comes along with a 3D model. If there is any flexibility, it is in the thickness of a layer, which can be varied. In the case of a change in the orientation of a part having unequal dimensions, the perimeter can be changed but is still decided by the model of the part and is therefore inflexible during fabrication.

In layerwise fabrication, it is the layer that is the basic building block—it is not because the layer is not made from smaller blocks but because unless a complete layer is fabricated, the fabrication does not progress to the next layer. There is no such possibility as to delay the fabrication of the fraction of a layer till the next two or three layers are formed.

One of the problems of layerwise fabrication is that the layer perimeter is always big—it does not imply that the method is only meant to make big parts—it implies that the layer size is always very big in comparison to the size of the tool that creates the layer. The size of the tool means anode size in this process, laser spot size in LPBF, e-beam spot size in EPBF, liquid drop size in IJP, or extrudate diameter in FDM.

What if the layer size is equal to the tool size? There will be no need for scanning, there will be no more problem finding the right overlap between adjacent scanned lines, and there will be no need to be careful to integrate every bit of such a big layer. There will also be no need to search for the right parameters to join such big layers. Thus, fabricating with a layer size equal to the tool size gives advantages over fabricating with a big layer size.

In a usual layer-tool setting, the layer, by virtue of being a layer, is big, while the tool, by virtue of being a tool, is small—these sizes are two extremes. Selecting a basic building block, smaller than and other than a layer, will provide a compromise between two extremes.

A layer has some finite thickness and is similar to a rectangular plate that can be considered as an assembly of many smaller cubes or volumes that are similar to a cube-like shape. These cubes make a layer and then make a 3D model. These cubes may not make a layer and then make a 3D model. A cube may look like a layer, and a layer may look like a cube. But the difference between a cube and a layer as a basic building block is build direction. An assembly of layers has only one fixed build direction that is perpendicular to all layers, since an assembly of layers can have only one normal passing through all layers.

An assembly of cubes has only one fixed build direction that is perpendicular to all cubes if a cube is a layer. An assembly of cubes can have many build directions. Since a cube has six faces, it can have a maximum of six directions moving away from six faces. Since fabrication happens on a substrate or a platform, there cannot be any build direction toward the substrate, resulting in a maximum of five build directions. It is not usual but also not impossible to have a case of six build directions where fabrication will take place without using a substrate—such as in two-photon polymerization, where fabrication happens in the middle of liquid and can proceed in any direction.

A 3D model, consisting of such cubes or cube-type elements also referred to as voxels, can have many build directions if the algorithm is well developed and will not be affected by the limitations of layerwise fabrication. Using voxels instead of layers in an example of pillar and arm, there will be no more need for orienting the cathodic plate or using more than a single CAD file to overcome the limitations of layerwise fabrication.

References

1. Zhang, H., Chen, K., Kang, C., & Liu, H. (2024). Atomic-scale understanding of microstructural evolution in electrochemical additive manufacturing of metallic nickel. *Materials & Design, 245*, 113288.
2. Uschpovskiy, D., Babchuk, R., Kotyk, M., et al. (2024). Electrochemical additive manufacturing of copper parts: Printed material properties vs. traditionally deposited. *Journal of Solid State Electrochemistry*, 1–9.
3. Habib, M. A., Gan, S. W., & Rahman, M. (2009). Fabrication of complex shape electrodes by localized electrochemical deposition. *Journal of Materials Processing Technology, 209*(9), 4453–4458.
4. Lin, J. C., Chang, T. K., Yang, J. H., et al. (2010). Localized electrochemical deposition of micrometer copper columns by pulse plating. *Electrochimica Acta, 55*(6), 888–1894.

5. Brant, A., & Sundaram, M. (2016). A novel electrochemical micro additive manufacturing method of overhanging metal parts without reliance on support structures. *Procedia Manufacturing, 5*, 928–943.
6. Manukyan, N., Kamaraj, A., & Sundaram, M. (2019). Localized electrochemical deposition using ultra-high frequency pulsed power. *Procedia Manufacturing, 34*, 197–204.
7. Singh, P., & Dutta, D. (2001). Multi-direction slicing for layered manufacturing. *Journal of Computing and Information Science in Engineering, 1*(2), 129–142.
8. Coupek, D., Friedrich, J., Battran, D., & Riedel, O. (2018). Reduction of support structures and building time by optimized path planning algorithms in multi-axis additive manufacturing. *Procedia CIRP, 67*, 221–226.

Chapter 57
EPBF (Electron Beam Powder Bed Fusion)

A schematic diagram of electron beam powder bed fusion (EPBF) is given in Fig. 57.1, which shows the following requirements to realize the process: (1) beam generation, (2) beam manipulation, (3) vacuum chamber, and (4) powder bed [1, 2].

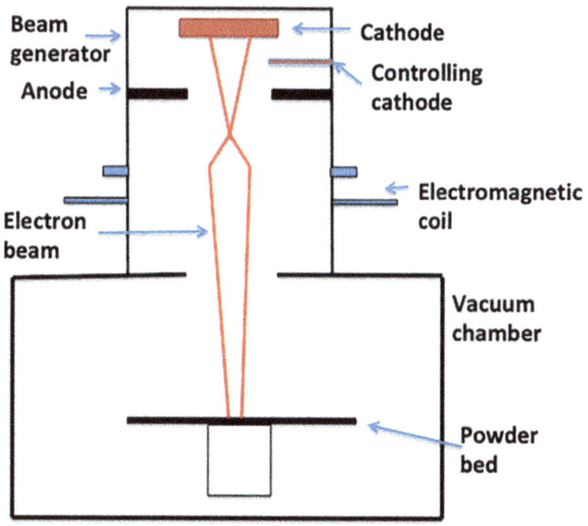

Fig. 57.1 Schematic diagram of EPBF

S. Kumar, *A Concise Encyclopedia of Additive Manufacturing*,
https://doi.org/10.1007/978-3-031-98394-8_57

References

1. Kahnert, M., Lutzmann, S., & Zaeh, M. F. (2007). Layer formations in electron beam sintering. In *18th solid freeform fabrication symposium, SFF 2007 - Austin, TX, United States* (pp. 88–99).
2. Schultz, H. (1993). *Electron beam welding* (Vol. 93). Woodhead Publishing.

Chapter 58
EPBF: Beam Generation

Electron beam (E-beam) is required to melt and sinter powders: melting for shaping while sintering for fixing the powder at a powder bed (PB) when required [1]. The beam is generated by electrically heating a cathode made from a material having a high melting point and low work function. Due to the high melting point, the cathode does not melt and degrade during electron generation, while due to the low work function, electrons detach from the cathode at a low applied voltage. Work function is a physics term that states how much voltage needs to be applied to a material to detach an electron from its surface. The lower the work function, the better the material for e-beam generation.

Tungsten (W) and lanthanum hexaboride (LaB_6) are used as a cathode material, each of which has a melting point and work function of 3422 °C, 4.5 eV, and 2210 °C, 2.5 eV, respectively. W is used because of its high melting point, while LaB_6 is used because of its lower work function and higher lifetime as a cathode [2]. Besides generating a beam by heating the cathode, which is called thermionic emission, it is generated by creating plasma between the cathode and the anode [3]. The beam, thus produced, is used in AM [4].

The generated electron (charge = -1.6×10^{-19} C, mass = 9.1×10^{-31} kg) moves toward the anode. The movement of electrons constitutes a beam. The higher the number of electrons, the higher the current of the beam. Thus, to have a high current, a large number of electrons need to be generated. To have a low current, either a smaller number of electrons needs to be generated or some electrons need to be stopped from reaching the anode. For stopping the electrons, placing another cathode just before the anode serves the purpose, as the cathode displaces some of the electrons before they reach the anode, which decreases the current. The anode is made perforated so the remaining electrons do not stop at the anode but pass through it so that they can be manipulated for processing PB.

The speed of an electron depends on the voltage applied between the cathode and anode. The higher the voltage, the higher the velocity of the electron, as per Eq. (58.1a):

S. Kumar, *A Concise Encyclopedia of Additive Manufacturing*,
https://doi.org/10.1007/978-3-031-98394-8_58

$$v = c \times \sqrt{1 - \frac{1}{\left(1 + \frac{e \times V}{m \times c^2}\right)^2}} \tag{58.1a}$$

where v, e, m are the velocity, charge, and mass of an electron, respectively. V is an applied voltage, and c is the speed of light [5]. For V of the order of kV, the speed of the electron is of the order of c, which brings relativistic effect in m, which can not be neglected if voltage is more than 100 kV [6]. Thus, m will increase by $\frac{1}{\sqrt{1 - \frac{v^2}{c^2}}}$.

The power of an e-beam having current I is given by Eq. (58.1b):

$$P = V \times I \tag{58.1b}$$

Thus, the power can be increased either by increasing the voltage or current. With an increase in the voltage, electrons move fast and go deep inside PB, while with an increase in the current, more electrons reach PB. Consequently, for the same value of power, there will be a number of different physical effects on PB depending upon the value of V and I. This is the reason why V and I (rather than P) are independent variables in EPBF.

To increase P, increasing V rather than I is a better option. Since, with an increase in I, the number of electrons increases, giving rise to an increased charging of PB, causing a disturbance on PB. Increasing V causes an increase in the momentum of the electron, but due to a huge difference in the mass of the electron and of the powder and a small beam-powder interaction time, the impinging electron is not able to displace the powder, causing no disturbance on PB. Though the high velocity of the electron causes an increase in the mass of the electron due to relativity, the increase is still minuscule to move the powder position.

References

1. Sun, S. H., Hagihara, K., Ishimoto, T., et al. (2021). Comparison of microstructure, crystallographic texture, and mechanical properties in Ti–15Mo–5Zr–3Al alloys fabricated via electron and laser beam powder bed fusion technologies. *Additive Manufacturing, 47*, 102329.
2. Edinger, R. (2018). Laser heated electron beam gun optimization to improve additive manufacturing. In *Solid freeform fabrication 2018: Proceedings of the 29th annual international* (pp. 2297–2304).
3. Bakeev, I. Y., Klimov, A. S., Oks, E. M., & Zenin, A. A. (2018). Generation of high density electron beams by a forevacuum- pressure plasma cathode electron source. *Plasma Sources Science and Technology, 27*, 075002.
4. Lee, H. J., Ahn, D. G., Song, J. G., et al. (2017). Fabrication of beads using a plasma electron beam and stellite 21 powders for additive manufacturing. *International Journal of Precision Engineering and Manufacturing-Green Technology, 4*, 453.
5. Sigl, M., Lutzmann, S., & Zaeh, M. F. (2006). Transient physical effects in electron beam sintering. In *2006 international solid freeform fabrication symposium* (pp. 464–477).
6. Krumeich, F. *Properties of electrons, their interactions with matter and applications in electron microscopy*. Laboratory of Inorganic Chemistry, ETH Zurich.

Chapter 59
EPBF: Beam Manipulation

An e-beam generated from a cathode and passing through a perforated anode will strike a limited area of powder bed (PB), i.e., an area equivalent to the beam spot size, which is on the order of a millimeter or less. It will not help make parts bigger than the spot size unless the e-beam generator moves relative to PB, or vice versa. Making parts by these relative movements can be helpful for investigating the effect of beam parameters on materials. But this has limited utility in a commercial application due to the lack of accuracy and speed.

Another option is to displace the e-beam from its original path so it will cover a wider area of PB and be able to make bigger parts. Since an e-beam consists of charged particles, it can be displaced by an applied electric field. Besides, the charged particle moves, which creates a magnetic field that can be displaced by an applied magnetic field. Hence, an e-beam can be displaced both by an applied electric and magnetic field, which is not possible for a laser beam as it does not consist of charged particles. These show that an electromagnetic field that has both electric and magnetic field components can be suitable for e-beam manipulation.

An electron of charge e moving with a velocity v will feel a force F, called Lorentz Force, in an electromagnetic field as per Eq. (59.1a):

$$F = e \times (E + v \times B) \tag{59.1a}$$

where E and B denote an electric field and a magnetic field, respectively. F acts normal to the plane made by v and B. Figure 59.1a shows x-, y-, and z-directions in Cartesian coordinates.

To comprehend the role of the magnetic field on a moving electron, the electric field is considered negligible, which changes Eqs. (59.1a) to (59.1b).

$$F = e \times (v \times B) \tag{59.1b}$$

S. Kumar, *A Concise Encyclopedia of Additive Manufacturing*,
https://doi.org/10.1007/978-3-031-98394-8_59

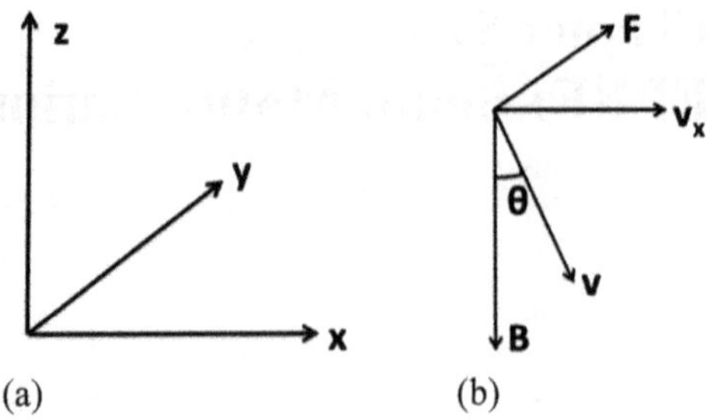

Fig. 59.1 (**a**) Cartesian coordinates, (**b**) Lorentz force *F* on an electron moving with velocity *v* in a magnetic field *B*

Expanding Eq. (59.1b) in terms of its components in the x-, y-, and z-directions, it can be shown by Eqs. (59.1c), (59.1d), and (59.1e).

$$F_x = e\left(v_y B_z - v_z B_y\right) \tag{59.1c}$$

$$F_y = e\left(v_z B_x - v_x B_z\right) \tag{59.1d}$$

$$F_z = e\left(v_x B_y - v_y B_x\right) \tag{59.1e}$$

To find a resulting force on an electron moving with a velocity *v* in the z-x plane and along a direction at an angle θ with *B* (which is in the z-direction) (Fig. 59.1b), the following information is used in Eqs. (59.1c), (59.1d) and (59.1e):

$B_y = 0$, $B_x = 0$ since *B* is in z-direction, $v_y = 0$ since *v* is along the z-x plane. It is found.

$F_y = -ev_xB_z$ while $F_x = 0$, $F_z = 0$.

Since, $v_x = vSin\theta$, $B = -B_z$

$$F = F_y = -ev_xB_z = evBSin\theta \tag{59.1f}$$

Equation 59.1f states the direction and magnitude of the Lorentz force acting on a moving electron. The force is in the y-direction and acting on the z-x plane (Fig. 59.1b).

The consequence of this force on a moving electron is known by observing the direction of the force. Since the magnetic field acts in a circular direction on a moving charged particle (electron), it acts in a clockwise direction on the electron to cause a resultant force in the y-direction. It is similar to the rotation of a screw that, if it rotates in a clockwise direction from v to B through a smaller angle in the z-x plane (Fig. 59.1b) causes an advancement in the positive y-direction. Vice versa, if there is a force in the y-direction, it means there was a rotation of electrons clockwise. Observing the direction of F in Fig. 59.1b, it is known whether electron direction is clockwise or counter-clockwise. Clockwise implies displacement of the electron toward the magnetic field, while counterclockwise implies away from the magnetic field.

In the present case, as per Fig. 59.1b, displacement of electrons is toward the magnetic field. How far an electron will displace depends upon the magnitude of v

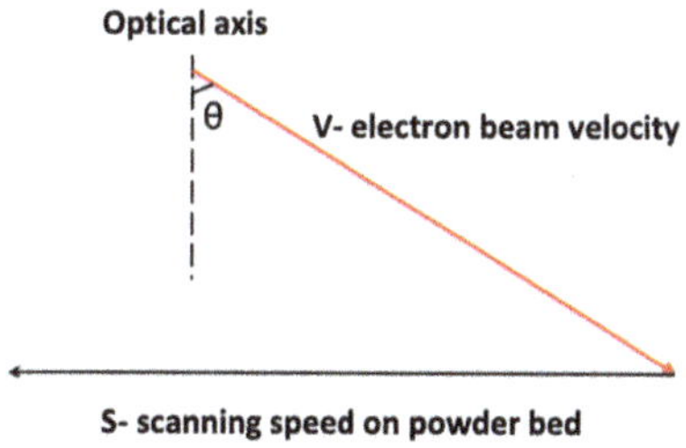

Fig. 59.2 An e-beam with velocity v scans powder bed with scan speed s

and B. If B is large or there is a strong magnetic field, the force will be large enough on the electron to align it with the magnetic field. With an alignment, angle θ will be zero, and, as per Eq. (59.1f), force will be zero, causing the electron to move indefinitely parallel to the magnetic field. This concept is utilized to focus the beam in EPBF.

Using $\theta = 0$ in Eq. (59.1f), $F = 0$ gives another inference that when a magnetic field is applied parallel to the beam, it has no effect on the electron, while for any other value of θ, the beam is deflected. It implies that the magnetic field can not be used to accelerate an electron along the optical axis; acceleration can only be accomplished by an applied voltage.

In case magnetic field B is reversed so $B = B_z$, it will change the equation to $F_y = -ev_xB$, which is acting in the negative y-direction normal to the z-x plane, implying a counterclockwise motion of vector v. It will make the angle θ bigger.

This shows that by reversing the magnetic field direction, electrons will be displaced away from the magnetic field. This will help the beam to move away from its original path and scan a wider length. By rapidly fluctuating magnetic field direction from positive to negative, the beam direction will change fast, while by changing the strength of the magnetic field, the beam velocity on PB (scan speed, Fig. 59.2) can be changed. This shows that there are two ways by which scan speed on PB can be changed—by changing either applied voltage or magnetic field. For scanning along more axes, more electromagnets (as sources of magnetic fields) are required. This concept is utilized in scanning the beam.

Figure 59.2 shows scanning of PB along a line using an e-beam moving with a velocity v. Scan speed s does not depend upon the beam velocity but depends upon the speed with which the beam is deflected toward the optical axis, which depends upon the magnitude of the magnetic field. If a high magnetic field is applied, the beam travels fast on PB, imparting fewer electrons on the bed. If the field is low, scanning speed will be slow, imparting a high number of electrons on the bed, resulting in a high effective current on the bed, and implying higher energy density applied.

For a constant applied voltage, changing applied current is a means to control the number of electrons on PB. If scan speed is set higher, resulting in a lower number of electrons, the applied current can be increased to compensate for a decrease in electron numbers, resulting in no change in the current density with an increase in scan speed. Thus, the scan speed and applied current can be adjusted to maintain the desired energy balance on the bed. Finding the right energy density by making an adjustment toward higher scan speed will result in a high production rate.

Chapter 60
EPBF: Beam Shape Control

1 Focusing of Beam

Focusing of an e-beam is done by applying a magnetic field parallel to an optical axis along which the beam is required to travel. The direction of the magnetic field, optical axis, and velocity of the electron is shown in Fig. 60.1a. An e-beam, after being accelerated by an applied voltage, does not exactly move in a straight line along the optical axis but moves in a curved path due to the presence of other electromagnetic fields (Fig. 60.1a). If an e-beam entering the magnetic field makes an angle with the optical axis, it will be deflected by the field and will become parallel with the optical axis (Fig. 60.1b).

Applying a magnetic field parallel to an optical axis helps achieve focusing of a beam, but it can also be achieved by applying the field perpendicular to the axis (Fig. 60.2). If the beam is deflected from the optical axis, it can be brought back to the same path by applying B perpendicular to it (Fig. 60.2a), in which case, the beam is making an angle θ wih B. The magnitude of the field should be set to a particular value so the resulting force will deflect it by an angle $\theta - 90°$ and align it with the axis, causing focusing of the beam.

To execute it, two magnets are attached on the periphery of an e-beam column—north pole of one magnet while the south pole of another magnet is facing the periphery (Fig. 60.2b), which generates magnetic fields. The direction of the field and of the force are perpendicular to each other (Fig. 60.2c). These magnetic poles displace the beam along the x-direction. To displace it along the y-direction, two more magnetic poles need to be set on the periphery normal to the fitted poles. These four magnetic poles will force alignment of an e-beam making any angle with the optical axis, causing focusing of the beam.

S. Kumar, *A Concise Encyclopedia of Additive Manufacturing*,
https://doi.org/10.1007/978-3-031-98394-8_60

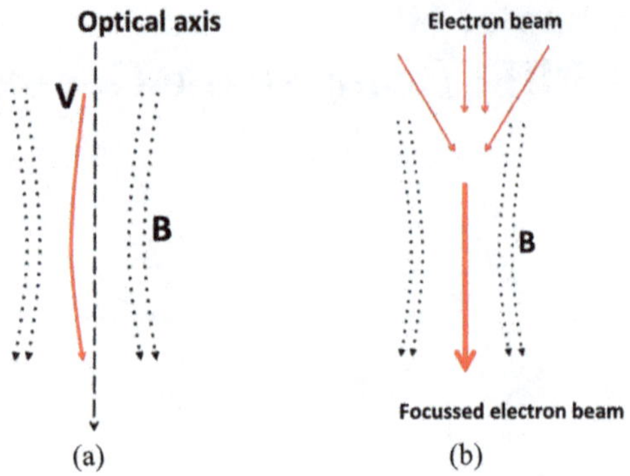

Fig. 60.1 An electron moving with velocity *v* in a magnetic field *B* (**a**) approximate beam path [1], (**b**) focusing of e-beams entering the field at different angles

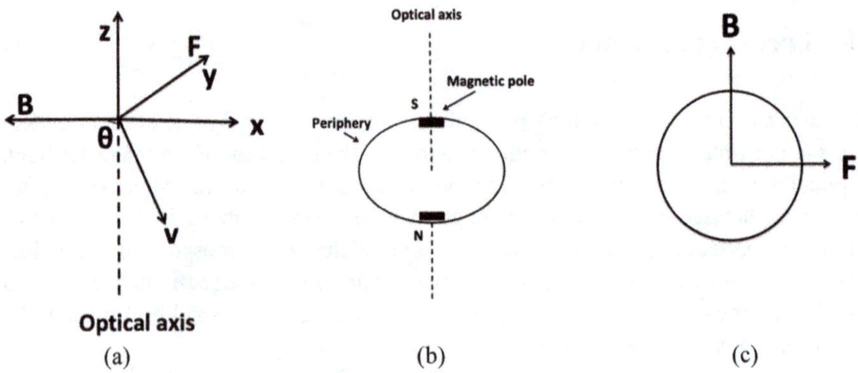

Fig. 60.2 Focusing of e-beam by deflection (**a**) directions of B, F and v in Cartesian coordinates, (**b**) position of electromagnet on the periphery of e-beam column, (**c**) directions of B and F

2 Beam Shape Control

Beam during scanning may not retain its original shape and deform to various shapes. Figure 60.3a shows a beam of original spherical shape that deforms, resulting in elongation along AB while compression along CD (Fig. 60.3b). The beam can be brought to its original shape if compressive force acts along AB and pulling force acts along CD (Fig. 60.3c). Electrons at point D of the beam need to be deflected outwards, while electrons at point A need to be deflected inwards to bring them back to their original position. As per Fig. 60.2a, the former deflection can be achieved by applying B along the x-axis, orthogonal to the optical axis (z-axis), while the latter inward deflection can be achieved by applying another magnetic field in the negative direction along the y-axis—this requires an arrangement of four magnetic poles.

This can be achieved by a quadrupole magnet (Fig. 60.3d), which applies the force exactly in the same direction as required [1]. The figure shows that four poles

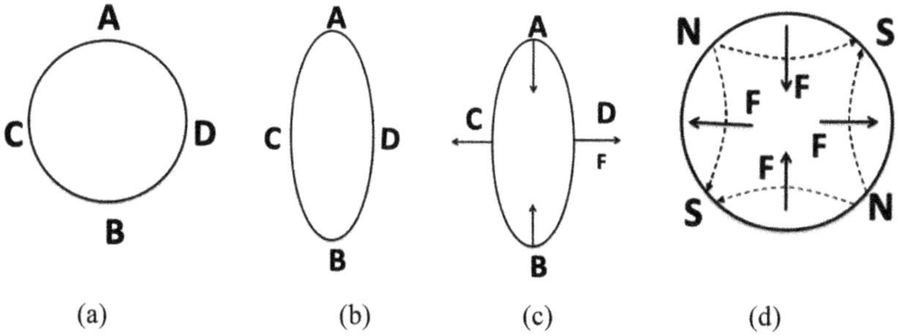

Fig. 60.3 (**a**) original beam shape, (**b**) deformed beam shape, (**c**) direction in which force needs to be applied for regaining the shape, (**d**) quadrupole magnet as a device to apply required force

are placed alternatively, magnetic lines are travelling from N to S, creating two opposite forces: one in the x-axis and the other in the y-axis. For a higher degree of deformation, a higher number of poles of equal number of similar poles are required.

Reference

1. Azhirnian A, Svensson D (2017) Modeling and analysis of aberrations in electron beam melting (EBM) systems. Master thesis, Chalmers university of technology.

Chapter 61
EPBF: Chamber

A vacuum chamber is an essential part in which e-beams travel and interact with a powder bed (PB). In the absence of a vacuum, electrons will interact with air molecules and ionize them, thus losing energy, speed, and direction during travel, causing difficulty in processing PB. In some EPBF systems, a vacuum of 10^{-5} mbar is maintained, which is sufficient to allow unhindered movement of e-beams. Besides, the vacuum enables reactive metals to be processed without getting reacted by oxygen or nitrogen gas [1].

Vacuum has disadvantages—it decreases the melting point of metals and increases their evaporation rate. For an alloy made up of two different metals of different melting points, the evaporation rate of the lower melting point metal will be higher, resulting in its higher loss, causing a change in alloy composition. This is noticed in Ti6Al4V alloy, where the higher loss of Al changes the composition.

Interaction of the e-beam with powder leads to the transfer of electrons to the powder, causing its charging. Accumulation of charges leads to coulombic repulsion between the same charges, which destabilizes powders' positions and brings disruption in a layer formation. When charges are high, the resulting repulsion is higher than the inertia of powders, leading to their displacement. The presence of helium gas (10^{-3} mbar partial pressure in Arcam systems) in the vacuum chamber mitigates this effect. Helium has a low atomic number, and at a pressure of 10^{-3} mbar, the number of helium atoms is not large enough to disturb the beam. But these numbers are big enough to take a substantial amount of charges away from charged

S. Kumar, *A Concise Encyclopedia of Additive Manufacturing*,
https://doi.org/10.1007/978-3-031-98394-8_61

powders by being present in the vicinity of these powders, causing mitigation of the disruption of the powder bed.

Reference

1. Korner, C. (2016). Additive manufacturing of metallic components by selective electron beam melting- a review. *International Materials Reviews, 61*(5), 361–377.

Chapter 62
EPBF: Processing

After layer formation, if powders get attached to a substrate or previously deposited layer, the incoming electrons will not be able to accumulate on the powder to disturb it. Attaching the powder can be done either by sintering or melting it. Melting or partial melting of powders cannot be an option, as all powders that are not melted by e-beams need to be detached and removed from the final part. As melting is the way for fabricating, the powders that are melted contribute to fabrication. Therefore, the powders are not supposed to be melted for any purpose other than contributing to the fabrication.

This leaves another option, i.e., sintering, to be exercised for making connections among powders and with the substrate [1]. Sintering implies that the powder will join another powder or substrate by forming a neck through atomic diffusion [2]. Formation of the neck increases electrical conductivity of the layer and enhances contacts between the substrate and powders, which helps dissipation of charges to the ground, resulting in removal of the cause (charges) responsible for dislodging of powders. A minimal size of neck obtained after some moment of sintering is sufficient for charge transfer and stabilization of PB. Since the role of the neck is to act as a conduit to transfer charge rather than to act as a fixture to hold a powder, a large powder [3], because of its higher inertia rather than a smaller powder, has a higher chance not to be displaced during charging. This makes the larger powder more suitable for EPBF.

Sintering is accomplished by the application of heat with the help of heaters and e-beams. A heater attached under a substrate is used to increase its temperature. When a layer comes in contact with it, powders of the layer are sintered. The low temperature of the heater increases the sintering time, which decreases the fabrication rate. While the high temperature increases the sintering of unmelted powders of previously deposited layers, resulting in difficulty in the removal of the powders from the part, which compromises fine features. This can be avoided by the use of a low heater temperature combined with heating the layer with the help of an e-beam. This will ensure no decrease in the fabrication rate and the nonoccurrence of

S. Kumar, *A Concise Encyclopedia of Additive Manufacturing*,
https://doi.org/10.1007/978-3-031-98394-8_62

oversintering. To increase the local temperature of the layer, high-power e-beams with high speed are scanned over the whole layer several times, which causes necessary sintering. An e-beam moving at high speed does not impart enough electrons at a point to cause disturbance on the layer.

E-beam for the purpose of sintering is used at several stages—on the substrate (as a substitute for the heater), on the deposited and processed layer (for helping the heater), and after processing all layers for annealing. Any of these stages increases the temperature of a processing chamber. A hot chamber gives advantages: decrease in thermal stress, cracks, and porosity, and disadvantages: decrease in strength in several alloys, grain size increase, etc.

References

1. Yan, W., Ma, W., & Shen, Y. (2020). Powder sintering mechanisms during the pre-heating procedure of electron beam additive manufacturing. *Materials Today Communications, 25*, 101579.
2. Rizza, G., Galati, M., Antonioni, P., & Iuliano, L. (2023). Effect of the sintering conditions on the neck growth during the powder bed fusion with electron beam (PBF-EB) process. *Journal Of Manufacturing And Materials Processing, 7*(2), 55.
3. Cordero, Z. C., Meyer, H. M., III, Nandwana, P., & Dehoff, R. R. (2017). Powder bed charging during electron-beam additive manufacturing. *Acta Materialia, 124*, 437–445.

Chapter 63
Electrophoretic Deposition

Deposition of metal ions using salt solutions does not allow many types of materials to be deposited. A colloidal solution provides versatility in materials' choice, as materials of any type, such as polymers, metals [1], or ceramics [2], can be deposited, as they are available in the form of colloids. These, unlike metal ions, can be of either charge, depending on the types of additives and solutions. These can be deposited on either electrode, which can be termed cathodic or anodic electrophoretic deposition [3]. Voltage is applied across electrodes to drive these colloidal charged particles, while for ionic solution-based AM, voltage is applied to ionize the solution and drive ions.

A solvent is chosen such that it will not ionize readily. If it ionizes, ions will change pH of the solution and disturb the stability of colloids, or ions will instead be deposited and interfere with the deposition of charged colloidal particles. The particles need to be smaller than one μm so their positions in the solution will not be affected by gravity. If particles are bigger, they will either sediment or be sedimenting.

If they are sedimenting, the solution will have fewer particles on the upper side of the solution while more particles are on its bottom side. It will result in a thinner deposition on the top and a thicker deposition on the bottom of a vertical electrode. Thus, using bigger particles will not provide a deposition of uniform but graded thickness. Since addition happens atom by atom, the process is slow, making it suitable for small parts.

Conductivity of an electrode is important for a uniform deposition to occur. If the conductivity is low, there will be a poor deposition on the electrode. This fact can be utilized to facilitate the deposition at a particular area on a low-conductive electrode by increasing the conductivity of that particular area. It can be increased by irradiating a photoconductive electrode with light [4].

If a photoconductive electrode plate is used, it can be irradiated on its back side to create deposition on its front side that is in contact with colloidal solution. If the irradiation makes a pattern, the deposition of charged particles will make a pattern

S. Kumar, *A Concise Encyclopedia of Additive Manufacturing*,
https://doi.org/10.1007/978-3-031-98394-8_63

on the front side. Thus, a 3D structure, though of limited geometry, can be created [5]. The technique can be applied to make an energy storage device [6, 7].

References

1. Nakazawa, K., Ozawa, S., & Iwata, F. (2021). Additive manufacturing of metal micro-ring and tube by laser-assisted electrophoretic deposition with Laguerre–Gaussian beam. *Nanomanufacturing And Metrology, 4*(4), 271–277.
2. Vogt, L., Schäfer, M., Kurth, D., & Raether, F. (2019). Usability of electrophoretic deposition for additive manufacturing of ceramics. *Ceramics International, 45*(11), 14214–14222.
3. Pikalova, E. Y., & Kalinina, E. G. (2019). Electrophoretic deposition in the solid oxide fuel cell technology: Fundamentals and recent advances. *Renewable and Sustainable Energy Reviews, 116*, 109440.
4. Dhungana, N., Nakazawa, K., & Iwata, F. (2024). Improve reproducibility of 3D microfabrication by laser-assisted electrophoretic deposition using feedback stage control. In *Proceedings of JSPE spring conference* (pp. 715–716). The Japan Society for Precision Engineering.
5. Mora, J., Dudoff, J. K., Moran, B. D., et al. (2018). Projection based light-directed electrophoretic deposition for additive manufacturing. *Additive Manufacturing, 22*, 330–333.
6. Chakrabarti, B. K., Gençten, M., Bree, G., et al. (2022). Modern practices in electrophoretic deposition to manufacture energy storage electrodes. *International Journal of Energy Research, 46*(10), 13205–13250.
7. Martinez, A. C., Maurel, A., Yelamanchi, B., et al. (2024). Combining 3D printing of copper current collectors and electrophoretic deposition of electrode materials for structural lithium-ion batteries. *Advances in Manufacturing, 13*, 1–14.

Chapter 64
Energy Saving

AM saves energy using the following means: repair, efficient parts, lightweighting, assembly, creating energy devices, and material recycling.

1 Repair

AM can remanufacture a damaged part and help it come back into use [1]. Otherwise, the energy spent on its fabrication will be lost. If it is not recycled, its material, along with the energy spent on producing the material, is lost. Thus, AM saves energy associated with the fabrication of a part and the production of its material.

2 Efficient Part

AM is able to make parts that last long. For example, a turbine blade having an internal cooling channel lasts longer, which leads to a decreased need for its production, which saves production energy. By making an effective insulation part, energy is saved from getting dissipated [2]. Another example is by making better powertrain components that do not let the energy be wasted due to the absence of better powertrain components [3].

3 Lightweighting

AM makes lightweight vehicle parts [4], which save fuel energy because, being lighter in weight, the vehicle needs less fuel [5, 6].

S. Kumar, *A Concise Encyclopedia of Additive Manufacturing*,
https://doi.org/10.1007/978-3-031-98394-8_64

Producing 1 kg of iron powder consumes approx. 7–15 MJ of energy, depending upon how it is produced [7]. Production will consume more energy for spherical powders having a narrow size distribution.

If AM makes a part weighing 1 kg, fabrication will require around 40 kg of powder. The actual amount depends upon the type of system and the material density. The remaining 39 kg of powder can develop some form of degradation because of being present in the vicinity of the part during fabrication. These degradations are in the form of a shape change, oxidized layers on powder, satellites on a powder surface, changes in the chemical composition, microstructure, and phase changes. Shape and size changes will change powder size distribution, which will change the powder flowability and spreadability during layer formation [8]. The used powder does not give the same part property. Depending on the material type, some are wasted, while others turn into waste after reuse.

A part lighter by 100 g saves around 10 MJ in the lifetime of an automobile vehicle. While used in an aircraft, the same weight difference saves around 10 GJ [9].

Consider an example of a part made by conventional manufacturing (CM), weighing 1.1 kg. If AM redesigns and fabricates it to weigh 1 kg, AM makes it lighter by 100 g. If a vehicle uses it, the vehicle becomes lighter by 100 g and saves maximum fuel energy, i.e., 10 MJ. For fabricating it, 1 kg of powder is consumed, while 39 kg of powder goes through some changes. If even 1 kg out of 39 kg is completely wasted, it does not have the advantage of energy saving. Since producing 1 kg of powder requires 10 MJ, while the part saves only 10 MJ in the lifetime of an automobile vehicle. Even if there are no other factors that increase the energy requirement of production, such as powder spillage, extra consumption of beam energy, and noise, AM does not have an advantage over CM for saving energy.

AM will have an advantage if there is a high difference between the weight of two parts (made each in CM and AM). If the parts are made big so the difference between them is not in grams but in kilograms, AM helps save energy substantially. The fabrication of a big AM part consumes high energy, causing both saving and wasting of energy—wasting because it is big while saving because it makes a vehicle substantially lighter.

The same weight difference, i.e., 100 g, saves energy of 10 GJ when an aircraft uses the part, implying no need to make a big part for aircraft to save energy. Substantial energy savings mean even if fabrication wastes powder, net energy is saved.

4 Assembly

If many sub-parts are fabricated and then assembled, the energy spent on assembling can be saved if a single part [10] instead of separate sub-parts is fabricated. The energy can only be saved if fabricating a single part requires less than the combined energy (energy for fabricating sub-parts plus energy for assembly). In AM, an assembled part can be made all at once, which is not convenient in CM. In CM, sub-parts are made by casting or machining and assembled by welding.

If it is an interlocked part, one of the metallic rings is made by machining, casting, or forging [11], and the other ring connects to it by cutting and welding, while it is made conveniently in AM. However, if the gap between two rings is small or the clearance in a joint is small, the resolution of an AM process limits fabrication, which does not give an advantage over CM for assembly but gives an advantage for design.

5 Devices for Creating Energy

AM helps develop energy materials and makes parts for devices used in energy production, such as battery components, and thus contributes to creating energy [12–14]. Though making devices for creating energy is not equal to energy saving. However, when it makes, its demerit of wasting energy is diluted.

6 Material Recycling

When AM uses recycled or waste material, it saves the energy that could have been employed to produce virgin materials in the absence of the lack of utilization of recycled or waste materials [15–18].

7 Metal Depletion

Use of machining depletes metal resources more than AM. Therefore, the use of AM helps save the energy that was used to mine the metal [19].

References

1. Walachowicz, F., Bernsdorf, I., Papenfuss, U., et al. (2017). Comparative energy, resource and recycling lifecycle analysis of the industrial repair process of gas turbine burners using conventional machining and additive manufacturing. *Journal of Industrial Ecology, 21*, S203–S215.
2. Valente, M., Sambucci, M., Chougan, M., & Ghaffar, S. H. (2023). Composite alkali-activated materials with waste tire rubber designed for additive manufacturing: An eco-sustainable and energy saving approach. *Journal of Materials Research and Technology, 24*, 3098–3117.
3. Schuhmann, D., Merkel, M., & Harrison, D. K. (2023). Highly efficient and resource-saving function-integrated additively manufactured components for the mobility of tomorrow. In *Quality analysis of additively manufactured metals* (pp. 781–813). Elsevier.

4. Zhao, N., Parthasarathy, M., Patil, S., et al. (2023). Direct additive manufacturing of metal parts for automotive applications. *Journal of Manufacturing Systems, 68*, 368–375.
5. Ingarao, G., Priarone, P. C., Deng, Y., & Paraskevas, D. (2018). Environmental modelling of aluminium based components manufacturing routes: Additive manufacturing versus machining versus forming. *Journal of Cleaner Production, 176*, 261–275.
6. Verhoef, L. A., Budde, B. W., Chockalingam, C., et al. (2018). The effect of additive manufacturing on global energy demand: An assessment using a bottom-up approach. *Energy Policy, 112*, 349–360.
7. Kruzhanov, V., & Arnhold, V. (2012). Energy consumption in powder metallurgical manufacturing. *Powder Metallurgy, 55*(1), 14–21.
8. Cordova, L., Campos, M., & Tinga, T. (2019). Revealing the effects of powder reuse for selective laser melting by powder characterization. *JOM, 71*, 1062–1072.
9. Hettesheimer, T., Hirzel, S., & Roß, H. B. (2018). Energy savings through additive manufacturing: An analysis of selective laser sintering for automotive and aircraft components. *Energy Efficiency, 11*, 1227–1245.
10. Boschetto, A., Bottini, L., Eugeni, M., et al. (2019). Selective laser melting of a 1U CubeSat structure. Design for Additive Manufacturing and assembly. *Acta Astronautica, 159*, 377–384.
11. Sun, M., Xu, B., Xie, B., et al. (2021). Leading manufacture of the large-scale weldless stainless steel forging ring: Innovative approach by the multilayer hot-compression bonding technology. *Journal of Materials Science and Technology, 71*, 84–86.
12. Zhakeyev, A., Wang, P., Zhang, L., et al. (2017). Additive manufacturing: Unlocking the evolution of energy materials. *Advanced Science, 4*, 1700187.
13. De, A., Ramasubramian, B., Ramakrishna, S., & Chellappan, V. (2024). Advances in additive manufacturing techniques for electrochemical energy storage. *Advanced Materials Technologies, 9*(4), 2301439.
14. Sahu, M., Hajra, S., Kim, H. G., et al. (2021). Additive manufacturing-based recycling of laboratory waste into energy harvesting device for self-powered applications. *Nano Energy, 88*, 106255.
15. Wu, H., Mehrabi, H., Karagiannidis, P., & Naveed, N. (2022). Additive manufacturing of recycled plastics: Strategies towards a more sustainable future. *Journal of Cleaner Production, 335*, 130236.
16. Fan, W., Wang, Y., Liu, R., et al. (2024). Textile production by additive manufacturing and textile waste recycling: A review. *Environmental Chemistry Letters, 22*(4), 1929–1987.
17. Stratiotou Efstratiadis, V., & Michailidis, N. (2022). Sustainable recovery, recycle of critical metals and rare earth elements from waste electric and electronic equipment (circuits, solar, wind) and their reusability in additive manufacturing applications: A review. *Metals, 12*(5), 794.
18. Dhiman, S., Joshi, R. S., Singh, S., et al. (2022). Recycling of Ti6Al4V machining swarf into additive manufacturing feedstock powder to realise sustainable recycling goals. *Journal of Cleaner Production, 348*, 131342.
19. Landi, D., Zefinetti, F. C., Spreafico, C., & Regazzoni, D. (2022). Comparative life cycle assessment of two different manufacturing technologies: Laser additive manufacturing and traditional technique. *Procedia CIRP, 105*, 700–705.

Chapter 65
Energy Saving: Paradox

One of the advantages of AM is that it gives freedom to save energy. One of the disadvantages is that it is an energy-intensive process, implying that it wastes energy compared to other processes. What does it cause: saving or wasting?

AM is an energy-intensive process because it starts from a tiny building block, whether it is making a small or a large part. The size of a part is always bigger than the building block. AM does not have an advantage compared to other processes to start with a big one to make a big part. Machining starts with a big workpiece, forming processes with big sheet metal, and sintering with a big die (though a big die is not a big building block, sintering does not have the restriction to use only a small die)—all have options on how not to waste energy. AM does not have options, which is more pronounced when a big part is made and compared. For making small parts, the difference between AM and other processes will not be visible unless many parts are made.

What if there are not many small parts? AM makes only one. Is AM energy-intensive? If AM is supposed to make only one part, AM wastes less energy compared to when it makes many small parts or one big part. But AM is still energy-intensive, not because it wastes much energy but because other processes do not waste the same amount of energy as AM does. AM being energy-intensive is the result of its comparison with other processes. Imagining a situation in which all processes waste more energy than AM does to make one or more parts, AM is then not energy-intensive.

What if a part made by AM saves energy when in use? Then, energy is saved. But it does not change the fact that energy is not saved while the part is made. If the energy saved is more than the energy wasted during fabrication, AM saves net energy; otherwise, AM does not. But what the part does is not the same as what AM does. What a part does come under the application of AM. If AM saves energy, it is because its applications save energy, not because the characteristics of a process change. Its applications are as follows: the fabrication of an innovative part, repair, lightweight parts, etc.

S. Kumar, *A Concise Encyclopedia of Additive Manufacturing*,
https://doi.org/10.1007/978-3-031-98394-8_65

It does not mean if none of its applications saves energy, there is no way to save energy. Energy can still be saved. For example, if many parts are made together, the energy per part will be less. If that part consumes more energy to fabricate in other processes, AM saves energy. But AM saves energy not because it becomes non-energy-intensive but because it consumes the energy that is wasted when many parts are not made.

For example, if a small part is made in a big machine, it takes more energy than when the same part is made in a small machine. It is not because the characteristic of a process changes in a big machine, but because running the big machine requires more energy. Therefore, the fabrication of the part in the big machine entails the wastage of energy. If many parts are made in the machine, there is no wastage of energy, as whatever energy is required to run the machine is utilized (consumed during fabrication of these parts). This leads to the saving of energy. This saving is due to the management of the wastage of the energy. This management is possible only when there is a wastage of energy. And the wastage of energy is possible only when there is an additional energy that is not required for the execution of a process. Therefore, if there exists only a small machine that does not allow more than one part, AM will have no opportunity to utilize the additional energy and defeat the visibility of its energy intensiveness.

Chapter 66
Energy-Intensive Process

AM makes parts by joining, which requires energy. If it were only joining, it might not have required much energy—all pieces must be brought together to apply heat on them, which happens in sintering. But AM joins in a selective way. Considering powder bed fusion, when heat is applied, some of the heat that gets squandered is not utilized to join powder.

The squandering is due to the lack of absorption of heat, over-absorption, absorption by the other area of the bed, etc.; it is the side effect of joining. AM, compared to sintering, thus requires more heat to do the same amount of joining, which makes AM an energy-intensive process. For example, AM consumes 97 MJ/kg, while sintering consumes 2.4 MJ/kg to make a steel part [1]. Therefore, AM requires more electricity than CM [2].

The best way to stop heat wastage is not to use it. For example, when metal powders are not joined by melting but glued using an adhesive binder, heat is not expended to change their metallurgical states to join them [3].

If parts made using the binder need to be strong, they are treated in a furnace to go through sintering, which requires energy [4]. Energy is thus required twice: for fabrication and for post-process sintering. If conventional sintering makes a part, it requires energy only during sintering. The amount of energy required in post-process sintering and conventional sintering is not the same. Nevertheless, the energy comparison conveys that taking help of post-processing does not require less energy to fabricate a high-strength part through the AM route. It is therefore not possible to bypass the energy requirement by avoiding the energy-intensive process and compensating for the loss in properties by post-processing.

If AM is energy-intensive, it does not imply that it is not more energy efficient than CM [5]. It only implies AM may require more energy than CM to process a given area or volume. For example, if making a part does not lead to the same area to be processed, i.e., the processable layer area in AM is less than the processable surface area in machining, the energy spent in AM can be less.

S. Kumar, *A Concise Encyclopedia of Additive Manufacturing*,
https://doi.org/10.1007/978-3-031-98394-8_66

However, a complex part when formed in AM requires SS; the total energy spent is more due to the energy spent in fabricating SS. Though this does not negate the fact that with an increase in complexity, while machining consumes more energy, AM does not.

References

1. Azevedo, J. M. C., Serrenho, A. C., & Allwood, J. M. (2018). Energy and material efficiency of steel powder metallurgy. *Powder Technology, 328*, 329–336.
2. Van Sice, C., & Faludi, J. (2021). Comparing environmental impacts of metal additive manufacturing to conventional manufacturing. *Proceedings of the Design Society, 1*, 671–680.
3. Mirzababaei, S., & Pasebani, S. (2019). A review on binder jet additive manufacturing of 316L stainless steel. *Journal Of Manufacturing And Materials Processing, 3*(3), 82.
4. Lupi, G., de la Vega, F. M., Menezes, d., et al. (2025). Investigation on the thermal cycle and mechanical properties of Al 6061 produced by binder jetting. *Materials Science and Engineering A, 929*, 148129.
5. Giudice, F., Barbagallo, R., & Fargione, G. (2021). A Design for Additive Manufacturing approach based on process energy efficiency: Electron beam melted components. *Journal of Cleaner Production, 290*, 125185.

Chapter 67
Extrusion-Based AM

Extrusion in the extrusion-based deposition process implies extrusion of material in a continuous shaped form. The shaping of the material after it is out from a nozzle and before it reaches a substrate gives the process its name.

The shaping is not solely due to a nozzle but due to materials, temperature, environmental conditions, applied force, etc. But the role of the nozzle has more ability to separate extrusion-based deposition from other AM processes.

Extrusion-based deposition using solid polymers is fused deposition modeling, fused filament fabrication [1], fused pellet modeling [2], powder melt extrusion, composite extrusion modeling, etc. Ceramic-based extrusion deposition is fused deposition of ceramics, robocasting [3], direct writing [4], etc. Besides a variant of inkjet printing [5], liquid metal using extrusion [6] is another AM technique that relies on extrusion.

References

1. Singh, S., Singh, G., Prakash, C., & Ramakrishna, S. (2020). Current status and future directions of fused filament fabrication. *Journal of Manufacturing Processes, 55*, 288–306.
2. Wang, Z., Liu, R., Sparks, T., & Liou, F. (2016). Large scale deposition system by an industrial robot (I): Design of fused pellet modeling system and extrusion process analysis. *3D Printing and Additive Manufacturing, 3*(1), 39–47.
3. Feilden, E., Blanca, E. G. T., Giuliani, F., et al. (2016). Robocasting of structural ceramic parts with hydrogel inks. *Journal of the European Ceramic Society, 36*(10), 2525–2533.
4. Mondal, D., & Willett, T. L. (2020). Mechanical properties of nanocomposite biomaterials improved by extrusion during direct ink writing. *Journal of the Mechanical Behavior of Biomedical Materials, 104*, 103653.

S. Kumar, *A Concise Encyclopedia of Additive Manufacturing*,
https://doi.org/10.1007/978-3-031-98394-8_67

5. Zheng, S., Zlatin, M., Selvaganapathy, P. R., & Brook, M. A. (2018). Multiple modulus silicone elastomers using 3D extrusion printing of low viscosity inks. *Additive Manufacturing, 24*, 86–92.
6. Gannarapu, A., & Arda Gozen, B. (2019). Micro-extrusion-based additive manufacturing with liquid metals and alloys: Flow and deposition driven by oxide skin mechanics. *Extreme Mechanics Letters, 33*, 100554.

Chapter 68
Extrusion-Based AM: Feedstock

Filaments, pellets, and powders are used as feedstock. Filament-based deposition is FDM (Fig. 68.1a), while fused pellet modeling uses pellets. Powder-based deposition is powder melt extrusion and composite extrusion modeling (Fig. 68.1b).

Filaments are widely used feedstock, and almost all commercial systems use them, while few systems use pellets. The advantage of a filament is that it works not only as a feedstock but also as a machine part of an FDM system, because a moving filament acts as a piston to push molten filaments through a nozzle.

A filament brings simplicity to AM system development. This is one of the reasons why filament-based systems are in abundance. But if the filament is to act as a piston, it must possess more properties than required by a feedstock that needs to act only as a source of materials. This is the disadvantage, which demands extra effort in its development and places extra conditions on a material to be suitable to become a filament.

A filament should have enough rigidity that it will not buckle when it pushes the molten material, but it should not have so much rigidity that it will not bend when it needs to be driven in a non-straight path from its source. It should not be too weak to break when it pushes the material through. Moreover, it should not be made so strong that making it stronger means increasing its melting point, which in turn will cause difficulty in melting. Thus, it needs to satisfy conflicting property requirements, which excludes many materials from being converted into filaments and then into parts.

If there is an extrusion-based deposition that uses feedstock other than filaments, the process must not be deprived of the advantage gained from extrusion besides being free from the disadvantage of using filaments. If feedstocks such as pellets and powders are used, there will be a difference in how they are brought to a print head. They need a different setup for storing the feedstock and a screw or piston for extruding. But there will be no difference after they are melted and extruded. Using them provides the opportunity to use various polymers and their mixtures that otherwise would not have been possible, as various combinations of materials do not

S. Kumar, *A Concise Encyclopedia of Additive Manufacturing*,
https://doi.org/10.1007/978-3-031-98394-8_68

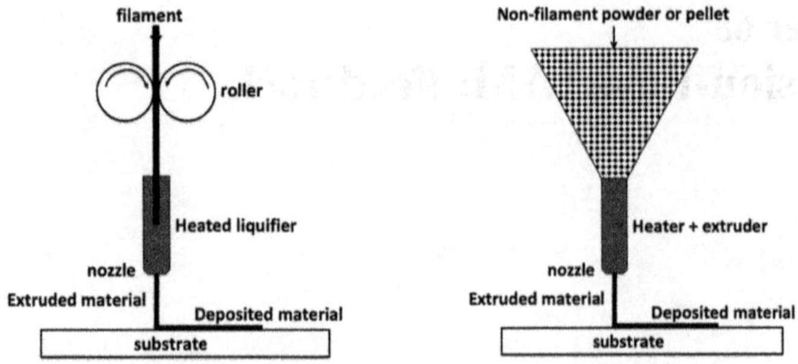

(a) filament type (b) non-filament type

Fig. 68.1 Schematic diagrams of extrusion-based deposition using: (**a**) filament, (**b**) non-filament

allow a filament to be drawn. This has facilitated the use of elastomeric, ceramic, and metallic materials in extrusion-based deposition. A metallic product can also be formed by using a composite feedstock made from metal and polymer and, in a downstream process, removing the polymer from the extruded part [1].

Reference

1. Singh, G., Missiaen, J. M., Bouvard, D., & Chaix, J. M. (2021). Additive manufacturing of 17–4 PH steel using metal injection molding feedstock: Analysis of 3D extrusion printing, debinding and sintering. *Additive Manufacturing, 47*, 102287.

Chapter 69
Four-Dimensional Printing

AM plus post-processing is as old as AM, but the aim of the post-processing has always been to improve properties. Four-dimensional printing (4DP) [1] is a relatively new technique that is also AM plus post-processing, but the aim of the post-processing is to change the shape rather than to change properties (Fig. 69.1). It does not mean that there will be no property changes in 4DP; properties will change, but changing the property is not the primary aim of 4DP. If there is no shape change in 4DP, it will not be called 4DP even if there is a change in properties.

4DP, as the name suggests, has four dimensions—three dimensions for space and an additional dimension for time. While an object of three dimensions is acquired by 3DP (or AM), neither the object nor its shape is final until the application of the fourth dimension is exercised. 4DP is a generic process that starts from various AM processes such as digital light processing [2, 3], stereolithography [4, 5], powder bed fusion [6], inkjet printing [7], and fused deposition modeling [8] and is thus not a variant of AM processes but is an application of AM. 4DP has been defined as "additive manufacturing of objects able to self-transform, in form or function when are exposed to a predetermined stimulus, including osmotic pressure, heat, current, ultraviolet, or other energy sources."

For post-processing, the part is kept in a certain environment where it changes shape due to the presence of pressure, heat, humidity, light, or chemicals. The part requires being made from a certain type of material that will respond to a particular stimulant when kept in the environment. Since two types of materials may change differently in an environment—if a part is made up of these materials, they will give rise to different shape changes. If they are treated in different environments for different lengths of time, the number of final shapes obtained will increase. Thus, 4DP is capable of providing a number of shapes obtained from a combination of materials, environments, designs, and time periods.

Shape change shows the degree of presence of a given parameter (temperature, pressure, etc.) in an environment. Thus, the shape change can work as an indicator to show how the environment has changed. This quality is utilized to develop

S. Kumar, *A Concise Encyclopedia of Additive Manufacturing*,
https://doi.org/10.1007/978-3-031-98394-8_69

Fig. 69.1 Schematic diagram of 4DP

sensors or actuators. The quality of shape change can also be utilized to trigger a shape change in a new environment so the shape can conform to the new environment. This quality is used in drug delivery, where the tablet shape of a drug changes to a new shape inside the body to enhance drug absorption in the body fluid [8]. AM can make complex tablets [9], but it has to make them before drug delivery, while 4DP is free from such compulsion [8]. If the final shape is so much important, then why not overcome the dependency on making the initial shape by AM and use other methods [10]?

References

1. Bodaghi, M., Wang, L., Zhang, F., et al. (2024). 4D printing roadmap. *Smart Materials and Structures, 33*(11), 113501.
2. Su, Y. T., Chiu, H. C., & Su, C. K. (2025). 4D-printed shape-programmable [H+]-responsive needles for determination of urea. *Talanta, 282*, 126998.
3. Wang, R., Yuan, C., Cheng, J., et al. (2024). Direct 4D printing of ceramics driven by hydrogel dehydration. *Nature Communications, 15*(1), 758.
4. Aydin, G., Abdullah, T., & Okay, O. (2025). 4D printing of self-healing and shape-memory hydrogels sensitive to body temperature. *European Polymer Journal, 223*, 113651.
5. Kong, D., Guo, A., Wu, H., et al. (2024). Four-dimensional printing of polymer-derived ceramics with high-resolution, reconfigurability, and shape memory effects. *Additive Manufacturing, 83*, 104050.
6. Shafranek, R. T., Millik, S. C., Smith, P. T., et al. (2018). Stimuli-responsive materials in additive manufacturing. *Progress in Polymer Science, 93*, 36–68.
7. Nachimuthu, M., & PK, R. (2023). Inkjet four-dimensional printing of shape memory polymers: A review. *Rapid Prototyping Journal, 29*(3), 437–446.
8. Panraksa, P., Hamdallah, S. I., Yilmaz, O., et al. (2025). 4D printing of shape-memory polymer-based floating tablets via fused deposition modelling: Transformable helical structure to tablet-like form. *Journal of Drug Delivery Science and Technology, 104*, 106534.
9. Afkhami, S., Abdi, M., & Baserinia, R. (2025). Effect of geometry on the dissolution behaviour of complex additively manufactured tablets. *Journal Of Manufacturing And Materials Processing, 9*(1), 11.
10. Tao, Y., Wang, S., Ji, J., et al. (2023). 4Doodle: 4D printing artifacts without 3D printers. In *Proceedings of the 2023 CHI conference on human factors in computing systems* (pp. 1–16).

Chapter 70
Freedom to Move

What AM can manufacture does not completely show what it is. What an AM system is can help show what AM is. If AM can be used on an office desk [1], it has the ability to be packaged in such a way that it can work in an office environment. That is, it must have some qualities as follows: it does not have chemicals that create fumes; or it does not consist of vibrating components that create disturbances; or its size along with its subsidiary equipment is not too big to be kept only in a workshop; or it does not require special facilities that can be found only in a workshop, such as pressurized gas; or it does not have delicate components that demand extra precaution, asking for an environment different than that of an office. Though, not all AM techniques have such qualities. For example, an AM system equipped with a high-power laser cannot be kept in an office environment, but an AM system that uses plastic filaments can be used. Similarly, an AM system, i.e., powder bed fusion, can be used in a remote area to make high-value metallic parts, while an AM system using filaments is not suitable for that purpose.

Thus, an AM system that is used in an office environment is not used to make all products that AM makes, but the system is still useful because the types of designs it can convert are not possible by any conventional system. This makes AM versatile, which allows it to be utilized at different places (remote or non-remote). This enables it to be part of factory planning [2] or supply chain [3]. This is a type of freedom—the freedom to move or to be located.

References

1. Olsson, A., Hellsing, M. S., & Rennie, A. R. (2017). New possibilities using additive manufacturing with materials that are difficult to process and with complex structures. *Physica Scripta, 92*(5), 053002.

S. Kumar, *A Concise Encyclopedia of Additive Manufacturing*, https://doi.org/10.1007/978-3-031-98394-8_70

2. Luft, A., Bremen, S., & Luft, N. (2023). A cost/benefit and flexibility evaluation framework for additive technologies in strategic factory planning. *Processes, 11*(7), 1968.
3. Naghshineh, B. (2024). Additive manufacturing technology adoption for supply chain agility: A systematic search and review. *International Journal of Production Research*, 1–33.

Chapter 71
Friction-Based and Fusion-Based Deposition

- **Small features**: In friction-based deposition, fabrication of a small feature is possible if it withstands tool pressure during fabrication, while in fusion-based deposition, there is no tool pressure, and the smallest features depend upon the minimum dimension of the beam or arc.
- **Material properties**: Properties in both cases can be obtained better than the conventionally manufactured materials in some aspects [1, 2].
- **Grain types**: Fusion-based deposition is versatile, which provides grains of various types. These grains can be columnar, single crystal, equiaxed of various sizes, and a mixture of all types within a given area. By changing parameters and the temperature of a substrate, different types can be obtained at various heights from the substrate.
- Friction-based deposition provides recrystallized grains that are due to the interaction between material, tool rotation rate, and travel velocity [3]. After recrystallization, the grains are fine and equiaxed [4]. Therefore, if the property of a part is not adequate because other processes could not furnish equiaxed grains, the part can be tried in this process.
- **Processing an inaccessible area**: A beam can reach farther away than a friction tool can. Therefore, fusion-based deposition is better when modification needs to be done or a small feature needs to be added in an inaccessible area.
- **Tool wear**: Only in friction-based deposition does a tool wear out, needing replacement.
- **Flexibility**: Fusion-based deposition can make features of various sizes, while friction-based deposition does not have such flexibility due to its dependency on tool sizes.
- **Retaining the original crystal structure**: If the aim is to retain the original crystal structure of feedstock, friction-based deposition, because of being a solid-state and low-temperature process, is a better alternative.

S. Kumar, *A Concise Encyclopedia of Additive Manufacturing*,
https://doi.org/10.1007/978-3-031-98394-8_71

- **Higher deposition rate and low accuracy:** Friction-based deposition has a higher deposition rate than fusion-based deposition. However, the former has a low accuracy, which requires CNC machining [5].
- **Feedstock external property**: Fusion-based deposition, for operational reasons, emphasizes external feedstock properties such as shape, size, surface roughness, and dimensional accuracy. For example, if a wire is not of uniform diameter, it cannot be fed at a consistent rate; if powder has a wide size distribution, it will not give a uniform amount of deposition.

Friction-based deposition is free from such stringent requirements. For example, powders of any size can be fed through a hollow cylindrical tool. This has led to even experimenting with using any size of chips as a feedstock [6]. The absence of the stringent requirement makes the feedstock inexpensive, which makes the process inexpensive [7].

References

1. Ren, X. P., Li, H. Q., Guo, H., et al. (2021). A comparative study on mechanical properties of Ti–6Al–4V alloy processed by additive manufacturing vs. traditional processing. *Materials Science and Engineering A, 817*, 141384.
2. Chen, H., Meng, X., Chen, J., et al. (2023). Wire-based friction stir additive manufacturing. *Additive Manufacturing, 70*, 103557.
3. Griffiths, R. J., Garcia, D., Song, J., et al. (2021). Solid-state additive manufacturing of aluminum and copper using additive friction stir deposition: Process-microstructure linkages. *Materialia, 15*, 100967.
4. Korganci, M., & Bozkurt, Y. (2024). Recent developments in additive friction stir deposition (AFSD). *Journal of Materials Research and Technology, 30*, 4572–4583.
5. Shao, J., Samaei, A., Xue, T., et al. (2023). Additive friction stir deposition of metallic materials: Process, structure and properties. *Materials & Design, 234*, 112356.
6. Jordon, J. B., Allison, P. G., Phillips, B. J., et al. (2020). Direct recycling of machine chips through a novel solid-state additive manufacturing process. *Materials & Design, 193*, 108850.
7. Yu, H. Z. (2022). Emerging processes–friction stir based. In F. G. Caballero (Ed.), *Encyclopedia of materials: Metals and alloys* (pp. 153–161). Elsevier.

Chapter 72
Friction Surfacing-Based AM

In friction surfacing-based AM (FSBAM), a rotating consumable rod is heated due to the heat generated by an inter-frictional force between a rod and a substrate (Fig. 72.1a). The heating causes the rod material to plasticize, which gets smeared on the substrate due to an existing applied axial pressure on the rod (Fig. 72.1b). Thus, the process is suitable for heat-treatable alloys [1]. In another variant, the tool is kept parallel instead of perpendicular to the substrate to utilize the lateral friction [2].

If instead of a rod, a substrate gets heated and plasticized, causing it to flow, the purpose of material deposition by means of the rod will not be served. Therefore, the substrate must satisfy the following conditions:

1. It should have a higher thermal conductivity so the heat is dissipated through it.
2. It should be colder. When the rod is hot, it is generally weaker than the substrate even if both are of the same material, e.g., stainless steel [3], mild steel, etc., which ensures that the same material is deposited on the same material (of the substrate), or the same material (of one layer) is deposited on the same material (of another layer) to begin multi-layer fabrication and transition from friction surfacing to AM. This simplifies the role of the substrate. The substrate will only be used for anchoring and supporting fabrication. It is no longer a problem that the material to be deposited is to be compatible with the material of the substrate.

A rod can be heated by frictional heating either by rotating it on some other locations of a substrate or on the surface of another plate [4]. Preheating is another option [5]. The hot rod is then brought to the area of the substrate where it needs to deposit (or to be deposited). Cooling the substrate is another method to create a temperature difference between it and the rod. Besides satisfying process requirements, cooling is used to change grain sizes [6].

When a consumable rod begins to deposit, it faces inertia of a substrate. When it exits the substrate, it needs to be detached—this leaves marks on the surface at both

S. Kumar, *A Concise Encyclopedia of Additive Manufacturing*,
https://doi.org/10.1007/978-3-031-98394-8_72

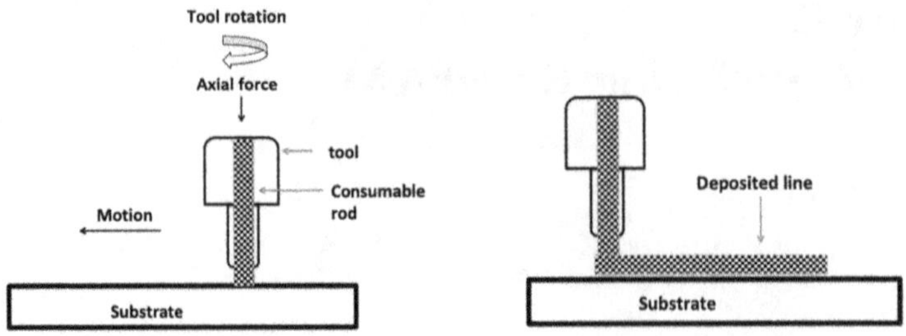

(a) Deposition (b) Deposited line over the substrate

Fig. 72.1 Schematic diagram of FSBAM. (**a**) Deposition. (**b**) Deposited line over the substrate

entry and exit points, which causes both surface and dimensional inaccuracies. This can be avoided if there is neither an entry nor an exit point, which is possible if processing is planned so that both points lie outside the design of a part and can be trimmed away after the fabrication. If machining is planned along with deposition for shaping a product, it will give rise to hybrid additive manufacturing [7].

The situation, similar to arising from entry and exit points, will arise again within the design of a part if the rod slows down, takes a turn, or deposits on a feature smaller than its diameter [8]. The bonding between a substrate and a layer or between two layers is a metallurgical bond that is achieved due to forging (axial pressure of rod) on a plasticized material. Thus, the properties of the resulting material are similar to a wrought material, while powder bed fusion gives better properties than a wrought material [9]. The speed of the process depends upon how fast the deposited layer cools down and becomes strong enough for the second layer to be deposited upon.

FSBAM, like additive friction surface deposition (AFSD), does not have the means to stir the deposited material, which in turn does not make the material as refined and homogenized as in AFSD. The advantage of the absence of stirring is that small features can be conveniently built without being potentially bent by the stirring action.

References

1. Sharifi, A., Khodabakhshi, F., Kashani-bozorg, S. F., & Gerlich, A. P. (2023). Microstructure and mechanical properties in additive manufacturing by friction surfacing of AA6061 alloy. *Materials Science and Engineering A, 884*, 145520.
2. Seidi, E., & Miller, S. F. (2022). Feasibility of multilayer solid-state deposition via lateral friction surfacing for metal additive manufacturing. *Journal of Materials Research and Technology, 20*, 1708–1725.
3. Suo, Y., Fan, S., Du, L., & Zheng, J. (2021). Experimental study on the mechanical properties of high-strength stainless steel at room and elevated temperatures. *Thin-Walled Structures, 165*, 107973.

4. Rao, K. P., Sankar, A., & Rafi, H. K. (2012). Friction surfacing on nonferrous substrate: A feasibility study. *The International Journal of Advanced Manufacturing Technology, 65*(5–8), 755–762.
5. Gandra, J., Krohn, H., Miranda, R. M., et al. (2014). Friction surfacing- a review. *Journal of Materials Processing Technology, 214*(5), 1062–1093.
6. Mishra, R. S., & Ma, Z. Y. (2005). Friction stir welding and processing. *Materials Science and Engineering: R: Reports, 50*(1–2), 1–78.
7. Abdelall, E. S., Al-Dwairi, A. F., Al-Raba'a, S. M., & Eldakroury, M. (2021). Printing functional metallic 3D parts using a hybrid friction-surfacing additive manufacturing process. *Progress in Additive Manufacturing, 6*, 731–741.
8. Kallien, Z., Rath, L., Roos, A., & Klusemann, B. (2024). Application of friction surfacing for solid state additive manufacturing of cylindrical shell structures. *Additive Manufacturing Letters, 8*, 100184.
9. Nagalingam, A. P., Gopasetty, S. K., Wang, J., et al. (2023). Comparative fatigue analysis of wrought and laser powder bed fused Ti-6Al-4V for aerospace repairs: Academic and industrial insights. *International Journal of Fatigue, 176*, 107879.

Chapter 73
Heater-Based Sintering

If a heater instead of a beam source is placed above a powder bed, the radiation from the heater will not converge to a point below the heater on the bed. With an increase in the separation between the heater and the bed, the heat energy density at the point will further decrease. Therefore, if the aim is to melt powder, the heater must be in proximity to the bed, which, however, can damage the heater if not shielded. Though, it can work well with a low-melting-point material such as polymer, for which less heat-induced risk is involved.

A heater needs either to move to cover the whole area or to be present everywhere so the whole area will be under its reach. If the heater needs to move, it reminds the movement of an inkjet printhead that covers the whole area without needing the quality of deflecting a beam. An inkjet printhead consists of a number of ink jets. Thus, if the printhead instead consists of a number of small heaters, it will serve the purpose. This is the concept behind selective heat sintering (Fig. 73.1). If the heater needs to be present everywhere, a number of small heaters need to be fitted in a cover or mask covering the whole area. This reminds the presence of a number of small mirrors in digital light processing. Selecting some of the heaters to be switched on while others are off will help print selectively. This is the concept behind microheater array powder sintering [1].

S. Kumar, *A Concise Encyclopedia of Additive Manufacturing*,
https://doi.org/10.1007/978-3-031-98394-8_73

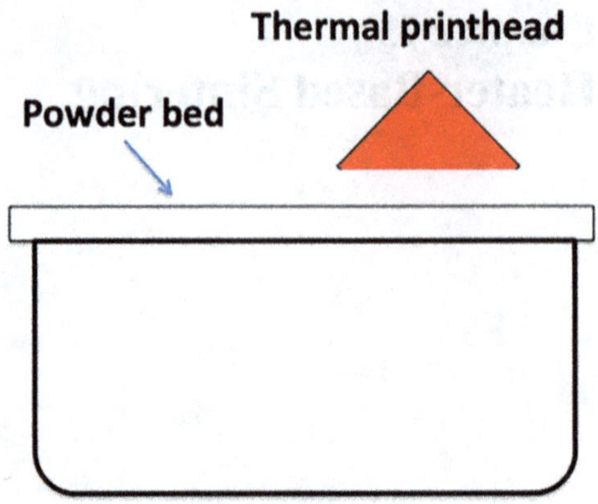

Fig. 73.1 Schematic diagram of selective heat sintering

Reference

1. Holt, N. (2018). *Microheater array powder sintering (MAPS) for printing flexible electronics.* University of Arkansas.

Chapter 74
High-Speed Sintering

High-speed sintering (HSS) is a non-beam powder bed fusion in which each layer is scanned twice [1]. First, a shape is defined on a layer by ink jetting using a printhead (Fig. 74.1a), and then the whole layer is irradiated by a thermal lamp to consolidate that shape (Fig. 74.1b).

The ink is a radiation-absorbing material, which causes the area on the layer where the ink is jetted to absorb more thermal radiation than other adjacent areas. The role of the printhead is to mark areas on a layer, which constitute a part. The thermal lamp does not mark any area and thus does not need the details of the CAD file of the part. The lamp is thus free from toolpath and covers the whole layer from one end to the other. Therefore, the speed of the lamp is relatively fast, which gives the name "high-speed sintering."

The speed of HSS is the summation of two steps: (1) time taken by the inkjet printhead for jetting on a layer, and (2) time taken by the lamp for moving over the layer (scanning the layer). As scanning by laser in laser powder bed fusion (LPBF) is slow while jetting by a printhead is fast, the speed is faster in HSS than in LPBF. Though for processing a layer in LPBF, only one step (laser scanning) is required, while for the same in HSS, two steps (jetting plus lamp scanning) are required. Since time taken by lamp scanning is insignificant in comparison to the time taken by jetting, the speed of the process is mostly influenced by the speed of the jetting.

However, if the size of a part is small, HSS will no longer have an advantage over LPBF in terms of speed. It is because there will not be a significant difference between the time taken by laser scanning and the time taken by inkjet printing, while the time taken by lamp scanning will not be insignificant. On the contrary, if the size of a part is big or if the combined size of many small parts is big, so processing to complete a layer will take more time, then HSS will have an advantage over LPBF in terms of speed, which will justify "speed" of high-speed sintering. It is assumed that LPBF is using only one beam; if it employs many beams (could be 26), the speed would be different [2].

S. Kumar, *A Concise Encyclopedia of Additive Manufacturing*,
https://doi.org/10.1007/978-3-031-98394-8_74

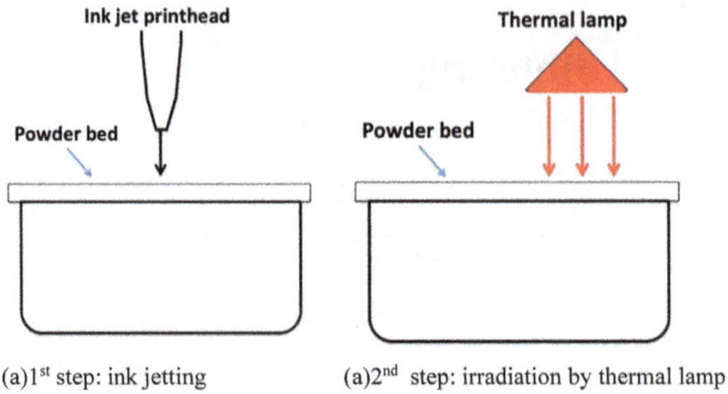

Fig. 74.1 Schematic diagram of HSS. (**a**) first step: ink jetting. (**b**) second step: irradiation by thermal lamp

In a variant of HSS named multi jet fusion, two types of inks are used—one for enhancing absorption of the radiation (the same as that used in HSS) while the second ink is an inhibitor-type ink used at boundaries of a scanned pattern. The role of the second ink is to decrease the diffusion of heat outside the boundary. If heat is diffused outside, it will cause some powder particles to join outside the boundary, resulting in a rough side surface. Using the second ink will make the side surface smooth and increase the definition of a scanned pattern [3]. The use of two types of inks is the same as two types of parameters in beam PBF—one for scanning at the boundary and the other at the non-boundary area [4].

The radiation-absorbing material is carbon black. The heat absorption by carbon must be enough to join nearby powders. Though this absorbed heat is not high, it is sufficient enough to join low melting point materials such as polyamide and elastomers. This heat will not be enough to join high melting point materials such as iron or titanium. The process, thus, has limited applications.

1 Energy Efficiency

HSS does not need a laser and is therefore free from expensive investment. Laser is itself not an efficient energy-conversion device; approximately only 5–70% of electrical energy is converted into laser energy [5, 6]. The most energy-efficient laser (of more than 40% efficiency) is the diode laser, but it has no widespread use in AM [7, 8] due to its poor beam quality [9]. Since HSS uses a thermal lamp that is almost 98% efficient (~2% is lost in nonthermal light energy), it is more energy-efficient than LPBF as far as the selection of heat source is concerned. In HSS, all powders of the layer get processed; the majority of them do not end up in a part. In LPBF, only those powders of the layer that end up in a part get irradiated.

In HSS, the majority of the powders that do not end up in a part still get processed and directly affected by thermal radiation, which may cause them to be degraded and to lose their ability to be recycled. It is argued that print speed is high and therefore exposure time is low, and there will be recyclability, but low exposure time is not better than no exposure time. In LPBF, there is no such process-induced degradation, risk, or exposure to powders. Besides, there is no such emergency to speed up the process at the cost of non-selection of powders in a layer. It does not mean powders are not at all degraded in LPBF; powders are degraded but not at such risk of getting as degraded as in HSS.

LPBF is more efficient than HSS in non-degradation of powders, and, therefore, HSS is more powder-inefficient than LPBF. Powder inefficiency of HSS will increase if the HSS system becomes bigger to speed up the production process and production, because the size of a layer (powder bed) will increase and therefore more powder will be at risk of degradation. Since production of powder is an energy-intensive process [10], HSS being powder-inefficient cannot be energy-efficient as far as utilization of powder is concerned. In this regard, LPBF is more energy-efficient. If LPBF is not considered energy-efficient, it is not because the process itself is inefficient but because the process uses an energy-inefficient device (laser).

References

1. Fahad, M., & Hopkinson, N. (2017). Evaluation and comparison of geometrical accuracy of parts produced by sintering-based additive manufacturing processes. *The International Journal of Advanced Manufacturing Technology, 88*, 3389–3394.
2. Yang, Y., Jiang, R., Han, C., et al. (2024). Frontiers in laser additive manufacturing technology. *Additive Manufacturing Frontiers, 3*, 200160.
3. Sillani, F., Kleijnen, R. G., Vetterli, M., et al. (2019). Selective laser sintering and multi jet fusion: Process-induced modification of the raw materials and analyses of parts performance. *Additive Manufacturing, 27*, 32–41.
4. Tian, Y., Tomus, D., Rometsch, P., & Wu, X. (2017). Influences of processing parameters on surface roughness of Hastelloy X produced by selective laser melting. *Additive Manufacturing, 13*, 103–112.
5. Hecht, J. (2018). *Understanding lasers: An entry-level guide*. John Wiley & Sons.
6. Li, L. (2000). The advances and characteristics of high-power diode laser materials processing. *Optics and Lasers in Engineering, 34*(4–6), 231–253.
7. Pinkerton, A. J. (2016). Lasers in additive manufacturing. *Optics & Laser Technology, 78A*, 25–32.
8. Arredondo, M. Z., Boone, N., Willmott, J., et al. (2017). Laser diode area melting for high speed additive manufacturing of metallic components. *Materials & Design, 117*, 305–315.
9. Santos, E. S., Shiomi, M., Osakada, K., & Laoui, T. (2006). Rapid manufacturing of metal components by laser forming. *International Journal of Machine Tools and Manufacture, 46*(12–13), 1459–1468.
10. Fredriksson, C. (2019). Sustainability of metal powder additive manufacturing. *Procedia Manufacturing, 33*, 139–144.

Chapter 75
How Material Is Wasted

1. Spillage of feedstock—when feedstock needs to be transferred to or from a machine [1].
2. Wastage of feedstock during fabrication [2, 3].
3. Wastage of processed feedstock during fabrication—when trial parts are formed.
4. Wastage of feedstock due to the lack of its processability—the ongoing fabrication needs to stop because of problems in layer consolidation [4] or some other reasons [5].
5. Wastage due to the formation of SS [6], sizing of oversized parts, finishing of rough parts [7], etc.
6. Wastage during separation from a substrate when the first few layers of a part are not able to be removed.

References

1. Scime, L., Wolf, S. D., Beuth, J., et al. (2018). Safety and workflow considerations for modern metal additive manufacturing facilities. *JOM, 70*, 1830–1834.
2. Graziosi, S., Faludi, J., Stanković, T., et al. (2024). A vision for sustainable additive manufacturing. *Nature Sustainability, 7*(6), 698–705.
3. Gibbons, D. W., Govender, P., & van der Merwe, A. F. (2024). Metal powder feedstock evaluation and management for powder bed fusion: A review of literature, standards, and practical guidelines. *Progress in Additive Manufacturing, 9*(4), 805–833.
4. Breuning, C., Renner, J., Markl, M., & Körner, C. (2025). Processing strategies for electron beam based powder bed fusion. In D. Drummer & M. Schmidt (Eds.), *Progress in powder based additive manufacturing* (Springer tracts in additive manufacturing). Springer.
5. Luzon, Y., & Khmelnitsky, E. (2019). Job sizing and sequencing in additive manufacturing to control process deterioration. *IISE Transactions, 51*(2), 181–191.

S. Kumar, *A Concise Encyclopedia of Additive Manufacturing*, https://doi.org/10.1007/978-3-031-98394-8_75

6. Jiang, J., Xu, X., & Stringer, J. (2019). Optimization of process planning for reducing material waste in extrusion based additive manufacturing. *Robotics and Computer-Integrated Manufacturing, 59*, 317–325.
7. Zhu, P., Zhang, G., Teng, X., et al. (2022). Investigation and process optimization for magnetic abrasive finishing additive manufacturing samples with different forming angles. *The International Journal of Advanced Manufacturing Technology*, 1–17.

Chapter 76
How to Make AM Sustainable

The following are the ways to make AM sustainable:

1. Saving energy for the creation of feedstock [1].
2. Using waste material as feedstock [2].
3. Using recycled materials to make big parts [3].
4. By recycling feedstock [4].
5. Using sustainable materials, e.g., geopolymer [5].
6. Optimizing product design [6].

References

1. Riyad, M. F., Han, P., Torabnia, S., & Hsu, K. (2025). Thermoacoustic consolidation of metal particles for energy-efficient metal powders recycling in metal additive manufacturing. *Journal of Manufacturing Science and Engineering, 147*(2), 021001.
2. Baruah, S., Hoekstra, J., Carter, M., et al. (2025, February). Towards achieving sustainability in additive friction stir deposition. In *TMS annual meeting & exhibition* (pp. 191–200). Springer Nature Switzerland.
3. Romani, A., & Levi, M. (2024). Large-format material extrusion additive manufacturing for circular economy practices: A focus on product applications with materials from recycled plastics and biomass waste. *Sustainability, 16*(18), 7966.
4. Jandaghi, M. R., & Moverare, J. (2024). Exploring the efficiency of powder reusing as a sustainable approach for powder bed additive manufacturing of 316L stainless steel. *Materials & Design, 244*, 113222.
5. Haydaraslan, K. S. (2024). Examining the relationship between sustainable material and building design in building construction with additive manufacturing. *International Journal of 3D Printing Technologies and Digital Industry, 8*(3), 361–369.
6. Nyamekye, P., Lakshmanan, R., Tepponen, V., & Westman, S. (2024). Sustainability aspects of additive manufacturing: Leveraging resource efficiency via product design optimization and laser powder bed fusion. *Heliyon, 10*(1), e23152.

S. Kumar, *A Concise Encyclopedia of Additive Manufacturing*,
https://doi.org/10.1007/978-3-031-98394-8_76

Chapter 77
How to Save Material

The following are the ways to save material:

1. Using one material type in one system to minimize the need for cleaning after each build. If the next build uses another material, the system needs cleaning to prevent mixing with trace amounts of earlier material.
2. Making parts simultaneously.
3. Development and use of small systems for investigation.
4. Monitoring, changing parameters [1], and stopping processing in case of any failure.
5. Changing the orientation to do away with the need for support structure (SS) [2].
6. Recycling of feedstock [3], processed feedstock (trial parts, SS), and end-of-life parts [4].
7. Reprocess metal powders to make new powders [5] or wires [6].
8. Making SS and parts not in different materials so as to eliminate possible waste due to mixing of different materials.
9. Optimizing design to make lightweight, hollow, or lattice structures [7].
10. Optimizing design to make lightweight SS [8].

References

1. Diao, Z., Yang, F., Chen, L., Wang, R., Zhang, Y., Sun, J., et al. (2023). Effects of deposition height stability of CuCrZr alloy based on arc voltage sensing: Influence of materials and energy saving on wire arc additive manufacturing. *Journal of Cleaner Production, 425*, 138665.
2. Jin, Y., Du, J., & He, Y. (2017). Optimization of process planning for reducing material consumption in additive manufacturing. *Journal of Manufacturing Systems, 44*, 65–78.
3. Punia, U., Dhull, S., Habeeb, A., et al. (2025). Adopting additive manufacturing: Spotlight on sustainability and material utilization. In *Green materials* (pp. 1–11).
4. Di, L., & Yang, Y. (2022). Towards closed-loop material flow in additive manufacturing: Recyclability analysis of thermoplastic waste. *Journal of Cleaner Production, 362*, 132427.

S. Kumar, *A Concise Encyclopedia of Additive Manufacturing*,
https://doi.org/10.1007/978-3-031-98394-8_77

5. Nagulin, K. Y., Terentev, A. A., Vasilev, I. S., et al. (2025). Recycling of used powder materials for additive manufacturing by processing in an inductively coupled plasma. *Powder Technology*, 453, 120586.
6. Riyad, M. F., Han, P., Torabnia, S., & Hsu, K. (2025). Thermoacoustic consolidation of metal particles for energy-efficient metal powders recycling in metal additive manufacturing. *Journal of Manufacturing Science and Engineering, 147*(2).
7. Gairola, S., & Jayaganthan, R. (2025). Lattice infill strategies for topology optimisation towards achieving lightweight designs for additive manufacturing: Structural integrity, and manufacturing consideration. *Journal of Manufacturing Processes, 139*, 224–238.
8. Kuo, Y. H., Cheng, C. C., Lin, Y. S., & San, C. H. (2018). Support structure design in additive manufacturing based on topology optimization. *Structural and Multidisciplinary Optimization, 57*(1), 183–195.

Chapter 78
Ion Deposition

There are two ion deposition processes: (1) electrolytic solution-based and (2) colloidal solution-based (or electrophoretic deposition).

When an external potential difference is applied across a solution using electrodes, charged particles or ions move through the solution—positive ions (cations) move toward the negative electrode (cathode), while negative ions (anions) move toward the positive electrode (anode).

Movement of ions toward electrodes leads to an accumulation of materials on electrodes, which is akin to deposition of materials on a substrate in AM—this is a method used to add materials to an ionic solution-based AM. This method is not new; it is already practiced in electroplating or electrophoretic deposition for making thin coatings that are 2D structures. The method is also applied in electroforming to make 3D structures [1].

In electroforming, materials are deposited on electrodes of various shapes; these electrodes are called tools or mandrels. On the removal of the electrode from the deposited material, the remaining hollow structure is a desired 3D structure. This structure is dependent upon tools [2]. The dependency limits the process to fabricating only those structures that conform to the shape of tools. The present AM process is free from such limitations.

Ionic solution implies a liquid either containing ions (e.g., NaCl solution) or having the potential to be ionized when an external voltage is applied. When potential is applied across the ionic solution, such as copper sulphate solution, copper ions move, giving rise to copper plating. This happens in electroplating, and the solution is called electrolytic solution. AM based on these types of solutions is electrolytic solution-based AM, also known as electrochemical AM [3].

A colloidal solution may also be an ionic solution, as it contains charged particles suspended in liquid. These charges remain present in the solution even in the absence of an external applied electric field. These charges move in the presence of an external field and give rise to a deposition—this happens in electrophoretic deposition, and the ionic solution is a colloidal solution. AM based on these types of

S. Kumar, *A Concise Encyclopedia of Additive Manufacturing*,
https://doi.org/10.1007/978-3-031-98394-8_78

solutions is colloidal solution-based AM, also known as electrophoretic deposition-based AM [4].

References

1. Castellano, P. M. H., Vega, A. N. B., Padilla, N. D., et al. (2017). Design and manufacture of structured surfaces by electroforming. *Procedia Manufacturing, 13*, 402–409.
2. Matsuzaki, R., Kanatani, T., & Todoroki, A. (2019). Multi-material additive manufacturing of polymers and metals using fused filament fabrication and electroforming. *Additive Manufacturing, 29*, 100812.
3. Kamraj, A., Lewis, S., & Sundaram, M. (2016). Numerical study of localized electrochemical deposition for micro electrochemical additive manufacturing. *Procedia CIRP, 42*, 788–792.
4. Mora, J., Dudoff, J. K., Moran, B. D., et al. (2018). Projection based light-directed electrophoretic deposition for additive manufacturing. *Additive Manufacturing, 22*, 330–333.

Chapter 79
Lack of Repeatability

Lack of repeatability means parts required to be the same are different due to differences in their surface roughness, dimension, strength, or other properties. They need to be the same when made in any of the following ways:

- Using the same process [1].
- Using the same feedstock, parameter, and machine.
- Using the same process but different machines [2, 3].
- Using the same machine (from the same company) but different models.
- By different operators in the same machine.
- At different locations inside a machine [4, 5].
- In two different batches.
- Using the different processes [6].

Repeatability can be ensured by online monitoring and control [7] or by optimizing parameters [8]. If it is limited only to dimension, parts can be post-processed [9]. Post-processing helps improve repeatability and facilitates industrial-scale production [10]. However, if material properties are different, post-processing will not always work.

1 Comparison with Mold-Based Process

In sintering, a mold is kept at the same temperature and pressure, which gives the same external condition to all powder. It may not ensure the same effect on all powder occupying different sections of the mold. But sintering does not bring different effects by applying several section-specific external conditions in the same way AM does by applying different parameters to make different features.

The application of an external condition creates bonds of various strengths among particles. The same condition may not create bonds of the same strength in

S. Kumar, *A Concise Encyclopedia of Additive Manufacturing*,
https://doi.org/10.1007/978-3-031-98394-8_79

two successive parts, causing variation in their properties and loss of repeatability. This can be due to unintended variations in temperature, pressure, or mold condition for the next part. But these unintended variations come due to only one external condition; therefore, if the external condition is controlled, there can be repeatability.

In powder bed fusion, all powders belonging to a part are spread as several layers. All lines in a layer are treated with the same parameters and are thus subject to the same external conditions. But the condition is applied separately, which does not guarantee that the effect of the same condition will not be different at different lines [11].

If the total effect of the same condition in a layer is equal to the total effect in another layer, it does not imply that the total effects on the first section of a part will be equal to the total effects on its next section, causing no certainty for repeatability. To ensure repeatability, every condition needs to be duplicated when the next part is formed. Since a part is made up of several different lines, there are more external conditions to be controlled than if it is made using one mold.

In IM, the confining space of a mold does not let material go astray. Otherwise, parameters need to be searched to confine the material. In AM, each object needs attention, monitoring, and changes in processing parameters if required. Hence, IM is better than AM to give repeatability [12] and expedite fabrication and is preferred when the same object needs to be fabricated [13].

2 Comparison with Machining

Machining makes a part by separating it from a block. This act of separation is confined to a surface, which can have some effects on the surface, i.e., low or high roughness, oxidation, etc. If the cutting is unoptimized, the damage can reach the bulk in the form of bending or heat-induced transformation. The bulk is the last to get damaged. If there is a lack of repeatability, it is not because machining changes the bulk property.

In AM, joining is not confined within the surfaces of powders or wires, but it does or does not transform the whole block (surface + bulk) from solid to liquid and then to solid again. In machining, there is no such transformation, and feedstock property continues after machining; the feedstock property is part property. In AM, feedstock transforms; the part property is not a feedstock property.

When there is no solid-liquid-solid transformation in AM, the feedstock property before and after joining can be the same or different. The part property is due to both feedstock and joining. Since joining is the net sum of the effects of many joinings corresponding to feedstock blocks, joining has many values depending upon the type of small joining. To have repeatability, small joining must have the same values in all parts, which will give the same properties in all parts.

References

1. Moylan, S., Slotwinski, J., & Cooke, A. (2012). Proposal for a standardized test artifact for additive manufacturing machines and processes. In *SFF proceedings* (pp. 902–920).
2. Vorkapic, N., Pjevic, M., Popovic, M., et al. (2020). An additive manufacturing benchmark artifact and deviation measurement method. *Journal of Mechanical Science and Technology, 34*, 3015–3026.
3. McGregor, D. J., Rylowicz, S., Brenzel, A., et al. (2021). Analyzing part accuracy and sources of variability for additively manufactured lattice parts made on multiple printers. *Additive Manufacturing, 40*, 101924.
4. Seifi, M., Christiansen, D., Beuth, J. L., et al. (2016). Process mapping, fracture and fatigue behavior of Ti-6Al-4V produced by EBM additive manufacturing. In *Proceedings of the 13th world conference on titanium* (pp. 1373–1377).
5. Galarraga, H., Lados, D. A., Dehoff, R. R., et al. (2016). Effects of the microstructure and porosity on properties of Ti-6Al-4V ELI alloy fabricated by electron beam melting (EBM). *Additive Manufacturing, 10*, 47–57.
6. Hopfensperger, L. J., Talmazov, G., Ammoun, R., et al. (2023). Accuracy of 3D printed scan bodies for dental implants using two additive manufacturing systems: An in vitro study. *PLoS One, 18*(4), e0283305.
7. Monu, M. C., Chekotu, J. C., & Brabazon, D. (2025). Eddy current testing and monitoring in metal additive manufacturing: A review. *Journal of Manufacturing Processes, 134*, 558–588.
8. Tien, C., Jean, C., Poupaud, L., et al. (2025). Enhancing dental model accuracy through optimized vat Photopolymerization additive manufacturing parameters. *Additive Manufacturing Letters, 13*, 100278.
9. Shah, P., Racasan, R., & Bills, P. (2016). Comparison of different additive manufacturing methods using computed tomography. *Case Studies in Nondestructive Testing and Evaluation, 6B*, 69–78.
10. Shipley, J., Magnusson, A., Karlsson, M., & Beamer, C. (2025). Lean additive manufacturing using HIP and how cost reduction involves the entire process chain. *Journal of the Japan Society of Powder and Powder Metallurgy, 72*(Supplement), S67–S74.
11. Dowling, L., Kennedy, J., O'Shaughnessy, S., & Trimble, D. (2020). A review of critical repeatability and reproducibility issues in powder bed fusion. *Materials and Design, 186*, 108346.
12. Wesemann, C., Spies, B. C., Schaefer, D., et al. (2021). Accuracy and its impact on fit of injection molded, milled and additively manufactured occlusal splints. *Journal of the Mechanical Behavior of Biomedical Materials, 114*, 104179.
13. Ituarte, I. F., Coatanea, E., Salmi, M., et al. (2015). Additive manufacturing in production: A study case applying technical requirements. *Physics Procedia, 78*, 357–366.

Chapter 80
Laminated Object Manufacturing

As per the Oxford dictionary, a sheet is a large rectangular piece of cotton, fabric, or paper. In AM, the word sheet is used for a rectangular piece of any material, such as metal, polymer, ceramic, paper, and glass. Sheet-upon-sheet fabrication gives the concept of layer-upon-layer fabrication an opportunity to showcase a material that resembles a layer. This opportunity is not available for other feedstocks, such as powder, wire, or hydrogel, as they have to go through processing before, they reach the status of a layer that the sheet has already achieved. This exactness in resemblance gives some advantages—if a big product needs to be fabricated, it could be better to start from the biggest available feedstock (sheet) rather than from a small feedstock such as powder. The sheet-upon-sheet phenomenon conveys a message that before another sheet is placed and joined, there is an opportunity to place electronics, sensors, or fiber between two sheets if joining does not damage them and make smart products [1–3].

Laminated object manufacturing (LOM) is a sheet-based AM process in which a 3D object is formed either by joining pre-cut sheets or by in-situ cutting and joining. As the name implies—it is about creating a laminated object, i.e., an object made from several laminae [4]. Lamina is another name for a layer. In this process, various types of laminae or sheets are used, such as paper, plastic [5], metals [6], ceramics [7], metallic glass [8], and sheets made from slurry [9]. Besides, various types of joining methods are adopted, such as laser welding, using glue, using binder [10], freezing [11], thermal pressing [5], cold pressing [12], etc.; various types of cutting methods are adopted, such as using a knife, milling [12], and laser cutting [11]. In one variant, named laser foil printing, metallic glass is processed by laser welding and laser cutting [8].

Commercial LOM systems require continuous feeding of sheets to the platform for building an object. Feeding is done by unfolding a roll that has several meters of sheet wrapped around it; unfolding the roll will enable fresh sheet to come over the platform, where part of it will be separated from the sheet and be joined on the platform. The sheet will move forward so a fresh part of the sheet will again come over

S. Kumar, *A Concise Encyclopedia of Additive Manufacturing*,
https://doi.org/10.1007/978-3-031-98394-8_80

the platform to be cut and joined—since the platform already has one sheet joined over it—the sheet will join on the sheet on the platform. This repetition of moving a fresh sheet and thereafter cutting and joining will give rise to sheet-upon-sheet manufacturing [13].

Unfolding the roll to supply a fresh sheet requires the used sheet to be folded on another roll on the other side of the platform. Thus, there are two rolls connected by a single sheet required to complete the feeding. The width of the sheet needs to be bigger than the width of the platform so when part of the sheet is cut, the cutting will not consume all the width of the sheet, i.e., the cutting will not break the connection between two rolls, as a continuity is required to ensure the feeding. This also means that for every fresh sheet there will be a used sheet—the process thus generates waste. The process is thus not better than machining in terms of the generation of waste.

Cutting of the sheet happens by a knife or a laser beam, while joining may happen by pressing the sheet at a high temperature so the glue coated on the sheet will be activated and will subsequently join an underlying sheet. There are two types by which fabrication can be accomplished: (1) bond-then-cut [14], and (2) cut-then-bond [15].

In bond-then-cut type (Fig. 80.1), a complete fresh sheet is first joined on a platform or a previous sheet, then the undesired part of the joined sheet is removed by cutting—this will leave the remaining part of the bonded sheet to remain bonded and to become part of an end product. After joining, removing by cutting may not be easy, especially if cutting is not at the boundary of a sheet. To facilitate removing by cutting, an anti-glue liquid or powder is sprayed on a selected area before joining, so the selected area is not strongly joined and can be removed without damaging an underlying sheet. Spraying thus requires extra time and information from a CAD file, which increases the fabrication rate. Another way to facilitate removing is to do cross-cutting or crosshatching on a selected area several times so to-be-cut area becomes weak and can be removed afterwards easily [16]. This type of removal, known as decubing, takes time and cutting energy, which makes the process energy-inefficient.

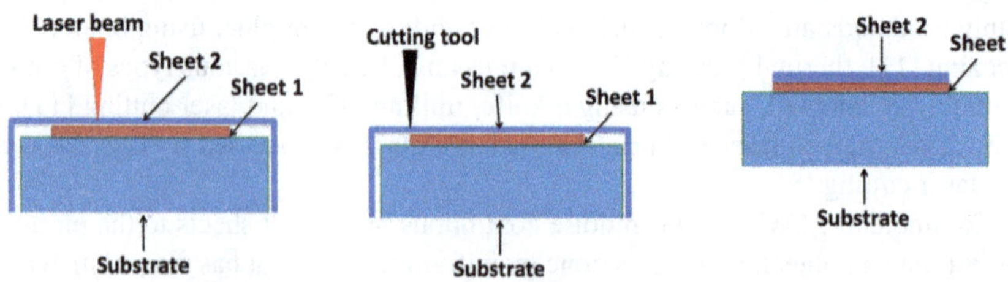

(a) bonding by laser beam (b) cutting of 2nd sheet (c) final part on a substrate

Fig. 80.1 Bond-then-cut type: (**a**) Sheet 2 is bonded by laser beam with sheet 1 which is attached to a substrate, (**b**) After bonding, excess material from sheet 2 is removed by a cutting tool, (**c**) Final part on a substrate

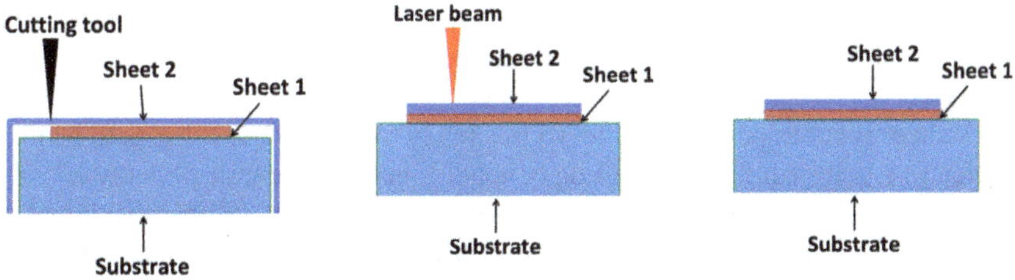

(a) cutting of 2nd sheet (b) bonding by laser beam (c) final part on a substrate

Fig. 80.2 Cut-then-bond type: (**a**) Excess material from sheet 2 is removed by a cutting tool. (**b**) After cutting, sheet 2 is bonded with sheet 1 attached on a substrate. (**c**) Final part on a substrate

In cut-then-bond type (Fig. 80.2), a sheet is cut or a pattern is made on a sheet as per the corresponding section of a CAD file. The sheet is not cut at the boundary, or it is not cut in such a way that it will be severed from a continuous roll. If a sheet is severed from the roll before the joining, it will lose alignment, or in the worst case, it will no longer be tensioned enough to be able to be cut. After cutting, the sheet is then joined on a platform or on the previous sheet, followed by cutting at the boundary and by detaching the sheet from the roll. In this type, there is no hassle for decubing. Therefore, this type can be preferred for its simplicity and energy efficiency. But the action of joining itself may cause misalignment. Therefore, this type is not preferred for accuracy. This type is better than the previous type when a small area needs to be bonded, while the previous type is better when a small area needs to be removed.

1 Why Cut-Then-Bond Type is a Subset of Bond-Then-Cut Type

In cut-then-bond type, a sheet is first cut and then joined to make a part, while in bond-then-cut type, the sheet is first joined and then cut. These two types seem to be different, but they are not much different.

LOM system is aligned for bonding, as the position of the bonding with reference to a platform is fixed. It is the cutting that changes its position to create different patterns. Since the position of the bonding is already determined, it will not change if the cutting happens either before or after the bonding. In a case when cutting precedes, then it is for some reason (simplicity), and in another case when cutting follows, then it is for another reason (accuracy). In both cases, it is the bonding that is decided first—a cutting that precedes bonding does not change the fact that the cutting is going to be decided as per the predetermined and prefixed position of bonding, and not vice versa. Therefore, cut-then-bond type is a special case of bond-then-cut type, in which case bonding is delayed for the sake of convenience.

It does not imply that all cut-then-bond types are subsets of bond-then-cut types. The following example shows why cut-then-bond type is different and is not a subset.

A sheet is cut as per a CAD file, so it is ready to be bonded. It needs to be transported to a platform where it will be bonded. It needs to be transported well, so its orientation should not change. If its orientation changes, it will be positioned and bonded differently, causing it to make a different part. If a sheet is complex, it requires being aligned on a platform with the help of fixtures; otherwise, it will again bond differently, as it does not have the luxury, as in a previous example, to have pre-aligned transportation and bonding. If the sheet is cut after bonding, it will be free from such problems of alignment during transportation and final positioning. In this example of cut-then-bond type, bonding of a sheet is not determined by the prefixed position of bonding, as cutting is done independently from bonding. If the cutting of a square ABCD is going to be done, no side (AB, BC, CD, AD) of the square is a preferred side. It is the orientation of ABCD that will decide which side will align with which side of the platform. While in the previous example, the ABCD portion of a continuous sheet is fixed with respect to the platform (Fig. 80.3a, b), the position of AB and CD cannot be changed arbitrarily unless the position of the roll is changed by disassembling the LOM system. Therefore, cut-then-bond type in this example (Fig. 80.3c) is different from cut-then-bond type in the previous example (Fig. 80.3a). Thus, cut-then-bond type and bond-then-cut type in this example are drastically different from those in the previous example.

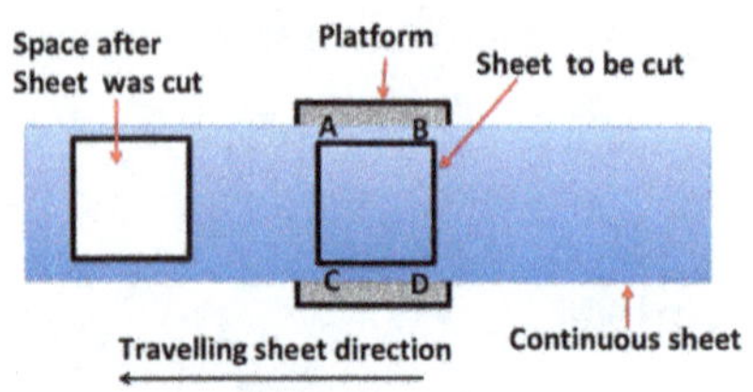

(a) Cut-then-bond type with one fixed orientation of a sheet to be cut

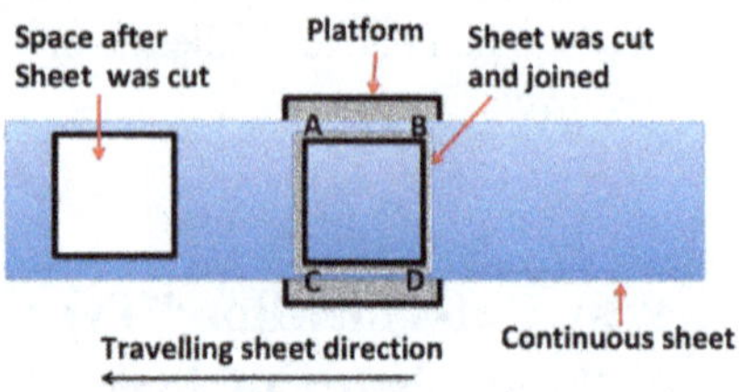

(b) Cut-then-bond type with one fixed orientation of a sheet

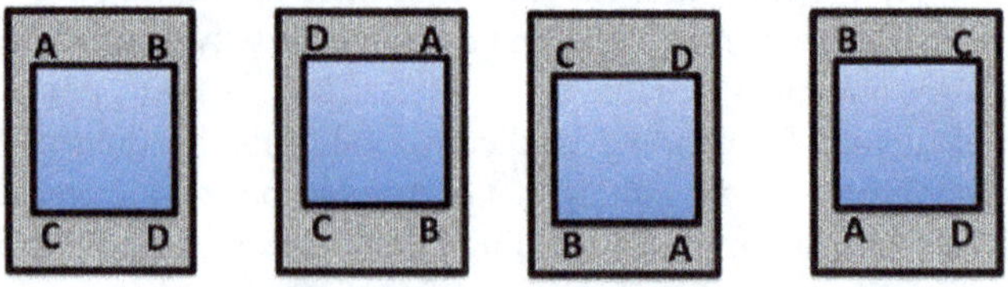

(c) Cut-then-bond type with many orientations of a sheet

Fig. 80.3 Cut-then-bond type: (**a**) with one orientation of a sheet to be cut, (**b**) with one orientation of a sheet cut and joined (**c**) with many orientations of a sheet

References

1. Hehr, A., Norfolk, M., Kominsky, D., et al. (2020). Smart build-plate for metal additive manufacturing processes. *Sensors, 20*(2), 360.
2. Hyer, H. C., Sweeney, D. C., & Petrie, C. M. (2022). Functional fiber-optic sensors embedded in stainless steel components using ultrasonic additive manufacturing for distributed temperature and strain measurements. *Additive Manufacturing, 52*, 102681.
3. Ramanathan, A. K., Gingerich, M. B., Headings, L. M., & Dapino, M. J. (2022). Metal structures embedded with piezoelectric PVDF sensors using ultrasonic additive manufacturing. *Manufacturing Letters, 31*, 96–100.
4. Wimpenny, D. I., Bryden, B., & Pashby, I. R. (2003). Rapid laminated tooling. *Journal of Materials Processing Technology, 138*(1–3), 214–218.
5. Zhang, Y., & Wang, J. (2017). Fabrication of functionally graded porous polymer structures using thermal bonding lamination techniques. *Procedia Manufacturing, 10*, 866–875.
6. Hung, C. H., Sutton, A., Li, Y., et al. (2019). Enhanced mechanical properties for 304L stainless steel parts fabricated by laser-foil-printing additive manufacturing. *Journal of Manufacturing Processes, 45*, 438–446.
7. Gomes, C., Travitzky, N., Greil, P., et al. (2011). Laminated object manufacturing of LZSA glass-ceramics. *Rapid Prototyping Journal, 17*(6), 424–428.
8. Li, Y., Shen, Y., Chen, C., et al. (2017). Building metallic glass structures on crystalline metal substrates by laser-foil-printing additive manufacturing. *Journal of Materials Processing Technology, 248*, 249–261.
9. Liu, S., Ye, F., Liu, L., & Liu, Q. (2015). Feasibility of preparing of silicon nitride ceramics components by aqueous tape casting in combination with laminated object manufacturing. *Materials and Design, 66A*, 331–335.
10. Bhatt, P. M., Kabir, A. M., Peralta, M., et al. (2019). A robotic cell for performing sheet lamination-based additive manufacturing. *Additive Manufacturing, 27*, 278–289.
11. Zhang, G., Chen, H., Yang, S., et al. (2018). Frozen slurry-based laminated object manufacturing to fabricate porous ceramic with oriented lamellar structure. *Journal of the European Ceramic Society, 38*, 4014–4019.
12. Schindler, K., & Roosen, A. (2009). Manufacture of 3D structures by cold low pressure lamination of ceramic green tapes. *Journal of the European Ceramic Society, 29*(5), 899–904.
13. Chiu, Y. Y., Liao, Y. S., & Hou, C. C. (2003). Automatic fabrication for bridged laminated object manufacturing (LOM) process. *Journal of Materials Processing Technology, 140*(1–3), 179–184.
14. Liao, Y. S., Chiu, L. C., & Chiu, Y. Y. (2003). A new approach of online waste removal process for laminated object manufacturing (LOM). *Journal of Materials Processing Technology, 140*(1–3), 136–140.
15. Thomas, C. L. (1996). Automating sheet-based fabrication: The conveyed-adherent process. In *1996 international solid freeform fabrication symposium* (pp. 281–290).
16. Dermeik, B., & Travitzky, N. (2020). Laminated object manufacturing of ceramic-based materials. *Advanced Engineering Materials, 22*(9), 2000256.

Chapter 81
Laser PBF (Powder Bed Fusion)

Laser PBF (LPBF), which is selective laser sintering (SLS) and selective laser melting (SLM), traces its origin back to 1986, when Carl Deckard, a master student at the University of Texas, filed a patent on this process, which was granted in 1989 [1]. The term "laser" in SLS and SLM implies that a laser is used for processing; the term "sintering" implies that powders are involved in the process. It infers that powder processing is done by the laser, which is used as a point heat source. "Selective" implies that all powders are not processed by the laser simultaneously, i.e., powders are processed selectively when and where required. In conventional sintering, all powders are processed simultaneously. "Melting" refers to a particular case of powder processing in which powders are completely melted. Thus, SLM refers to a case in which full melting occurs.

1 Description

In LPBF, the main aim is to make a layer of predefined geometry by fusing powders using an l-beam (laser beam). The process follows the following sequence: (1) A substrate is lowered down to a depth equal to the layer thickness, (2) a powder layer is spread on the substrate, and (3) the deposited powder layer is scanned by the l-beam to fuse powders at selected areas. The sequence (1), (2), (3) is repeated until the fabrication is complete (Fig. 81.1).

In an initial stage of the process, powders are placed in a powder container and are protruded from the container by an adjoining piston. Adjacent to the container, a scraper is placed, which carries powders toward the substrate. The substrate is placed over a piston so its vertical position can be changed by adjusting the piston. A scanning mirror is used to scan the deposited layer on the substrate using an l-beam coming from a laser source.

S. Kumar, *A Concise Encyclopedia of Additive Manufacturing*,
https://doi.org/10.1007/978-3-031-98394-8_81

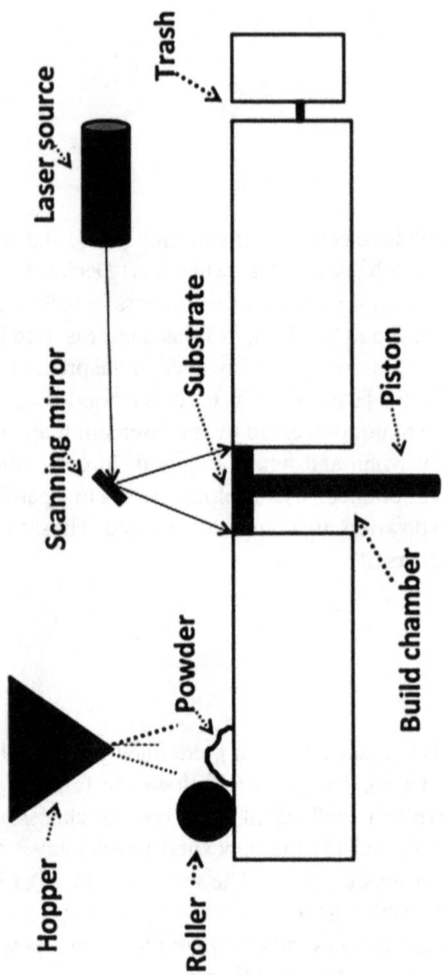

Fig. 81.1 Schematic diagram of laser powder bed fusion

In step 1 of the sequence, the piston of the powder container moves upward, and the piston of the substrate container moves downward. The step gives requisite powders to be carried away by the scraper and creates space on a substrate container for the powder to be deposited. In step 2 of the sequence, powders are deposited over the substrate, and the position of the scraper changes to the right of the substrate. In the last step (step 3), deposited powders are scanned by an l-beam.

The above description shows the necessities of the process: (1) formation of powder bed (PB), (2) consolidation of powders by an l-beam, and (3) a mechanism to repeat the above-listed first and second points.

Instead of using a substrate over a build platform, PB made directly over the platform can also act as a substrate. In this case, the final part does not need to be cut off from the substrate, as it is not attached to the substrate. PB can also be formed by using a counter-rotating roller instead of a scraper/blade, and the powder feeding can also be accomplished by using a hopper instead of a powder container. From the hopper, a predetermined amount of powder falls in front of the roller, which then carries it away for deposition. The excess powder carried away falls into a trash that is equipped at the other side of the build chamber [2].

References

1. Yadroitsev, I., Yadroitsava, I., Du Plessis, A., & MacDonald, E. (Eds.). (2021). *Fundamentals of laser powder bed fusion of metals*. Elsevier.
2. Chowdhury, S., Yadaiah, N., Prakash, C., et al. (2022). Laser powder bed fusion: A state-of-the-art review of the technology, materials, properties & defects, and numerical modelling. *Journal of Materials Research and Technology, 20*, 2109–2172.

Chapter 82
Laser Powder Deposition: Types

In laser powder deposition, an l-beam coming from a nozzle acts as a heat source. The beam melts blown powder (supplied by the same or different nozzle) or wire (fed by the different nozzle). By moving the nozzle parallel to a fixed substrate (or the substrate parallel to a fixed nozzle), a melting point becomes a line. By creating such lines as per the design of a first layer, a physical layer is made.

By moving the nozzle up (or the substrate down), space is created to accommodate second to-be-fabricated layer—this step, followed by melting, is repeated to fabricate the second layer and so on. The process has variations due to variations in nozzle, feedstock, position of the nozzle, and laser-material interaction.

Coaxial continuous, coaxial discrete, and off-axial powder deposition types are practiced.

- **Coaxial continuous:** In this type, powder is deposited around the axis of an l-beam (Fig. 82.1a). The beam comes from the central area of a nozzle, while powders come from a continuous annular orifice (Fig. 82.1d) made at the periphery of the nozzle. The powder stream takes a conical shape around the beam—the axis of the cone is the same as that of the I-beam, giving the name coaxial deposition. Since the powder is injected from all points at the annular space, the coaxial deposition is coaxial continuous, which is used for fabricating structures with high resolution [1]. As the deposition has no directional dependence, it is suitable to make multilayer depositions [2].
- **Coaxial discrete:** In this type, powder is deposited not from all points at the annular space but from selected points (Fig. 82.1b). Figure 82.1e shows three discrete points from where powder is injected, forming a conical shape (having the same axis as that of the beam) around the beam. For less than three discrete points, a conical shape will not form. Discretization helps achieve a higher deposition rate but has lower symmetry of powders' position around the beam in comparison to the former type. Increasing the number of discrete points increases the number of powder jets and symmetry. Using nozzles to create powder jets

S. Kumar, *A Concise Encyclopedia of Additive Manufacturing*,
https://doi.org/10.1007/978-3-031-98394-8_82

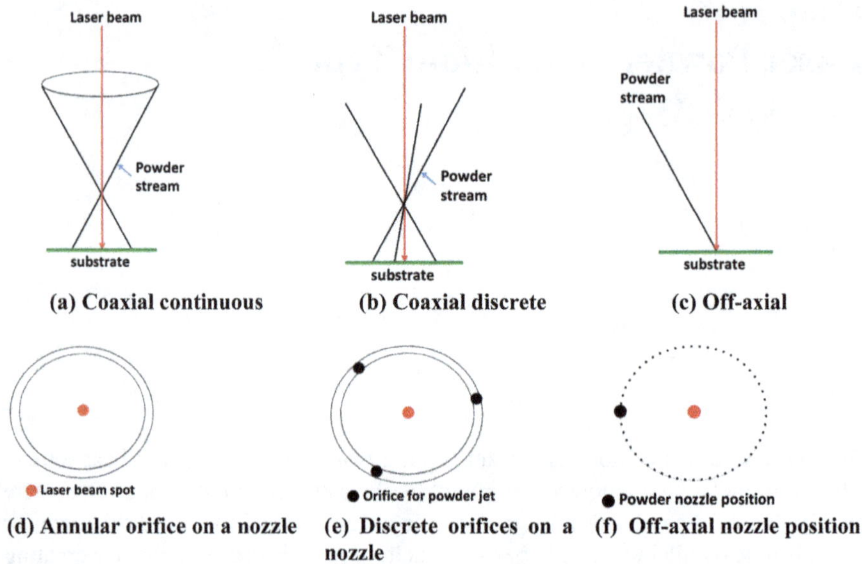

Fig. 82.1 Schematic diagrams for powder deposition: (**a**) coaxial continuous, (**b**) coaxial discrete, (**c**) off-axial; top views of: (**d**) annular orifice, (**e**) discrete orifices, (**f**) off-axial position

will provide the same effect. This has higher powder divergence than the coaxial continuous type, causing a smaller standoff distance from the substrate. If divergence in coaxial continuous is 8°, it is 15° in this type [3].

- **Off-axial:** In this type, the axis of a powder stream makes an angle between 0° and 90° with the l-beam axis (Fig. 82.1c, f). At 90°, the stream will not impinge upon a substrate unless either the substrate is tilted or the stream loses its speed and falls. At 0°, the powder jet will no longer be off-axial but coaxial. The powder nozzle is not associated with a beam nozzle, which gives flexibility in setting up an off-axis angle and a stand-off distance for the powder nozzle. This deposition type is used for its simplicity and flexibility [4].

However, this has directional dependence. The relative motion between the nozzle and the substrate may change the deposited powder amount and the deposition site (Fig. 82.2). The powder stream does not follow a straight line (Fig. 82.1c) but is bent due to gravity (Fig. 82.2a and b) [5]. Powders after leaving the nozzle are no longer tied to the nozzle. When the substrate is moving toward them, powders strike the substrate at a nearer location (Fig. 82.2a), while when the substrate is moving away, powders will strike at a farther location (Fig. 82.2b). This causes the intended strike position to be negatively or positively missed. Changes from the intended position can happen in any deposition process [6].

This situation can be avoided in two cases: (1) the powder jet is injected with a higher speed, which has the disadvantage that it may cause ricochet from a solid substrate and create ripples in the melt pool; (2) the substrate moves slowly

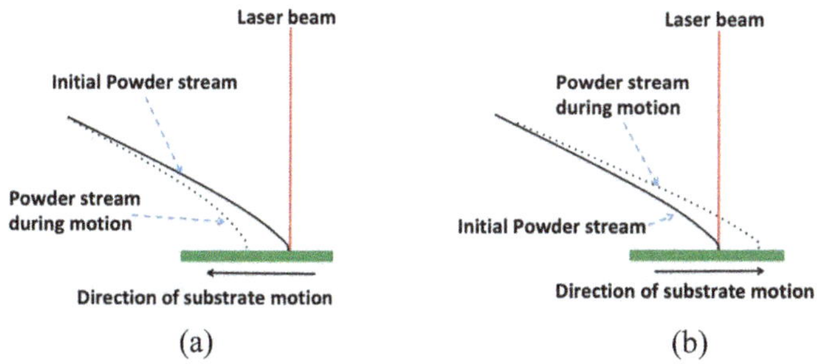

Fig. 82.2 Directional dependence in off-axial powder deposition, motion in: (**a**) opposite direction, (**b**) same direction

compared to the feed rate; and (3) the powder nozzle is kept closer to the substrate, which may damage the nozzle. If the time for a powder to reach the substrate is small (milliseconds), it will not be affected by gravity or drag force [7].

Implications of directional dependence depend on other parameters. For example, if an l-beam spot size is far smaller than a powder stream spot size, and with displacement of the powder stream (due to any motion), the l-beam spot size is always covered by the powder stream spot size, there will be no implication. However, if the l-beam spot is under—or overflowed with powders, these will cause non-uniform deposition.

Employing two powder jets in opposite directions will help—in one direction, one jet will cause deprivation while the other will cause overflowing, while in another direction, the reverse will happen. Thus, the net effect will be zero.

References

1. Oliveira, U. D., Ocelík, V., & De Hosson, J. T. M. (2005). Analysis of coaxial laser cladding processing conditions. *Surface and Coatings Technology, 197*(2–3), 127–136.
2. Eisenbarth, D., Esteves, P. M. B., Wirth, F., & Wegener, K. (2019). Spatial powder flow measurement and efficiency prediction for laser direct metal deposition. *Surface and Coatings Technology, 362*, 397–408.
3. Gao, X., Yao, X. X., Niu, F. Y., & Zhang, Z. (2022). The influence of nozzle geometry on powder flow behaviors in directed energy deposition additive manufacturing. *Advanced Powder Technology, 33*(3), 103487.
4. Vilar, R. (1999). Laser cladding. *Journal of Laser Applications, 11*(2), 64–79.
5. Turichin, G., Zemlyakov, E., Klimova, O., & Babkin, K. (2016). Hydrodynamic instability in high-speed direct laser deposition for additive manufacturing. *Physics Procedia, 83*, 674–683.
6. Yuan, L., Pan, Z., Ding, D., et al. (2021). Fabrication of metallic parts with overhanging structures using the robotic wire arc additive manufacturing. *Journal of Manufacturing Processes, 63*, 24–34.
7. Wu, J., Zhao, P., Wei, H., et al. (2018). Development of powder distribution model of discontinuous coaxial powder stream in laser direct metal deposition. *Powder Technology, 340*, 449–458.

Chapter 83
Laser Powder Deposition: Laser-Powder Interaction

The timing of interaction between an l-beam and a powder stream influences the process. There are three possibilities as follows:

1 L-Beam and Powder Stream Interacts Midway

If an l-beam and a powder stream meet on the way before they reach a substrate, it can affect the process. If the powder comes on the way, it will obstruct the l-beam from reaching the substrate. Powder will get heated, and the heated powder will reach the substrate; this shadowing effect will bring variation in a melt pool. In a coaxial continuous deposition, the degree of interaction depends upon the geometry of the l-beam cone and the cone made by the stream [1] (Fig. 83.1).

In case the l-beam cone is thinner than the surrounding powder stream envelope, the beam will intersect the envelope in a smaller zone (Fig. 83.1a). Otherwise, the intersection zone will be big (Fig. 83.1b), which will obstruct the beam more, resulting in more heating of powders. If an off-axial deposition is used, this can be minimized by depositing either at an angle (Fig. 83.2a) or perpendicular to the substrate in combination with an inclined beam (Fig. 83.2b).

It can also be minimized in a usual setup (Fig. 83.2c)—inclined l-beam [2], though it will increase the dissipation of laser energy by increased reflection of the beam. Increasing the angle between the l-beam path and powder stream path (Fig. 83.2a) will have a minimum shadowing effect, but this decreases both energy efficiency and powder catchment efficiency (this is the percentage of blown powders that ended up in a solidified layer).

If powders, instead of just getting heated, get melted, the creation of the melt pool on the substrate is no longer required as a means to melt powders, and problems associated with creating and controlling the melt pool will be over, but there are the following reasons why powders are not melted:

S. Kumar, *A Concise Encyclopedia of Additive Manufacturing*,
https://doi.org/10.1007/978-3-031-98394-8_83

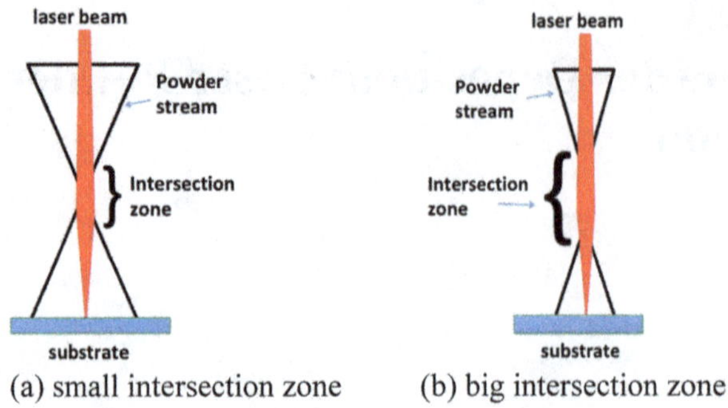

Fig. 83.1 Intersection between l-beam and powder stream in coaxial continuous deposition: (**a**) small zone, (**b**) big zone

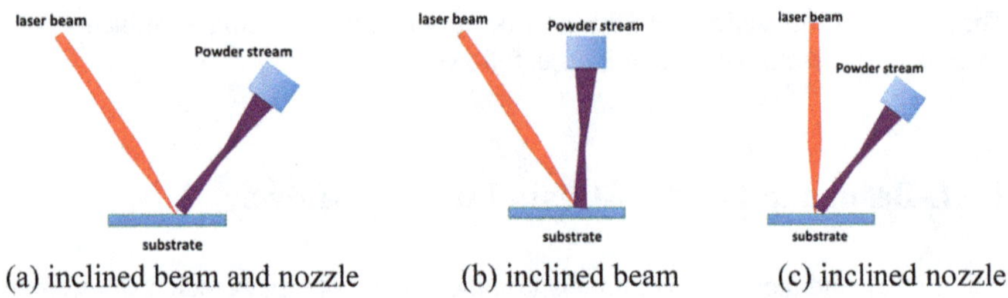

Fig. 83.2 Minimizing shadowing effect: (**a**) inclined l-beam and inclined off-axial nozzle, (**b**) inclined beam and perpendicular nozzle, (**c**) perpendicular beam and inclined nozzle

1. They are not small enough—powders used are from the size of about 20 μm and above; these are not small enough to be melted during small laser-powder interaction time. If these are of sub-micron size, they can be melted. But using them is not possible, as these cannot be fluidized [3] and, therefore, cannot be accelerated using a carrier gas. During acceleration, they agglomerate and no longer remain the size of an isolated small powder small enough to be melted.
2. If they are small (depending upon the experimental condition), they vaporize and can form plasma that attenuates the beam. If they are not small, they vaporize partially, which gives rise to recoil pressure to the powder, which changes its direction of flow [4].
3. Laser energy is not sufficient—focal length of the beam is small, which does not let them have sufficient interaction with the beam to gain sufficient energy for melting. This can be overcome by increasing the focal length as well as the laser power. Increasing the focal length, though possible by optics, will bring an additional problem of having a big focused powder stream matching that big length. Increasing the laser power may vaporize the substrate in the course of melting the powder.

Though the problem can be overcome, liquid drops (by melting powders) will fall on the substrate, which, on being continued, will give rise to a structure. But the aim of the process is not to facilitate the complete melting on the way but to do the opposite, i.e., to prevent even an isolated case of powder being melted on the way. Besides, there are more economic and energy-efficient ways to create liquid drops than using expensive powder and lasers.

2 L-Beam Follows Powder Stream

Powders are blown on a substrate to deposit them, but they will be rebounded from the substrate, causing it not to have the intended amount of powders. With an increase in the feed rate and the powder size, rebound force [5] will increase, which will make it increasingly difficult to place a powder layer on the substrate to replicate a powder deposition type familiar in powder bed fusion (PBF).

This difficulty is also a source of difference between two processes: bed type and deposition type. Imagine a state where the difficulty disappears—the nozzle is used to additionally create a powder layer that acts as a support to make overhangs on it, and there is no need to maneuver the axes of a 5-axis CNC machine to circumvent the need for a support structure. Though the difficulty will not disappear, it can be minimized with the help of an additional nozzle fitted with the sole purpose of placing powders [6].

Though it is impossible to make a powder layer, it is possible to have some powder on a substrate when powders are blown on it. These few powders count when the aim is to create a thin melt pool (of μm thickness), which is desired when minimum dilution (it is a measure of how much substrate material is becoming part of a build) is required. There are a number of strategies (such as decreasing laser power, increasing scan speed, or decreasing feed) to create a thin pool, and these strategies are better than how a thin pool results due to the presence of stray powders. But the point is, what will be the consequence if the beam reaches the substrate after the powder has reached it?

3 Powder Stream Follows L-Beam

When an l-beam strikes a substrate, it is supposed to create a melt pool so that when the powder stream reaches the substrate, powders will be melted in the pool and the deposition of material will start. How long these powders remain unmelted in the pool determines the surface roughness and build quality. If they remain unmelted for long, there will be no more space for other powders to be accommodated in the pool [7]. If incoming powders cannot be accommodated, the build rate will decrease.

Though powders attenuate the coming beam depending on the configuration of the beam and powder stream, this type of creation of a melt pool and subsequent deposition is practiced in multilayer deposition. The size of the pool will increase with an increase in the injection of powders, and its temperature will increase with an increase in the injection of hot powders [8].

References

1. Li, L., & Huang, Y. (2018). Interaction of l-beam, powder stream and molten pool in laser deposition processing with coaxial nozzle. *Journal of Physics Conference Series, 1063*, 012078.
2. Meacock, C., & Vilar, R. (2008). Laser powder microdeposition of CP2 titanium. *Materials and Design, 29*, 353–361.
3. Geldart, D. (1973). Types of gas fluidization. *Powder Technology, 7*, 285–292.
4. Sergachev, D. V., Kovalev, O. B., Grachev, G. N., et al. (2020). Diagnostics of powder particle parameters under laser radiation in direct material deposition. *Optics & Laser Technology, 121*, 105842.
5. McLaskey, G. C., & Glaser, S. D. (2010). Hertzian impact: Experimental study of the force pulse and resulting stress waves. *Journal of the Accoustical Society of America, 128*(3), 1087–1096.
6. Kumar, S. (2020). Future additive manufacturing processes. In *Additive manufacturing processes* (pp. 187–202). Springer.
7. Haley, J. C., Schoenung, J. M., & Lavernia, E. J. (2019). Modelling particle impact on the melt pool and wettability effects in laser directed energy deposition additive manufacturing. *Materials Science and Engineering A, 761*, 138052.
8. Pirch, N., Linnenbrink, S., Gasser, A., & Schleifenbaum, H. (2019). Laser-aided directed energy deposition of metal powder along edges. *International Journal of Heat and Mass Transfer, 143*, 118464.

Chapter 84
Layer

A layer is the shortest unit of a design that is built in AM. It is usually defined by its thickness, shape, and area. The definition needs to be revisited.

When a design is divided into a number of layers, a layer is not meant to be different from its projection on a horizontal plane. If the design is simple, e.g., a rectangular block, the layer of any thickness gives the same projection, which does not create the difference between a layer of any thickness and its projection. If the design is not such a simple one, e.g., a sphere, the layer of any thickness does not give the same projection. For such a design, if a thick layer is selected, it gives problems because it has many projections. If the differences in projections are ignored and the layer is selected to manufacture a sphere, the result will not be a sphere. Therefore, a layer of any thickness can be selected as long as the selection does not create more than one projection. If the design is a sphere, the thickness of the layer should be selected to a minimum. So even if the selection creates more projections, their differences can be ignored. This will allow to make an approximate sphere. That is a sphere manufactured in AM. Can AM ignore the differences in order to manufacture? This is what AM is. It ignores it because it has limitations. Its limitation is that it manufactures a layer through its single projection. It selects a single projection and consolidates through it. If a layer has two projections, it does not select the upper projection and consolidate through it to meet the lower projection. Because it has no ability to give a direction to the consolidation.

In some cases, consolidation gets a direction when the directional solidification takes place [1]. But it has only one direction to take, and that is the direction that happens by default in AM. If it takes another direction than the default direction, it is due to the lack of control of process parameters and not because AM has the ability to change the direction as per the vertical design of a layer.

Therefore, a layer is required to be defined by a single projection. It does not limit the thickness of a layer it can manufacture. Thus, a rectangular block can have a layer equal to the rectangular block, but a sphere cannot have a layer equal to the

S. Kumar, *A Concise Encyclopedia of Additive Manufacturing*,
https://doi.org/10.1007/978-3-031-98394-8_84

sphere. The block can have layers of various thicknesses without affecting its size and shape (Fig. 84.1).

If a layer is a projection, which 3D shapes can a layer have? It can take only those 3D shapes that are the extension of the projection. A layer can be a circular disk—an extension of a circular area, a ring—an extension of a circular line, a rectangular block—an extension of a rectangular area, a square block—an extension of a square area, a star-shaped disk—an extension of a star-shaped area, etc., but it cannot be a pyramid, a trapezium, or a rhombohedron, etc.

Figure 84.2 shows the projections of various 3D layers on horizontal planes. When the shape is either a rectangle or a circular disk, the projection is only one area (Fig. 84.2a, b). This is because the upper and the lower surface of the layer have the same area. For a trapezium layer, its upper and lower surfaces have different areas, which gives two projections, as shown by a dotted and a solid line (Fig. 84.2c). AM requires only one projection, assuming the consolidation will extend normal to the projection (Fig. 84.3). The downward consolidation gives the depth of the consolidation, which is a layer thickness. For a trapezium layer (Fig. 84.3c), when there are two projections, it is required that AM consolidates through the upper projection and reaches the lower projection, i.e., the extension of the consolidation should not be according to what AM does but what the design of the layer dictates. But the extension will be normal to the projection, making a rectangular layer rather than a trapezium layer.

Therefore, a layer is what its upper projection is, i.e., a single projection defines a layer in AM. What is the use of a trapezium layer when it acts like a rectangular layer? What is the use of its two projections when one of the projections is not recognized by AM? A 3D shape of a layer does not exist in AM unless it is made by the extension of the 2D shape of a layer. When a trapezium layer is selected, the result will be a rectangular layer, which will create a gap that is the difference between the solid area of the rectangular area and that of the trapezium (Fig. 84.3c).

A gap is generally created when a layer selected does not match the curve of the design of a part; this gap is created because AM does not meet the requirement of the design—this is a fact. When a layer is selected that has more than one projection and the gap is created, it is not because AM does not meet the requirement but because the 3D design of the layer does not meet the requirement of AM—this is a fact. Both facts say the same fact: if a layer does not match the design, there is no solution for it, as the layer cannot be stretched to match.

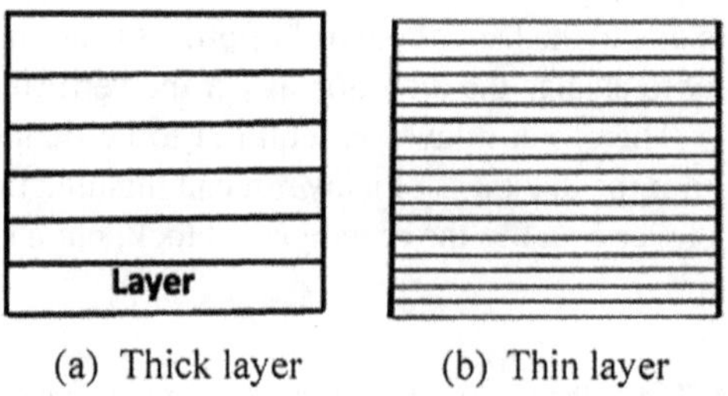

Fig. 84.1 No change in rectangular block. (**a**) Thick layer. (**b**) Thin layer

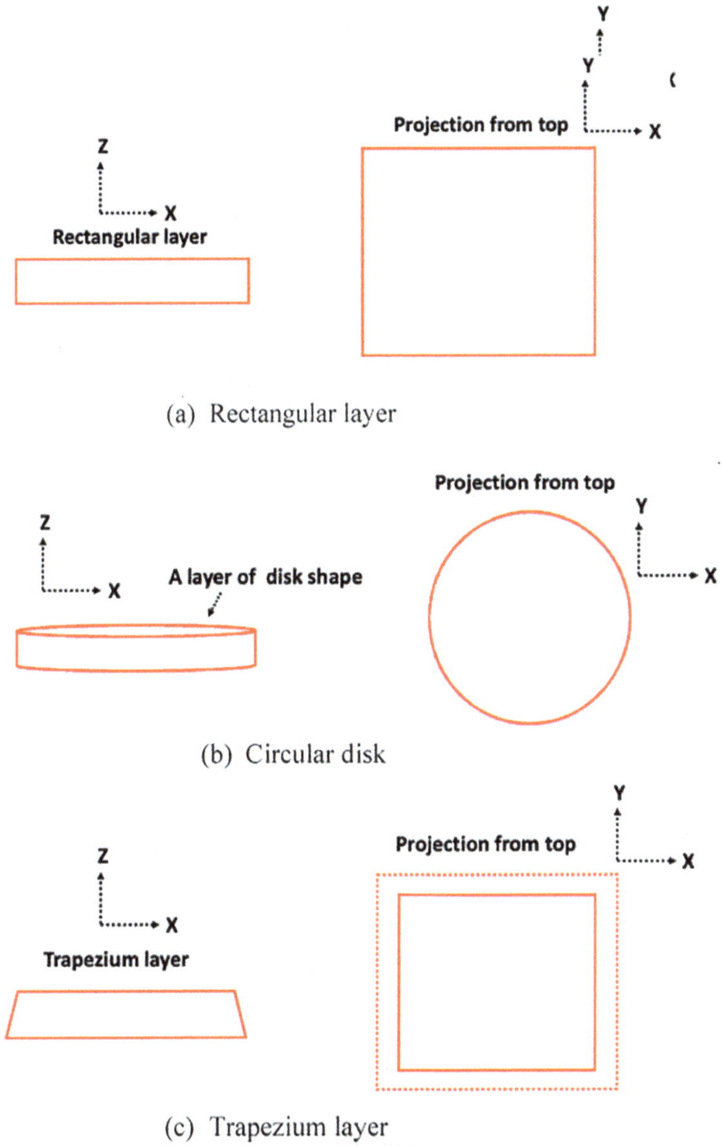

Fig. 84.2 Projection of layer of various 3D shapes. (**a**) Rectangular layer. (**b**) Circular disk. (**c**) Trapezium layer

The thickness, shape, and area of a layer are not sufficient to define a layer, as this information does not help say a difference between two designs where the layers of the same thickness, shape, and area are located differently. For example, this information will not be able to say how a rectangular block is different from a staircase made from the layers of the same thickness, shape, and area (Fig. 84.4a, b).

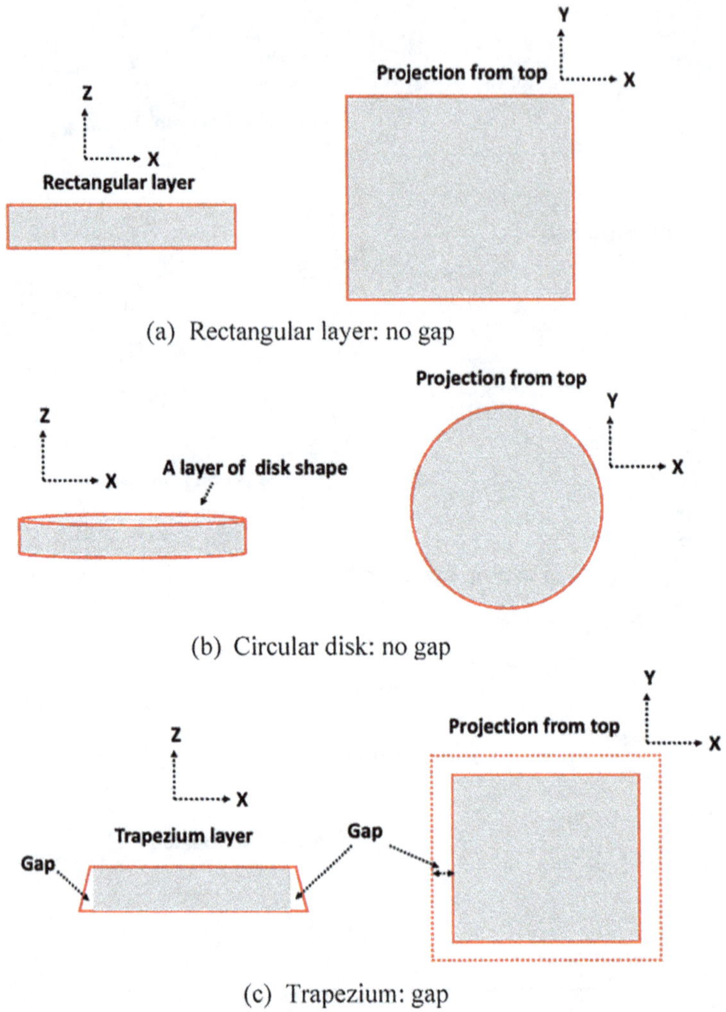

Fig. 84.3 Layer consolidation of various 3D shape. (**a**) Rectangular layer-no gap. (**b**) Circular disk- no gap. (**c**) Trapezium layer- gap

Thus, besides the thickness, shape, and area of a layer, its location is required to define it completely.

(a) Rectangular block

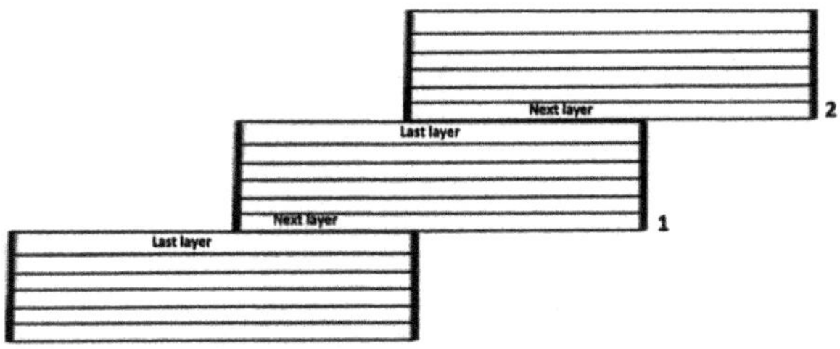

(b) Three steps of stair-case

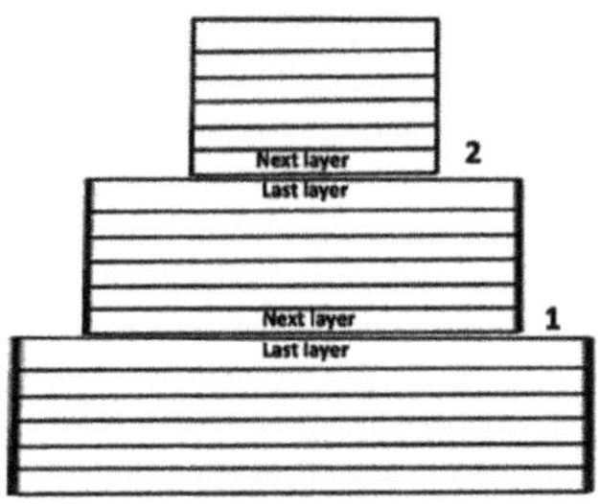

(c) Three sub-blocks of rectangular block

Fig. 84.4 Difference between parts because of changes in next layer. (**a**) Rectangular block. (**b**) Three steps of stair-case. (**c**) Three sub-blocks of rectangular block

Reference

1. Fortuna, S. V., Gurianov, D. A., Kalashnikov, K. N., et al. (2021). Directional solidification of a nickel-based Superalloy product structure fabricated on stainless steel substrate by electron beam additive manufacturing. *Metallurgical and Materials Transactions A, 52*, 857–870.

Chapter 85
Layer Deposition and Deposition Process

Layer deposition is an action that informs that a specific layer is deposited. The layer is specific because it is part of the design of a part. AM makes a part when layer deposition takes place for all layers of the design of the part.

Processes and techniques are used interchangeably in AM. The meaning of the process is not strictly a set of steps. Otherwise, any part that is made in an AM process with a variation in any step will not belong to the same process.

Deposition process is a family of AM processes or techniques [1] that restrict how layer deposition can take place. It can take place only when there is a free space to be filled. It cannot take place when the material it deposits finds the space occupied.

When the deposition process requires that the space should not be occupied, this requirement informs that other techniques are possible that can facilitate the layer deposition without providing a free space. One of the examples of other techniques is powder bed fusion (PBF), where the layer deposition does not require material to travel from one point to another to fill up the required space [1]. Because the material exists everywhere on a plane. However, if the material exists everywhere, this does not make a layer. A layer means the material should exist only where the design suggests it to exist. This requirement demands that the unnecessary material be removed from the plane so that the resulting layer will be part of the layer deposition. If it is not removed immediately, a promise should be given that this unnecessary material should not be part of the final product. PBF is not about immediacy. It is about the promise. It gives promise by selective consolidation of the bed. The selective consolidation distinguishes unnecessary materials from necessary materials, so the unnecessary material can be removed. This is the selective consolidation that allows the layer deposition to take place in PBF. How the selective consolidation will take place changes from one technique to another. In one technique, it can happen by using a laser beam [2], while in another technique, it can happen by using an electron beam [3]. The selective consolidation can comprise either complete melting [4] or partial melting [1, 5].

S. Kumar, *A Concise Encyclopedia of Additive Manufacturing*,
https://doi.org/10.1007/978-3-031-98394-8_85

When unnecessary material is removed, this reminds of machining. For making a part, as the machining relies on removing unnecessary material, PBF is also not free from the reliance on removing unnecessary material. For example, powder as unnecessary material is removed in PBF. Although the degree of removal force required in machining and PBF does not match—a removal force is required in machining, while in PBF, the removal force can be so small that it can be negligible. In the deposition process, material is removed as well, but it is not because material removal is required as part of the process. But because when the material could not be deposited well while executing the process, the inaccurate deposition needs to be corrected. For example, in the powder deposition process, powder is not supposed to be removed because only that amount of powder needs to be deposited that becomes part of an ongoing build, leaving no powder required to be removed. Thus, in an ideal case, there is no need to remove powder in the powder deposition process. While in PBF, there is a need to remove powder even in an ideal case. Therefore, out of the deposition process and PBF, it is the PBF that demands the removal of unnecessary material, making it closer to resembling machining. The detailed comparison between machining and AM can be found elsewhere [6, 7].

When it is said that the material exists everywhere on a plane in PBF, it means bringing material to exist everywhere on a plane is not the same as bringing material to exist on specified points on the plane. Bringing material to exist everywhere on a plane is the formation of a bed in PBF, while bringing material to exist on specified points on the plane is the deposition of a layer in a deposition process. In PBF, a bed is not formed by bringing material to exist on specified points on a plane [8].

References

1. Kumar, S. (2022). *Additive manufacturing classification* (Synthesis lectures on engineering, science, and technology). Springer.
2. Azarmi, F., & Amiri, A. (2019). Microstructural evolution during fabrication of alumina via laser stereolithography technique. *Ceramics International, 45*(1), 271–278.
3. Yang, Z., Liu, Y., Li, H., et al. (2024). Optimizing processing parameters in electron beam powder bed fusion (EB-PBF) for CuNiFeSnTi/diamond composites. *International Journal of Refractory Metals and Hard Materials, 128*, 107027.
4. Yang, X., Gibbons, G. J., Tanner, D. A., et al. (2023). Scan strategy induced microstructure and consolidation variation in the laser-powder bed fusion (L-PBF) additive manufacturing of low alloy 20MnCr5 steel. *Materials & Design, 232*, 112160.
5. Grossin, D., Montón, A., Navarrete-Segado, P., et al. (2021). A review of additive manufacturing of ceramics by powder bed selective laser processing (sintering/ melting): Calcium phosphate, silicon carbide, zirconia, alumina, and their composites. *Open Ceramics, 5*, 100073.
6. Kumar, S. (2021). Comparison. In *Additive manufacturing solutions* (pp. 57–92). Springer.
7. Kumar, S. (2023). Advantage and Disadvantage. In *Additive manufacturing advantage, synthesis lectures on engineering, science, and technology* (pp. 1–60). Springer.
8. Yim, S., Sun, J., Minowa, K., et al. (2023). In-situ observation of powder spreading in powder bed fusion metal additive manufacturing process using particle image velocimetry. *Additive Manufacturing, 78*, 103823.

Chapter 86
Layer Deposition and Layer Arrangement

Layer deposition is the deposition of a single layer or many layers. Layer arrangement is an arrangement of layers. A design is an arrangement of layers. A part is an arrangement of layers. AM is an arrangement of layers.

A design can be called an arrangement of layers because it does not work unless it is divided into many layers. These layers are arranged in the design as per the design. If a design is different, the arrangement of layers in the design is different. This is how different parts are made. However, if the orientation of a design changes, it changes how layers are arranged [1]. Therefore, for a design, there will be a number of arrangements as per the orientation (Fig. 86.1). The orientation of a design is more important than the design itself, because this is the orientation that informs which arrangements can allow a design into a part conveniently. None of the arrangements (Fig. 86.1a, b, c, d) accurately fits in the design, showing the limitation of converting a design through layers. This limitation shows different printing speeds and surface finishes for different orientations [2]. Optimizing the build orientation will decrease manufacturing cost and support structure usage [3].

A part can also be called an arrangement of layers because it is made from a design that is made up of many layers.

When it is said AM is an arrangement of layers, it informs what AM does when it is used—it deposits one layer after another, giving rise to an arrangement of layers. But AM has a restriction that it does not decide about the arrangement. The decision is taken when a design is made. AM is to make the replica of the arrangement that is in the design. This allows AM to be understood from the design point of view but not from the layer arrangement point of view. But it is the layer arrangement point of view that is more important because there is no dearth of designs to be converted, but there is a dearth of manufacturing techniques to convert designs.

S. Kumar, *A Concise Encyclopedia of Additive Manufacturing*,
https://doi.org/10.1007/978-3-031-98394-8_86

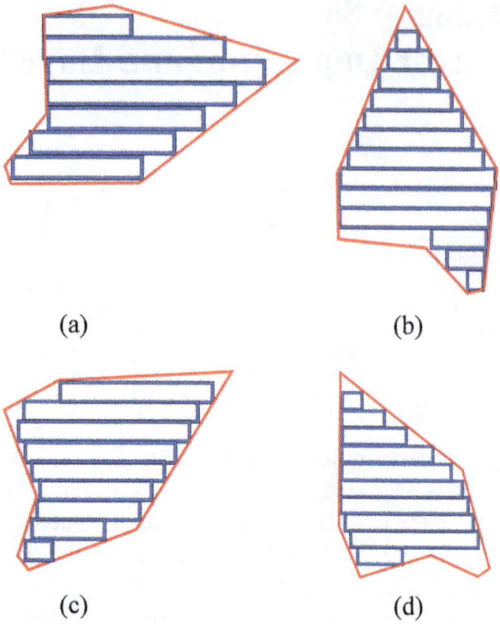

Fig. 86.1 Different layer arrangements for the same design. (**a**) 1st type. (**b**) 2nd type. (**c**) 3rd type. (**d**) 4th type

References

1. Zwier, M. P., & Wits, W. W. (2016). Design for Additive Manufacturing: Automated build orientation selection and optimization. *Procedia CIRP, 55*, 128–133.
2. Ben Amor, S., Zongo, F., Eltaief, A., et al. (2023). A new method to select optimal part building orientation for additive manufacturing processes based on geometric complexity and heat shrinkage. *Progress in Additive Manufacturing, 8*(2), 211–222.
3. Mathur, J., Miller, S. R., Simpson, T. W., & Meisel, N. A. (2024). Using virtual reality to orient parts for additive manufacturing and its effects on manufacturability and experiential outcomes. *Additive Manufacturing, 94*, 104421.

Chapter 87
Layerless FDM (Fused Deposition Modeling)

In fused deposition modeling (FDM), during the formation of a layer, the height of a nozzle from the layer does not change. The nozzle moves up after the last line of the layer is deposited, i.e., it gets the opportunity to move up only after the completion of each layer.

What if a nozzle does not wait for the completion of a layer before it moves up? In a layerless FDM, the nozzle constantly moves up while it deposits a line (Fig. 87.1). Thus, simultaneously moving up and forward by the nozzle makes a deposition path spiral [1]. Moving up in such a way does not let a layer to be made. If there are no layers, there does not exist any problem associated with joining them.

A constantly upward-moving nozzle means there will always be an empty space, i.e., the process does not have the capability to fill up an empty space that is made by a spiral path. For example, for making a thin cylinder by a spiral path, there will always be space at the center, i.e., the formation of the cylinder will necessarily create a cylindrical space at the center of the cylinder along its height.

A solid thin cylinder can be made if extruded material has low viscosity, so the material is not able to retain itself within a deposited line and will move inwards to fill up a gap at the center. But that material will also move outwards, increasing inaccuracy. A solid, inaccurate, thin cylinder will thus be made, which is not due to the merit but the demerit of the process.

To make a big solid cylinder, the nozzle has again to go down and start moving up; the movement of the nozzle will be obstructed by the structure it has already made. The process is thus not able to make a solid part but only hollow parts. Layerwise process has disadvantages, but unlike this process, it provides a route to make a solid part.

If a process makes only hollow objects such as empty cylinders, the process is not different from FDM that already makes such cylinders. Without support structures, a layerless FDM can make empty cylinders with an increasing or decreasing diameter, resulting in conical or inverted conical-type structures, respectively. Thus,

S. Kumar, *A Concise Encyclopedia of Additive Manufacturing*, https://doi.org/10.1007/978-3-031-98394-8_87

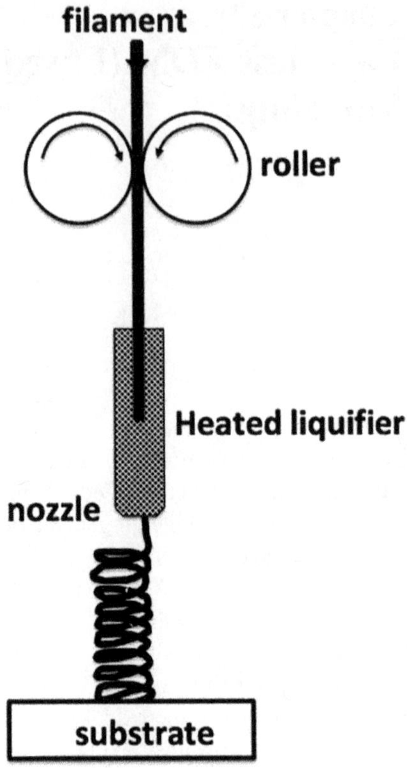

Fig. 87.1 Schematic diagram of layerless FDM

it has more scope than FDM to change diameters and vertical angles of conical structures.

In FDM, a deposited line is partially supported by the previously deposited line. Depending upon a change in diameter, the next deposited line will slip, which is prevented by creating support structures. In a layerless FDM, there is a continuous deposited spiral line that provides lateral support and prevents the structure from falling. The process thus demonstrates a method to avoid support structures [2].

References

1. O'Dowd, P., Hoskins, S., Geisow, A., & Walters, P. (2015). Modulated extrusion for textured 3D printing. In *NIP & digital fabrication conference* (Vol. 1, pp. 173–178).
2. Kanada, Y. (2015). Support-less horizontal filament stacking by layer-less FDM. In *International SFF symposium 2015* (pp. 56–70).

Chapter 88
Layerless FDM and Layerless PBF

Layerless fused deposition modeling (FDM) is a process that is an application of FDM to create spiral structures. Layerless FDM and FDM are different because they have different process steps. The process step that is different is how extrudates are deposited. In layerless FDM, it is deposited layerless, while in FDM, it is layerwise, i.e., when FDM is applied to create a structure for which the same process steps are not followed, a new process is formed.

Thus, a layerwise process, FDM, gives rise to another process, layerless FDM. For creating a new process, what FDM does is not something that cannot be done by other processes. Any deposition process can deposit material in a spiral way, and a new process will be ready. It has already been done by other processes, e.g., WAAM [1] and solvent-cast direct writing (DW) [2] (in DW, solvent of extrudate ink is evaporated, causing the extrudate to retain its shape).

There is no need for a nozzle moving in a circular path to go up if the platform is going down. There is also no need for the nozzle either to move or to go up if the platform is simultaneously rotating and moving down. Creating a relative spiral movement between the platform and the nozzle gives a new process belonging to a different process group and a spiral structure.

This happens in powder bed fusion (PBF) when the platform rotates and goes down [3], and in photopolymerization (in inverse orientation) when the platform rotates and goes up [4].

In PBF, there is no nozzle to deposit materials, and therefore the material needs to be arranged without the help of a nozzle that is moving along a tool path. Thus, a material hopper is required to spread the material and create the bed. Thus, both processes (PBF and deposition) require material to be deposited. One process does it with a nozzle, while another process does it with a hopper. When a spiral structure is made, both processes look to be closer to each other, blurring the division between them. Though powder is deposited in both cases, there is a difference. One process requires just the deposition with a nozzle. Another process requires deposition plus spreading. Though a fraction of spreading can be enough to make a part in PBF,

S. Kumar, *A Concise Encyclopedia of Additive Manufacturing*,
https://doi.org/10.1007/978-3-031-98394-8_88

there is no absence of spreading in PBF. That's why instead of a nozzle, a powder hopper is required, which is having an attached coater with it. Therefore, creating a layerless structure in PBF does not convert PBF into a deposition process.

Why is a spiral structure made? It is made because for the fabrication of certain geometries, support structures are not needed when a part is made spiral way. Support structures are needed more in a deposition process than in PBF because of the lack of a bed in the deposition process [5]. Therefore, fabrication through spiral structure has more utility in a deposition process when a part necessarily requires a support structure in either process, and a bed in PBF provides the necessary support.

What if the bed in PBF does not provide the necessary support? It means when a part is made in either process, it requires the creation of special support structures in both processes. Even then, a spiral deposition process is preferred to a spiral PBF when a spiral-based process needs to be developed only to overcome the need for support structures. It is because the working of the spiral deposition process does not require any modification in the existing deposition system, provided the system has enough degrees of freedom for its nozzle to move in both horizontal and vertical planes. While existing PBF, especially commercial systems, are not fit for a spiral-based process and require extensive modification in terms of platform movements and material feedings.

However, for photopolymer, a spiral deposition process does not have much edge over a spiral PBF, as there exist systems for inverse stereolithography that can be modified for the spiral PBF.

References

1. Gu, X., Hou, Z., Xu, J., et al. (2017). A novel additive manufacturing method for spiral parts. In *2017 IEEE 7th annual international conference on CYBER Technology in Automation, control, and intelligent systems* (pp. 791–796).
2. Guo, S., Gosselin, F., Guerin, N., et al. (2015). Solvent-cast three-dimensional printing of multifunctional microsystems. *Small, 9*(24), 4118–4122.
3. Hauser, C., Sutcliffe, C., Egan, M., & Fox, P. (2005). Spiral growth manufacturing (SGM)—A continuous additive manufacturing technology for processing metal powder by selective laser melting. In *SFF symposium proceedings*.
4. Dudley, K. (2015). 3D printing using spiral buildup. US patent US20140265034A1.
5. Kumar, S. (2022). Fabrication strategy. In *Additive manufacturing solutions* (pp. 111–144). Springer.

Chapter 89
Layerless PBF

Powder bed fusion (PBF) makes a layer when a powder bed is formed. This layer is of constant thickness because the layer is made by spreading on a planar platform that is positioned parallel to the direction of the spreading. The thickness of the layer will increase when the platform goes down, increasing the separation between the build plane and the platform. Thus, the thickness is changed by changing the separation.

But this changing of the layer thickness does not change the fact that there will always be a fixed thickness. Though there will be a number of thicknesses that can be obtained when a part is made—this is how a part property is optimized by combining a number of thicknesses. But for a given layer when the processing is happening, the thickness does not change. This limitation of PBF to not being able to change the thickness unless the processing of a layer is completed gives a uniform thickness of a powder layer.

What will happen if the thickness for a particular layer does not remain constant but changes? Then the solidified layer that is obtained after processing will not be planar but will be of varied thicknesses. This means there will no longer be one thickness but many thicknesses of a layer, implying how a layer thickness is perceived presently will not be applied in this case.

In order to change the thickness of a layer many times during processing, it is necessary to know how a particular layer thickness is obtained and how it can be changed.

A particular layer thickness is obtained because when a platform goes down, a scraper moves from one side of the platform to the other side, filling a gap with powder on the platform. This movement of the scraper from one side to the other side helps a constant thickness to be obtained on the platform. Unless the scraper has completed moving and reaching to the other side, it is not possible any more to decrease the platform position or index. But after the scraper has reached the other side, the platform position is decreased. The next movement of the scraper again will give another layer when the scraper reaches the other side. This is the scraper's

S. Kumar, *A Concise Encyclopedia of Additive Manufacturing*,
https://doi.org/10.1007/978-3-031-98394-8_89

to-and-fro movement, which gives one layer of one thickness for its to movement and another layer of another thickness for its from movement.

What if when the scraper starts to move, the platform starts to go down—the platform does not go down fast, and the scraper brings enough powder to fill the increasing gap on the platform (Fig. 89.1a). At the end, there is not a layer thickness of constant size but of increasing size. What will happen if a laser beam or a binder jet follows the scraper when the scraper is moving—the laser beam or the binder does not wait to let the scraper reach the end of the platform to start processing or solidifying; instead, the processing follows the scraper. It means there is processing, then there is no processing, then there is processing—the cycle goes on before the scraper reaches the end. Each processing means a fraction of the layer is solidified, giving a small solidified bed. Each no processing means a fraction of a layer is going to be spread, but at a slightly different height.

This cycle reminds of the usual cycle that happens in PBF when there is processing of a layer followed by no processing during the spreading of the layer—the cycle requires the completion of a layer.

While this cycle happens without the completion of a layer. At the end of this cycle, there will be a triangular solid block consisting of many steps (Fig. 89.1b). The number of steps is equal to the number of times of the processing. Formation of the step means the downward movement of the platform was not too fast to disturb the planarity of the step. At the beginning, the step has the minimum height corresponding to the initial layer thickness while at the end, the step has the maximum height equal to the layer thickness achieved at the end. It is feasible if the processing

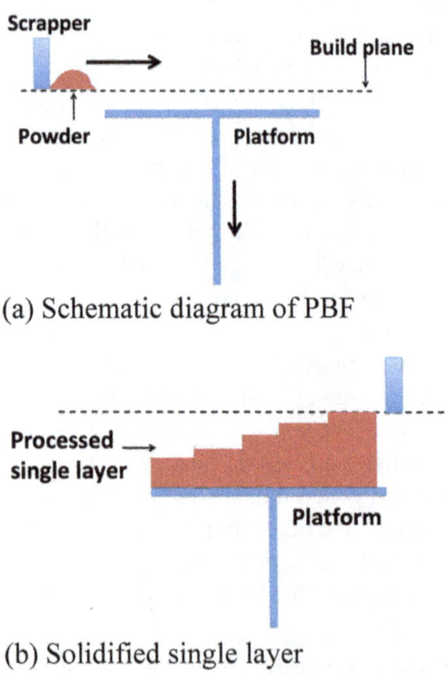

Fig. 89.1 Formation of an inclined layer: (**a**) Schematic diagram of PBF. (**b**) by decreasing the platform while moving the scraper

is possible from minimum to maximum layer thickness. If there is not much difference between the minimum and the maximum thickness, the height of the block will be small.

The shape of the layer thus obtained is not part of state-of-the-art AM. The layer must be planar, so a CAD model is sliced parallelly to give planar layers so that the difficulty of processing a layer will not change from one location to another location within a layer due to variations in the thickness.

When AM is called additive layer manufacturing (ALM), the name conveys that the layer is planar in practice. The reason is that ALM is not advanced enough to accommodate all types of layers, whether they are nonplanar [1–3], curved [4, 5], spiral, or of increasing or decreasing thickness. This is the reason why ALM in the case of PBF is not free from 'bed.'

Thus, in the state-of-the-art PBF, there is an unwritten pledge between a platform and a scraper that the platform will remain faithful and will not move down unless the scraper completes its journey and reaches the other side of the platform.

Formation of a bed in PBF requires cooperation from the platform. The platform provides constant support, so a layer is formed, comprising a myriad of powders having unmovable positions. These positions of the powder can move during processing, but they will not move during processing for the reason that the platform fails to provide support. But when a layer is formed by constantly moving down the platform, it generates a layer having a slope, which does not guarantee that the powders will not move and change the position, leading to a possible collapse of the inclined layer.

Thus, processing of a layer after its formation, as happens in PBF, will not work in an inclined layer. The waiting that happens in PBF, that unless the scraper does not reach the other end, processing of a layer will not start, will not work in the case of an inclined layer. There needs to fix a layer before the powder of the inclined layer starts to unsettle because the slope starts to become steep with a constant moving down of the platform. Before the layer becomes steep, the layer is not steep—this means within some tolerance, the inclined layer can be considered a layer resembling a bed. This means, within a fraction of the distance from the start of the movement of the scraper, there is no difference between an inclined layer and a non-inclined layer (if formed). As the processing of the inclined layer within the fraction of the distance does not bring a non-uniformity in the property that could be expected to be obtained when processing an inclined layer. Thus, the inclined layer is not different from a powder bed when it is processed not far late from its time of formation.

Consequently, processing an inclined layer, if done within a certain time after it starts to get steep, is not different from processing a number of small powder beds arranged in increasing steps comprising the layer.

References

1. Ahlers, D., Wasserfall, F., Hendrich, N., & Zhang, J. (2019). 3D printing of nonplanar layers for smooth surface generation. In *IEEE 15th international conference on automation science and engineering (CASE)* (pp. 1737–1743).
2. Wulle, F., Gorke, O., Schmidt, S., et al. (2022). Multi-axis 3D printing of gelatin methacryloyl hydrogels on a non-planar surface obtained from magnetic resonance imaging. *Additive Manufacturing, 50*, 102566.
3. Ruan, J., Tang, L., Liou, F. W., & Landers, R. G. (2010). Direct three-dimensional layer metal deposition. *Journal of Manufacturing Science and Engineering, 132*(6), 064502.
4. Tan, W. S., Juhari, M. A. B., Shi, Q., et al. (2020). Development of a new additive manufacturing platform for direct freeform 3D printing of intrinsically curved flexible membranes. *Additive Manufacturing, 36*, 101563.
5. Emon, O. F., Alkadi, F., Kiki, M., & Choi, J. W. (2022). Conformal 3D printing of a polymeric tactile sensor. *Additive Manufacturing Letters, 2*, 100027.

Chapter 90
Layerless PBF: Part Formation

Formation of an inclined layer and its processing leads to the creation of a block having steps—this is a processed layer. Since a part consists of many layers—a part must be conceptualized in the form of an aggregation of such layers.

Consider a part that consists of more than one layer. For fabrication of the second layer, processing must happen on the first layer that is a triangular block having one end at the level of the build plane while the other end is nearest to the platform surface (Fig. 90.1). Powder spreading followed by stepwise processing can start at any of the two ends of the platform. At one end, where the upper surface of the block is at the level of the build plane, the platform needs to go down to accommodate the powder and start the processing. But, after that, the platform does not need to go down because there already exists enough space above the sloped surface to be filled up and processed.

Though powder spreading due to the scraper is required to level the surface. Processing, though, happens through small powder bed by small powder bed, as happened while making the first layer. But the processing needs to be done on a surface full of steps, which was not the case while processing the first layer. This requires overcoming extra difficulty in processing. There would not be any such difficulty should there be a planar layer. This is the reason why layerwise manufacturing is planar layerwise manufacturing. In the case of a planar layer, when the scraper returns from the other end to the first end while making the next layer, the scraper does not find different thicknesses or gaps to fill. But in the case of an inclined layer, the problem starts when the scraper returns to make the next layer as the processed layer is stepwise decreasing. There was no such problem when the first layer was getting processed. During the processing of the first layer, there is not much difference in difficulty in making either a planar layer or an inclined layer.

What will happen if only the first inclined layer is formed? This is not possible because a useful part consists of more than one layer. The problem starts when the scraper returns to make the next layer. Is it possible that the scrapper does not need to return? It is possible when the scraper moves in a circular path, and with a

S. Kumar, *A Concise Encyclopedia of Additive Manufacturing*,
https://doi.org/10.1007/978-3-031-98394-8_90

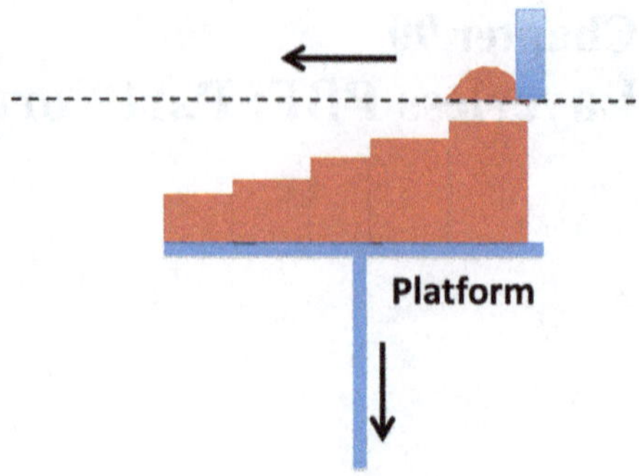

Fig. 90.1 For making a second layer on an inclined layer

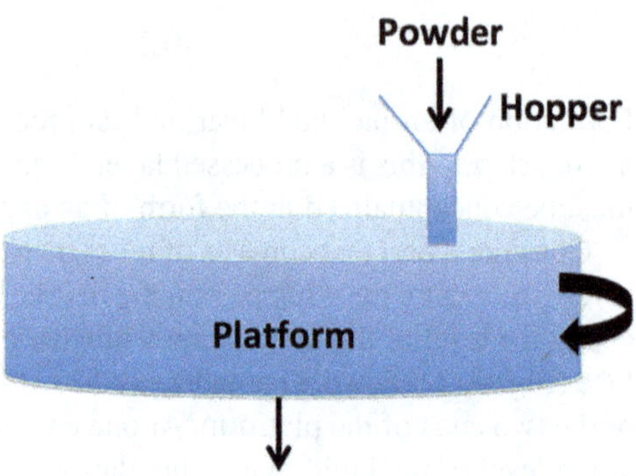

Fig. 90.2 A circular platform rotating and going down for making spiral parts

constant downward movement of the platform, the scraper acquires a spiral path, making it not to encounter the same point of the path [1].

Thus, the scraper always makes small powder beds. How small the small powder bed depends upon the speed of the downward movement of the platform. The scraper does not need to move when the platform is not big and can rotate (Fig. 90.2). It is the relative movement between the platform and the scraper that gives the required powder bed.

This type of powder spreading and consolidation will lead to a cylindrical part having the transition between two successive layers that will not be as visible as found in layerwise, which is an advantage. If a circular track is big, parts of any shape (not only the cylindrical) can be made by utilizing a small section of the track.

Reference

1. Hauser, C., Sutcliffe, C., Egan, M., & Fox, P. (2005). Spiral growth manufacturing (SGM)—A continuous additive manufacturing technology for processing metal powder by selective laser melting. In *SFF symposium proceedings, Texas, USA*.

Chapter 91
Layerless PBF: Why Layerless

A spiral path does not have a clear transition from the first track to the second track, while the circular path will not have any transition to the second track unless there is a clear transition. This is the difference between layerless and layerwise powder bed fusion (PBF). In layerwise (circular track), there comes a point when the index moves up from the first track to the second track (Fig. 91.1a); this point comes after completion of the first track. This index shows a discontinuity in fabrication, which makes it layerwise. Without this discontinuity, the ongoing fabrication will not resume.

In layerless (spiral track), there never comes a point for the index to move up from the first track to the second track, but the transition takes place—it means the index starts to move up before the end of the first track and continues moving up after the end of the first track (Fig. 91.1b). This upward movement of the index does not consume a point in the space but consumes a line. This line consists of the end section of the first track and the start section of the second track.

The track due to this curved line is not different from an inclined track due to a linear line made in PBF, as both tracks consist of steps. The steps of the inclined track give an indication of how the index moves up stepwise after the formation of a small powder bed each time.

Thus, spiral tracks can be understood to be a mixture of various lengths of powder bed. During transition, the powder beds are small, while at other locations of the track, the powder bed can be long. Depending upon the angle of spiral with the vertical direction, the lengths of the powder bed at various locations will vary.

In circular tracks, there is one length of a powder bed—the length of the circular track. Both circular and spiral tracks have one common thing—both make powder beds. This is the reason why both are PBF. But if bed means a layer, then why are both not layerwise processes?

A layerwise process is understood to be those processes that make products as per the slice of a CAD model, where the slice is a planar layer in practice. This planar layer does not represent the myriad types of powder beds formed in spiral

S. Kumar, *A Concise Encyclopedia of Additive Manufacturing*,
https://doi.org/10.1007/978-3-031-98394-8_91

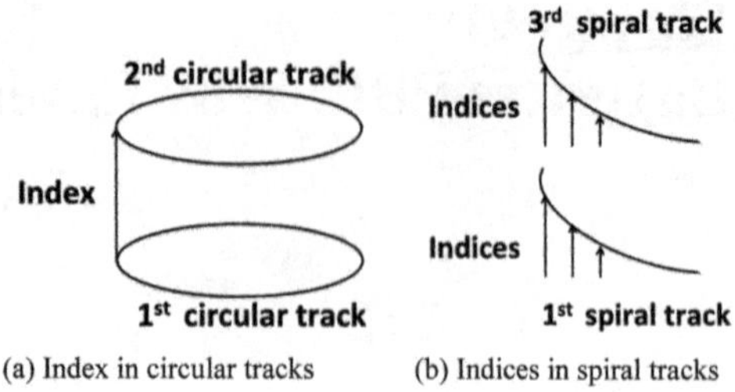

Fig. 91.1 Index in circular (**a**) and spiral (**b**) tracks

tracks. This planar layer is not meant to represent the composition of a particular track but whether a particular track is planar or not. As a spiral track is not represented by a planar layer, it does not come under layerwise.

But making a spiral part does not necessarily require a CAD model to be sliced into spiral layers. A CAD model sliced in planar layers can also make spiral tracks and then a spiral part. What is required in this case is the incorporation of the downward movement of the platform during the formation of a planar layer—the result will be a spiral track. Whether this is layerwise or layerless? It should not be considered layerwise because the spiral track that is formed does not represent the planar layer. Besides, the formation of spiral tracks is due to not following the right set of steps of a process. The right process steps do not include the downward movement of a platform during the formation of a layer. The transformation of planar layers into spiral tracks demonstrates either the realization of another process or a defect in the original process. Though this process defect is not visible because it comes as a manufacturing strategy, it shows how the limitation of layerwise can be overcome.

Chapter 92
Layerless 3D Printing

Layerless 3D printing is the same as additive layerless manufacturing [1].

Reference

1. Moayedi, S., Zamani, J., & Salehi, M. (2025). A review on layerless 3D printing: Challenges and applications. *Rapid Prototyping Journal, 31*(2), 409–431.

S. Kumar, *A Concise Encyclopedia of Additive Manufacturing*,
https://doi.org/10.1007/978-3-031-98394-8_92

Chapter 93
Liquid Deposition Process

Inkjet printing (IJP) [1], digital inkjet printing, direct inkjet printing, polymer jetting, photopolymer jetting [2] (Fig. 93.1), rapid freeze prototyping (RFP), etc., are liquid deposition processes.

If the material is not liquid, it is melted in order to be deposited. If it has a high melting point, e.g., ceramic, it is mixed in a carrier liquid. If it is a high molecular weight polymer, it is dispersed in a carrier liquid to make it low-viscous so that it can be ejected through a nozzle.

The liquid should not be solidified either in the nozzle or after leaving it and should reach a substrate in the state of liquid. Besides, the liquid composition should not change due to chemical reaction, physical segregation, or evaporation. It can reach in the form of either a continuous stream or drops. But the continuous stream has a propensity to break before it reaches, therefore, the liquid deposition is a dropwise deposition. The process can thus be controlled by controlling the ejection of drops.

Drops are detached from the bulk of liquid through a nozzle by the application of an electric field [3], variable magnetic field [4, 5], vibration, sound waves, heating, vapor pressure, or force using physical objects such as pistons or screws, etc. The number of drops ejected is synchronized with the speed of the nozzle relative to the substrate so that the speed should not be either high, causing a gap, or low, causing an overlap between two deposited drops.

When a drop reaches a substrate, the condition of the drop depends upon the impact [6]. If it has high impact, the drop will be flat, and its shape will be elliptical. The high impact will cause a splash, leading to a loss of material from the drop. The elliptical drop may retract if it has high surface tension [7]. If it has low surface tension, the drop will expand.

When successive drops reach the substrate, the drops coalesce and then lose their shapes to make a line. Solidification of drops or lines depends upon the phase transformation, polymerization (Fig. 93.1b), and gelling or vaporization of carrier liquid.

S. Kumar, *A Concise Encyclopedia of Additive Manufacturing*,
https://doi.org/10.1007/978-3-031-98394-8_93

(a) Photopolymer drop is deposited

(b) Deposited photopolymer is solidified by curing

Fig. 93.1 Schematic diagram of photopolymer jetting: (**a**) material jetted, (**b**) material solidified

It is facilitated by changing the environment on the substrate by localized heating or curing using a beam.

If there are many nozzles, many such adjacent lines can be made to make a layer, and thus the fabrication can be expedited. If there are many nozzles using many materials, there will be many lines made from different materials, which can provide a technique to make a multi-material part. If different materials are intended to serve different functions, the part can work as a multifunctional part.

Making a multi-material part requires compatibility between different materials so their ejection from the nozzles will not cause mismatch in ejection times and drop sizes. Besides, there should not be a large difference between inter-material (drops from different nozzles) coalescence and intra-material (drops coming from the same nozzle) coalescence. If drops coming from different nozzles are of different colors, a multicolor part can be made [8].

References

1. Zhu, Z., Gong, Z., Qu, P., et al. (2021). Additive manufacturing of thin electrolyte layers via inkjet printing of highly-stable ceramic inks. *Journal of Advanced Ceramics, 10*, 279–290.
2. Patpatiya, P., Chaudhary, K., & Kapoor, V. (2022). Reverse manufacturing and 3D inspection of mechanical fasteners fabricated using photopolymer jetting technology. *Mapan, 37*(4), 753–763.
3. Ball, A. K., Das, R., Das, D., et al. (2018). Design, development and experimental investigation of E-jet based additive manufacturing process. *Materials Today: Proceedings, 5*, 7355–7362.
4. Jayabal, D. K. K., Zope, K., & Cormier, D. (2018). Fabrication of support-less engineered lattice structures via jetting of molten aluminum droplets. In *Proceedings of the 29th annual international solid freeform fabrication symposium* (pp. 757–764).
5. Simonelli, M., Aboulkhair, N., Rasa, M., et al. (2019). Towards digital metal additive manufacturing via high-temperature drop-on-demand jetting. *Additive Manufacturing, 30*, 100930.
6. Wijshoff, H. (2018). Drop dynamics in the inkjet printing process. *Current Opinion in Colloid & Interface Science, 36*, 20–27.
7. Derby, B. (2015). Additive manufacturing of ceramic components by ink jet printing. *Engineering, 1*(1), 113–123.
8. Meisel, N., Dillard, D., & Williams, C. (2018). Impact of material concentration and distribution on composite parts manufactured via multi-material jetting. *Rapid Prototyping Journal, 24*(5), 872–879.

Chapter 94
Machining Tool: Problem

There are several reasons why a machining tool does not help make a complex part; some of the reasons are as follows:

- The tool cannot be made thin enough to make thin features. In extreme cases, if it is made, it can break while making the feature; or if it is not made thin, it may break the feature.
- A sharp corner depends upon the sharpness of a tool. If a sharp tool cannot be made and operated, there is no tool available to make the sharp corner.
- Making a complex part requires a number of tools that need to be arranged in a sequence of operations, i.e., a decision needs to be made about which tool will come after the use of which tool. This does not allow for making all types of parts.
- For example, a small milling tool can make a fine feature, but it cannot make the same at the inside base of a hollow cylinder because it cannot access the base. Hence, the sequence of operations, i.e., to make a hollow cylinder using a big milling tool and then using the small milling tool to carve a feature, will not work. The sequence of operations in the reverse order will not work either because a cylinder is made by removing the material (i.e., clearing the feature). Therefore, the complex part cannot be made because each tool does not have the ability to make a shape. But because when the operations of the two tools are combined, the result of the operation of one tool does not allow another tool to work.
- Making a complex part requires excessive material removal. If the part is formed from a difficult-to-cut material, the tool wear will further increase [1, 2]. In the absence of a suitable tool for such materials, the part will not be formed.

S. Kumar, *A Concise Encyclopedia of Additive Manufacturing*,
https://doi.org/10.1007/978-3-031-98394-8_94

References

1. Pramanik, A. (2014). Problems and solutions in machining of titanium alloys. *The International Journal of Advanced Manufacturing Technology, 70*, 919–928.
2. Singh, P., Pungotra, H., & Kalsi, N. S. (2016). On the complexities in machining titanium alloys. In *CAD/CAM, robotics and factories of the future: Proceedings of the 28th international conference on CARs & FoF 2016* (pp. 499–507). Springer India.

Chapter 95
Marking

AM can make a part that can be used as a tag to identify other parts [1]. What if an AM part contains a QR code to trace it in a supply chain or to distinguish it from other parts [2, 3]? What if an AM part contains a hidden spot of chemicals to verify its authenticity [4]? AM provides opportunities to engrave QR codes on parts' various surfaces during manufacturing.

If an identifying mark gives information about the design orientation, it will help repair [5].

References

1. Desavale, S., Ameri, F., Trueba, L., & Igveh, O. (2022). Direct part marking (DPM) supported by additively manufactured tags to improve the traceability of castings. In *IFIP international conference on advances in production management systems* (pp. 244–251). Cham.
2. Matuszczyk, D., & Weichert, F. (2023). Reading direct-part marking data matrix code in the context of polymer-based additive manufacturing. *Sensors, 23*(3), 1619.
3. Jahnke, U., Bornefeld, P. A., & Koch, R. (2016). Production-integrated markings for traceability of AM parts in the context of industry 4.0.
4. Sola, A., Sai, Y., Trinchi, A., et al. (2021). How can we provide additively manufactured parts with a fingerprint? A review of tagging strategies in additive manufacturing. *Materials, 15*(1), 85.
5. Kumar, S. (2023). A concept for refabrication. In *Additive manufacturing advantage, synthesis lectures on engineering, science, and technology* (pp. 82–98). Springer.

S. Kumar, *A Concise Encyclopedia of Additive Manufacturing*,
https://doi.org/10.1007/978-3-031-98394-8_95

Chapter 96
Material Extrusion

Material extrusion (MEX) is a type of AM technique in which filaments or pellets are melted enough to be extruded to make parts. Other names of the technique are fused deposition modeling [1], fused filament fabrication [2], fused pellet modeling [3], etc. [4]. The technique requires the feedstock in the form of filaments or pellets to be converted into materials that must have extrudability. The conversion of the feedstock in the form of an extrudable material is part of the technique. This technique is not applicable to those techniques that rely on material extrusion to make parts without prior conversion. For example, if there is a feedstock in the form of extrudable slurry, paste, gel, or bioink, material extrusion to make a part will not come under material extrusion (MEX). Though it is technically material extrusion. Extrusion of an extrudable ink comes under ink jet printing, ink writing, extrusion printing, extrusion-based 3D printing, etc. [5–9]. These extrusion-based 3D printing do not include MEX [9–12].

References

1. Siemiński, P. (2021). Introduction to fused deposition modeling. In *Additive manufacturing* (pp. 217–275). Elsevier.
2. Dey, A., Roan Eagle, I. N., & Yodo, N. (2021). A review on filament materials for fused filament fabrication. *Journal Of Manufacturing And Materials Processing, 5*(3), 69.
3. de Castro Gonçalves, C., Malavolta, A. T., Barbosa, G. F., et al. (2024). Influence of bed temperature on mechanical resistance, surface and dimensional quality of 3D printed parts by robotized fused pellet modeling. *Progress in Additive Manufacturing*, 1–15.
4. Oleff, A., Küster, B., Stonis, M., & Overmeyer, L. (2021). Process monitoring for material extrusion additive manufacturing: A state-of-the-art review. *Progress in Additive Manufacturing, 6*(4), 705–730.
5. Chung, J. H., Naficy, S., Yue, Z., et al. (2013). Bio-ink properties and printability for extrusion printing living cells. *Biomaterials Science, 1*(7), 763–773.

S. Kumar, *A Concise Encyclopedia of Additive Manufacturing*,
https://doi.org/10.1007/978-3-031-98394-8_96

6. Wu, Q., Song, K., Zhang, D., et al. (2022). Embedded extrusion printing in yield-stress-fluid baths. *Matter, 5*(11), 3775–3806.
7. Yuk, H., & Zhao, X. (2018). A new 3D printing strategy by harnessing deformation, instability, and fracture of viscoelastic inks. *Advanced Materials, 30*(6), 1704028.
8. Zhang, D., Hsu, Y. C., & Dunand, D. C. (2024). Ink-extrusion 3D printing and silicide coating of HfNbTaTiZr refractory high-entropy alloy for extreme temperature applications. *Advanced Science, 11*(17), 2309693.
9. Hossain, S. S., Jang, S., Park, S., & Bae, C. J. (2023). Understanding ink design and printing dynamics of extrusion-based 3D printing: Defect-free dense piezoelectric ceramics. *Journal of Manufacturing Processes, 92*, 1–11.
10. Outrequin, T. C. R., Gamonpilas, C., Siriwatwechakul, W., & Sreearunothai, P. (2023). Extrusion-based 3D printing of food biopolymers: A highlight on the important rheological parameters to reach printability. *Journal of Food Engineering, 342*, 111371.
11. Zhang, B., Belton, P., Teoh, X. Y., et al. (2024). An investigation into the effects of ink formulations of semi-solid extrusion 3D printing on the performance of printed solid dosage forms. *Journal of Materials Chemistry B, 12*(1), 131–144.
12. Agarwal, T., Costantini, M., & Maiti, T. K. (2021). Extrusion 3D printing with pectin-based ink formulations: Recent trends in tissue engineering and food manufacturing. *Biomedical Engineering Advances, 2*, 100018.

Chapter 97
Material Jetting

Material jetting (MJT) is a type of AM in which a product is formed only from the material deposited as per the design of the product [1]. The material is deposited using a nozzle or nozzle-like device. The only source of the material of the product is the material deposited from the nozzle or nozzle-like device. This type of AM is known by other names as well, such as direct ink writing [2, 3], inkjet 3D printing [4], etc.

MJT does not include binder jetting (BJT), as BJT is a different AM technique [5]. In BJT, as the name implies, binder is jetted to make a product. The purpose of the binder, as the name implies, is to bind materials to make a product. Thus, two types of materials are required in BJT to make a product: one is a binder, and the other is a building material to be bound. There is a restriction on how the building material will be sourced. The building material comes only from a powder bed. Therefore, BJT requires a binder to be jetted only on a powder bed. This means BJT requires two steps to complete: the first step is to make a powder bed, and the second step is to jet the binder on it. How the powder of the powder bed is sourced does not change whether this technique will be called BJT or not. The powder for the bed can be sourced either from the container kept under the platform, so the container will move up at one side of the platform to make the powder available to be spread on the platform. Or the powder can be sourced from the container above the platform, so the powder will be transferred to one side of the platform to spread on the platform. If the powder is transferred using a nozzle or nozzle-like device, this transfer of the material can also be called jetting of the material. Thus, in BJT, jetting can be the reason how both binder and building material can be sourced. But jetting the building material is not as per the design of the model of a product, while jetting of the binder is as per the design of the model.

This jetting of the building material makes a difference between MJT and BJT. In MJT, jetting is the only method to source the material. Besides, jetting of the material should be as per the design of the model of a product, because this is the only way a product is made. In MJT, jetting of a material means jetting of a building

S. Kumar, *A Concise Encyclopedia of Additive Manufacturing*,
https://doi.org/10.1007/978-3-031-98394-8_97

material. Although there is no need to distinguish between a material and a building material, because whatever is deposited is considered a building material, as there is no concept of the deposition of a binder in MJT.

What will happen if, in MJT, a building material is deposited by one nozzle and a binder is deposited from another nozzle? The aim of the binder deposited from the second nozzle is to bind the material deposited from the first nozzle. Whether this will be MJT, because the building material is jetted as per the design of a model, satisfying the condition of MJT? Whether this will be BJT because the binder material is jetting as per the design of a model? Whether this will be hybrid AM, because the product is made by the combination of the jetting of material (material jetting) and the jetting of binder (binder jetting)?

Above will be MJT because the basic difference between MJT and BJT is the presence of a powder bed. In MJT, there is an absence of a powder bed or a bed from any materials; the product is made only by sourcing the materials using an action, i.e., jetting, where the jetting does not help form a bed. MJT does not identify the material, only ascertaining that the constituent of the product will not be different from this material. Thus, MJT is silent on the role of the material.

A binder is a material. Therefore, jetting of the binder is jetting of a material. Therefore, binder jetting is technically a material jetting. But binder jetting as an AM technique (BJT) is not technically material jetting as an AM technique (MJT). Moreover, BJT and MJT do not have enough relations, as their names imply. If BJT has a relation with MJT because both use jetting techniques, then BJT also has a relation with powder bed fusion (PBF) because both use powder beds. If BJT has a relation with MJT because both use jetting techniques, then MJT also has a relation with (powder-based) directed energy deposition (DED) because both use different degrees of jetting. Knowing MJT does not intuitively lead to knowing BJT, and vice versa. MJT can only be known in the presence of pre-knowledge of BJT, and vice versa.

Where does this lead to? Let us consider an example. A material that is a glue or glue-like is taken to make a product using MJT. Since MJT is silent about how the jetted material is consolidated, it will not exclude a glue being taken and getting consolidated among themselves by using its property of gluing. Product formation by this jetting of binder cannot be binder jetting (BJT), because jetting of binder is not defined as binder jetting in AM.

References

1. Gülcan, O., Günaydın, K., & Tamer, A. (2021). The state of the art of material jetting—A critical review. *Polymers, 13*(16), 2829.
2. Saadi, M. A. S. R., Maguire, A., Pottackal, et al. (2022). Direct ink writing: A 3D printing technology for diverse materials. *Advanced Materials, 34*(28), 2108855.
3. Tagliaferri, S., Panagiotopoulos, A., & Mattevi, C. J. M. A. (2021). Direct ink writing of energy materials. *Materials Advances, 2*(2), 540–563.

4. Liu, Z., Huang, J., Li, P., et al. (2024). High-precision inkjet 3D printing of curved multi-material structures: Morphology evolution and optimization. *Additive Manufacturing, 94*, 104497.
5. Fang, X., Zu, Y., Ma, Q., & Hu, J. (2023). State of the art of metal powder bonded binder jetting printing technology. *Discover Materials, 3*(1), 15.

Chapter 98
Material Saving

In additive processes, such as sintering, casting, etc., materials are not needed more than what constitutes a part. Therefore, these processes should have the same advantage as AM has, but they do not have it because they do not have the design capability to save material. AM can change design to change the need where material is added—leading to lightweighting [1–2] or a strategy to build [3].

Material is saved because a part with a new design consists of less material than a part with an old design. This is due to an improvement in the old design, which is possible due to AM. When both parts weigh, the new part will be lighter than the old one, concluding that AM saves material. The conclusion would not have required any scrutiny if AM had not needed more material than that required by the new design. Or if all techniques other than AM had wasted much material while making the old design, it was obvious that AM saved material. But AM uses more material both before and during fabrication. To save material by converting a new design into a lightweight part, AM uses more material. This use of more material may or may not lead to the wastage of material. If the use leads to the wastage of material, AM wastes material in order to save material by converting a new design into a part.

1 Amount of Material

In powder bed fusion (PBF), the whole bed needs to be filled up with material irrespective of the size of a part fabricated. This amount of material is higher than the amount that becomes the part.

If the extra material involved in PBF is again utilized, the material will not be wasted. But not all materials that help develop a part have their properties unchanged. The deterioration rate depends upon the material type, the size of a part, scanning time, and the time the part spent in bed. If no material deteriorates in bed, even then material can be lost. Putting the material in a system, recouping after the job is done,

S. Kumar, *A Concise Encyclopedia of Additive Manufacturing*,
https://doi.org/10.1007/978-3-031-98394-8_98

handling the part and removing it from the system, and cleaning the part [4] and the system—all contribute to varying extents of spilling.

If deposition process (DP) wastes material, it is because the amount deposited is more than required. It also wastes when a support structure (SS) is made, but making an SS is not the basic requirement of DP in the same way as forming a bed in PBF is.

What if all three (PBF, DP, and machining) are developed so no material is wasted? It means, in PBF, no material is either spilled or deteriorated in bed. While in DP, whatever material is deposited ends up in a part. And in machining, all chips are recycled. Then, which one of the three is the best? DP is the best because all material becomes a part. PBF and machining involve more material—PBF for forming a bed and machining due to the difference between an original block and a final part.

2 Better than Machining

Consider an example of a titanium part fabricated by machining where only 10% of the material ends up in a part [5], and the remaining 90% is wasted as chips [6]. Fabricating it in AM saves material.

However, in machining, when 50% of the material ends up in a part and the remaining 50% is wasted, AM is not better to save material [7]. Then, why fabricate it in AM? Because even that small amount of machining is difficult to carry out, as the material is ceramic or a composite that has less machinability. What if the material has high machinability? AM can still be used if the design is too complex to be machined.

What if design is not complex? AM can still be preferred if the part is not one-off but a member of a group to be fabricated—it is for mass customization or a small batch of production in which the decision to fabricate is not taken solely on the basis of the merit of AM over CM but due to logistic reasons (such as utilizing machine space). In this case, machine space is shared by all members of the group, which minimizes the material loss per part.

What if the fabrication of a part does not accompany a group? The part is one-off; still, it can be fabricated if better property is required. Since machining does not improve properties but AM does.

3 By Multi-Material Part

AM gives an opportunity to make a multi-material part, which saves material by allowing to choose where within the part not to use the material. For example, to make a wear-resistant part, when AM uses a wear-resistant material to make the surface and another material to make all of the part except its surface. It saves the wear-resistant material by not using it to make sections other than the surface. Instead of wear resistance, it can also be corrosion [8], fatigue, oxidation, or thermal resistance [9].

What the process allows to save cannot be saved if there is no developed machine to execute the process steps well. If mixing or spilling wastes the material, the savings due to the formation of a multi-material part will provide less total savings.

References

1. Zhu, L., Li, N., & Childs, P. R. N. (2018). Light-weighting in aerospace component and system design. *Propulsion and Power Research, 7*(2), 103–119.
2. Fais, C., Yasko, I., Ali, M., et al. (2024). Experimental performance evaluation of a lightweight additively manufactured hydrodynamic thrust bearing. *International Journal of Lightweight Materials and Manufacture, 8*, 285–299.
3. Wang, X., Zhou, C., Luo, M., & Liu, F. (2022). Fused plus wire arc additive manufacturing materials and energy saving in variable-width thin-walled. *Journal of Cleaner Production, 373*, 133765.
4. Verhaagen, B., Zanderink, T., & Rivas, D. F. (2016). Ultrasonic cleaning of 3D printed objects and cleaning challenge devices. *Applied Acoustics, 103B*, 172–181.
5. Dominguez, L. A., Xu, F., Shokrani, A., et al. (2020). Guidelines when considering pre & post processing of large metal additive manufactured parts. *Procedia Manufacturing, 51*, 684–691.
6. Razumov, N. G., Masaylo, D. V., & Silin, A. O. (2021). Investigation of additive manufacturing from the heat-resistant steel powder produced by recycling of the machining chips. *Journal of Manufacturing Processes, 64*, 1070–1076.
7. Ingarao, G., Priarone, P. C., Deng, Y., & Paraskevas, D. (2018). Environmental modelling of aluminium based components manufacturing routes: Additive manufacturing versus machining versus forming. *Journal of Cleaner Production, 176*, 261–275.
8. Bartolomeu, F., Costa, M. M., Alves, N., et al. (2020). Additive manufacturing of NiTi-Ti6Al4V multi-material cellular structures targeting orthopedic implants. *Optics and Lasers in Engineering, 134*, 106208.
9. Wits, W. W., & Amsterdam, E. (2021). Graded structures by multi-material mixing in laser powder bed fusion. *CIRP Annals, 70*(1), 159–162.

Chapter 99
Material Saving: Why It Is Saving

AM saves material in two ways: (1) by default and (2) by design.

- **By default**: Material is added where it is needed in AM. It means more material than what is added is not required. In machining, more materials are required, i.e., a block bigger than a part is always required. This "more material" is saved when AM is used. This is material saving by default. In machining, "more material" will be saved only when the design of a part is similar to the design of a block, i.e., only when a finished product that does not require machining is brought to be machined. Therefore, for a machining process, the saving of a material and the action of machining do not go together. If it is said that some machining saves material, it only implies that some machining wastes less material.

Saving by default is not an advantage due to what AM does but due to what AM is. There is no effort done by AM to have this advantage. This saving is not an outcome over which AM has control, i.e., AM does not have an option to change it. Thus, there is no freedom to save in this case. As freedom occurs when it is left up to users to exercise it. For example, if material is intended to be saved by changing the design of a part, then it depends on the users to change the design either in one way and save or in another way and not save. For a design that is already selected and given to users, there is no more freedom left to select a design and enjoy the saving that follows the selection. In the absence of the selection, AM is helpless to act to save materials. If materials are still saved, AM has not done an effort by changing the design to save them.

What if this saving is considered a disadvantage? What if AM is required not to save this material because, for example, there is no facility or desire to store it? Even then, AM cannot work differently to consume more material. AM works only in one way, i.e., by adding the material where it is needed. If the way AM works is not considered an advantage, it will not be an advantage. Then, why is it considered an advantage?

S. Kumar, *A Concise Encyclopedia of Additive Manufacturing*,
https://doi.org/10.1007/978-3-031-98394-8_99

It is considered an advantage because it is observed from the perspective of other techniques. For example, it is an advantage only because the amount of material machining consumes is considered normal. If it were considered waste rather than normal, the amount of material AM consumes would have been considered normal; then there would not have been any material saving. Therefore, if material is saved by default in AM, it is not the advantage that is gained by the application of AM as a manufacturing tool.

It can be said that the material saving by default is not due to science but due to history. The history of manufacturing is dominated by the history of manufacturing techniques that do not save materials. Therefore, when machining cannot machine without wasting material, the waste is not called a waste but a necessity. This has been considered normal. This normalization is the lack of acceptance of what manufacturing should be. Manufacturing as a tool, if ideal, should not waste material, i.e., it should transfer faithfully the material from a raw state to a product state without consuming the material during transfer. If AM in an ideal case fits in the ideal state of manufacturing, it does not do any favor by not wasting the material. It does not save the material either. Because for saving the material, the amount of raw material should be lower than the amount of material in the product, which is not the case. Thus, if it is said AM saves material, it is not supported by science. But if it is still said AM saves material, a comparison is drawn with conventional techniques that, like AM, have failed to save material by default.

- **By design**: Material can be saved if a part consists of less material than what it previously required. For this, the design of the part needs to be improved so that it will consist of less material when converted into a part. For conversion, there requires a technique that is able to do it. When AM does it [1, 2], AM has the ability to save material more than it saves by default. This ability is freedom to save material.

What if machining also has the ability to convert this improved design? Then machining will also help make a part using this improved design. But it does not imply that machining will also be able to save material. Because machining does not work on a principle that says making a thinner section means less material addition. On the contrary, machining will remove more material if a thinner section is required. It is possible that an improved design leads to a design that has a smaller dimension, which leads to a smaller workpiece to start with. This will lead to material savings because a smaller workpiece means less material loss after machining. But this material saving is due to the requirement of a smaller workpiece and not due to the requirement of less material to machine a thinner section.

If AM is not able to convert an improved design, AM has no ability to save material by exercising its quality of converting an improved design. If machining is not able to convert the improved design as well, machining also has no ability to save material by exercising its quality of converting the improved design. Then AM is the same as machining, as both are not able to convert the improved design. But AM can still save material by using an old design instead of an improved design. This material saving, i.e., by default, shows that there can still be material saved even if AM fails to be better than machining. This material saving, thus happened, shows not the advantage but the disadvantage of AM. This advantage of AM (i.e., this material saving) is the result of the main disadvantage of AM (i.e., the inability to convert an improved design).

When AM has the ability to convert an improved design, i.e., when AM has the ability to make a part that is called complex when compared with other techniques, this ability is freedom to design. When AM has the ability to convert an improved design, this can or cannot lead to material savings. When this leads to material savings, this is freedom to save material. Thus, the freedom to save material is not an independent freedom but depends on the freedom to design. It means the freedom to save material follows the freedom to design. Thus, AM cannot save material without making a complex part unless the material it saves is saved by default.

References

1. Meng, L., Zhang, W., Quan, D., et al. (2020). From topology optimization design to additive manufacturing: Today's success and tomorrow's roadmap. *Archives of Computational Methods in Engineering, 27*, 805–830.
2. Mikulikova, A., Mesicek, J., Karger, J., et al. (2023). Topology optimization of the clutch lever manufactured by additive manufacturing. *Materials, 16*(9), 3510.

Chapter 100
Mold and Pattern: Mold

AM makes molds and patterns. They are both temporary and permanent. The permanent patterns or molds are those that are not meant to be used only once. The permanent patterns are further used for making molds and temporary patterns.

1 Permanent Mold

AM makes metallic or ceramic molds used to make cast parts (Fig. 100.1). AM allows making lattice structures in a mold wall for stress management and cooling channels for thermal management during die casting [1]. Extrusion-based AM can make a large-sized (several centimeters) compression mold using fiber-reinforced composites [2].

S. Kumar, *A Concise Encyclopedia of Additive Manufacturing*,
https://doi.org/10.1007/978-3-031-98394-8_100

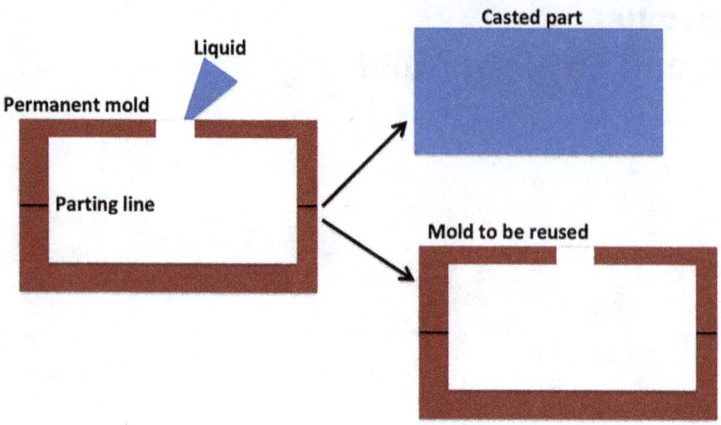

Fig. 100.1 Permanent mold

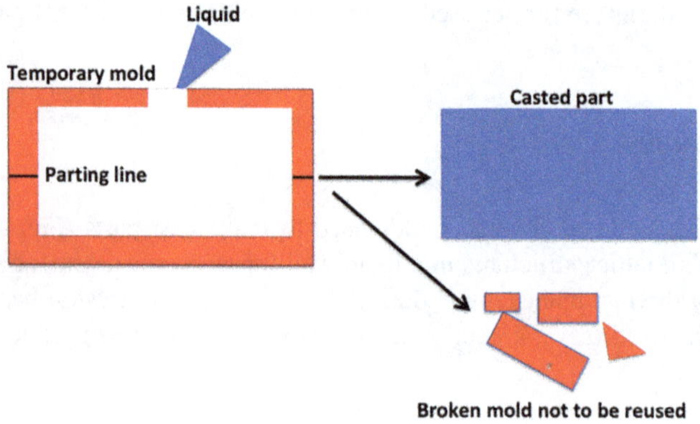

Fig. 100.2 Temporary mold

2 Temporary Mold

A complex part is difficult to be removed from a mold, which requires the mold to be broken to let the part out. AM makes temporary molds, e.g., a sand mold [3] for this purpose (Fig. 100.2).

The sand mold provides a higher surface finish, better mechanical properties, and higher saving of sands than that fabricated by CM [4].

References

1. Brøtan, V., Berg, O. A., & Sørby, K. (2016). Additive manufacturing for enhanced performance of molds. *Procedia CIRP, 54*, 186–190.
2. Yeole, P., Kim, S., Hassen, A. A., et al. (2021). Large-scale additive manufacturing tooling for extrusion-compression molds. *Additive Manufacturing Letters, 1*, 100007.
3. Mitra, S., Castro, A. R. D., & El Mansori, M. (2019). On the rapid manufacturing process of functional 3D printed sand molds. *Journal of Manufacturing Processes, 42*, 202–212.
4. Hawaldar, N., & Zhang, J. (2018). A comparative study of fabrication of sand casting mold using additive manufacturing and conventional process. *International Journal of Advanced Manufacturing Technology, 97*, 1037–1045.

Chapter 101
Mold and Pattern: Permanent Pattern

AM makes patterns that are used to make permanent or temporary molds. Figure 101.1 shows a permanent pattern that is inside a mold box where liquid metal or ceramic slurry is poured over it. The liquid after drying takes the external shape of the pattern—this is a mold (from the pattern). Thus, the shape and surface roughness of the internal cavity of the mold are due to the pattern, while its external shape is due to the mold box. The external shape thus formed does not affect the shape and surface finish of a part cast.

AM makes plastic patterns used to make permanent or temporary molds. These molds are used to make wax patterns. The wax pattern makes a sand mold used for one-off casting. A temporary pattern is made from a permanent pattern through a temporary mold (Fig. 101.2).

S. Kumar, *A Concise Encyclopedia of Additive Manufacturing*,
https://doi.org/10.1007/978-3-031-98394-8_101

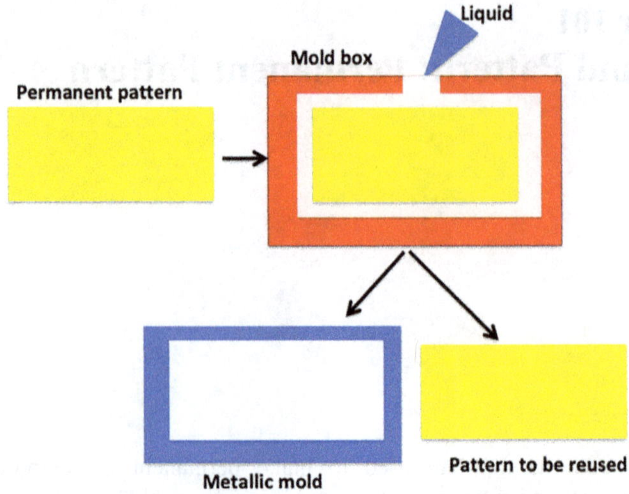

Fig. 101.1 Permanent pattern

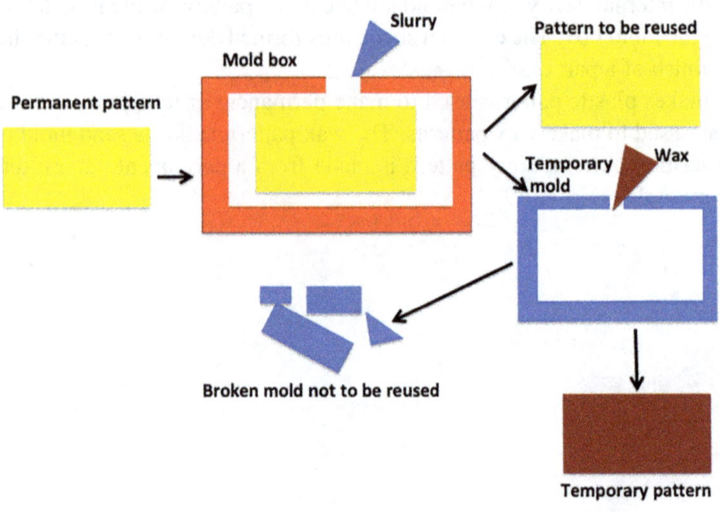

Fig. 101.2 From permanent pattern to temporary pattern

Chapter 102
Mold and Pattern: Temporary Pattern

If only one part is needed, AM makes temporary (expendable) patterns using acrylonitrile butadiene styrene [1], polymethyl methacrylate, wax, epoxy, etc. The patterns are used to make temporary [1] or permanent molds. Figure 102.1 shows a route to make a mold using a pattern.

The pattern expands while being burnt during its removal. The expansion will break the ceramic shell (mold) cast around it. The breaking is prevented in two ways. First, the ceramic shell is made stronger; second, the pattern is made weaker.

The shell is made stronger by using a stronger material. Thus, a shell made stronger by replacing ceramic with ceramic plus fiber has less probability to break. But if it loses its thermal conductivity after the improvement, it is not good for its next use. It causes the metal cast inside it to crack. Therefore, making a shell stronger will help its own fabrication but may not serve the purpose.

The second way is to weaken the pattern so its expansion will not cause the shell to break. Weakening can happen in two ways: material change and design change. When the material contains a low-melting-point component material, the component melts before the pattern starts to expand. This causes the pattern to collapse before it pushes the shell out. When the pattern has a design of lattice structures, the lattice gap accommodates the expansion during heating, which causes a weaker expansion outwards, resulting in a weaker stress on the shell.

AM makes such lattices and hollow structures. They consume less material, which minimizes material waste [2]. Instead of making a lattice pattern to be covered by shell, the lattice pattern can be made to be infiltrated with ceramic. After the removal of the pattern, the lattice ceramic shell is used to make a metallic lattice structure after casting [3].

S. Kumar, *A Concise Encyclopedia of Additive Manufacturing*,
https://doi.org/10.1007/978-3-031-98394-8_102

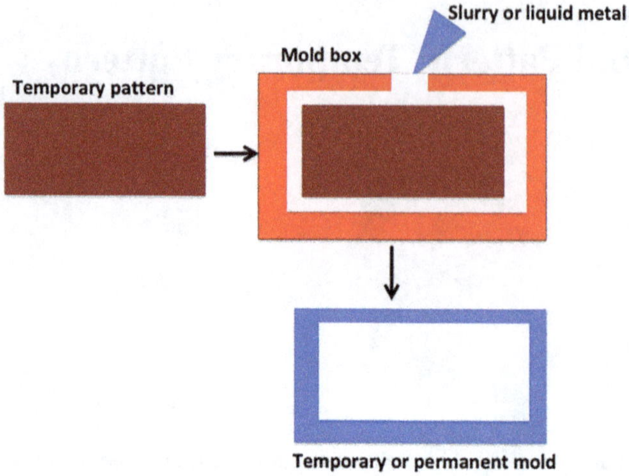

Fig. 102.1 From temporary pattern to mold

References

1. Xie, J., Xu, Y., Meng, Z., et al. (2025). Impact behavior of auxetic cementitious cellular composites (ACCCs) architected through additive manufacturing (AM) assisted casting: Experiment and modelling. *Construction and Building Materials, 471*, 140692.
2. Druschitz, A., Williams, C., Snelling, D., & Seals, M. (2014). Additive manufacturing supports the production of complex castings. In M. Tiryakioğlu, J. Campbell, & G. Byczynski (Eds.), *Shape casting: 5th international symposium* (pp. 51–57). Springer.
3. Carneiro, V. H., Rawson, S. D., Puga, H., et al. (2020). Additive manufacturing assisted investment casting: A low-cost method to fabricate periodic metallic cellular lattices. *Additive Manufacturing, 33*, 101085.

Chapter 103
Multi-Material and Composite

AM makes both multi-material [1, 2] and composite parts [3, 4]. A multi-material part consists of many materials, which does not inform how one material influences another. A composite consists of multiple materials or one material of different types. As a multi-material part, its properties and functions are due to the combination of materials. If it consists of two materials, one material influences another, which gives rise to a combined property. The individual property of each material is no longer applied when it is used in applications. This is the difference between a composite and a non-composite multi-material part.

For example, for a multi-material part as a transition part [5] in a furnace wall, one side of the part is made of ceramic to face high temperature inside the furnace, and another side is made of plastic to face room temperature outside the furnace. Thus, ceramic uses its individual property when used inside the furnace, while plastic uses its individual property outside the furnace. The transition part is not intended to consist of ceramic and plastic in such a way that they will give a combined property and not show their individual property when applied. In the application (furnace), it is the individual property of a material that provides functionality. Thus, in the transition part, there are two materials, two individual properties, and two functions.

While in a composite, e.g., made up of ceramic and plastic to make a structural part, the part has one strength, one function, and one material (composite). The part, in an application, uses its one strength to apply in all of its vicinity. It is possible that the strength will not be the same in all of its vicinity, but the part does not have the freedom to use plastic to apply one strength in one vicinity and ceramic to apply another in another vicinity.

S. Kumar, *A Concise Encyclopedia of Additive Manufacturing*,
https://doi.org/10.1007/978-3-031-98394-8_103

References

1. Hasanov, S., Alkunte, S., Rajeshirke, M., et al. (2021). Review on additive manufacturing of multi-material parts: Progress and challenges. *Journal Of Manufacturing And Materials Processing, 6*(1), 4.
2. Han, D., & Lee, H. (2020). Recent advances in multi-material additive manufacturing: Methods and applications. *Current Opinion in Chemical Engineering, 28*, 158–166.
3. Wang, Y., Ding, Y., Yu, K., & Dong, G. (2024). Innovative polymer-based composite materials in additive manufacturing: A review of methods, materials, and applications. *Polymer Composites, 45*(17), 15389–15420.
4. Hu, Z., He, J., Chen, W., et al. (2024). High-performance carbon fiber reinforced polyether-ether-ketone composite pellets 3D-printed via screw-extrusion additive manufacturing. *Composites Science and Technology, 246*, 110362.
5. Hu, Z., Meng, S., Li, J., et al. (2020). Continuous gradient ceramic/polymer composite for application in large temperature gradient connection by a polymer-derived ceramic route. *Composites Part A: Applied Science and Manufacturing, 132*, 105799.

Chapter 104
Non-beam PBF

Beam powder bed fusion (PBF) makes complex parts but is expensive because the beam is expensive. If the beam is replaced with an alternative heat source, the process will not be expensive. A resistance heater can do the job, but the heat from it is not as directed as a beam is. The beam has an advantage that it can be deflected and the whole powder bed (PB) can be raster scanned. The heater does not give such convenience; it needs to be moved relative to PB or vice versa. Alternatively, a number of small heaters need to be arranged all over close to PB so by switching them on and off, the whole PB can be selectively heated, which will lead to a process that is free from a beam.

Since the aim of applying heat is to join powders, spraying them with glue (or ink or binder) can fulfill the aim as well. Thus, there is another process where both the beam and heater are done away with by an ink-jetting or a binder-jetting device.

Beam PBF is widely researched because a beam provides high energy density and resolution. The high energy density allows processing metals and ceramics, while the resolution allows introducing details in a part. But the process is slow and requires much energy. These demerits drive the search for other processes not based on high-energy beams.

Non-beam heat sources are infrared lamps, concentrated microwave energy, and resistance-based heaters. Microwave energy controlled by a localized microwave heating applicator is not considered a beam because it, unlike the beam, cannot be deflected. Besides, the applicator needs to be placed near PB surface, while e-beam or l-beam sources do not have such compulsion.

S. Kumar, *A Concise Encyclopedia of Additive Manufacturing*,
https://doi.org/10.1007/978-3-031-98394-8_104

Chapter 105
Non-beam PBF: Localized Microwave Heating-Based AM

Microwave is an electromagnetic radiation of wavelength from 1 mm to 1 m, while laser used in AM is of the order of μm [1]. Microwave is used in sintering [2] as well as in post-processing of AM parts [3, 4] because it saves time and energy for processing.

While other radiation, such as laser, is absorbed on the surface of a powder, giving rise to surface absorption, microwave gives rise to volumetric absorption. Due to the nonlinear effect of the absorption, the temperature of powders increases drastically after some time, giving rise to a high temperature at low input microwave energy. This high rise in the temperature, which is called the thermal-runaway effect, makes microwaves a potential thermal source for high-temperature processing. Though it has unpredictable behavior at different conditions, which needs to be controlled.

In microwave-assisted selective laser melting, microwaves are studied as a complementary heat source to the l-beam, so in its presence, low laser power scanning will take advantage of the thermal-runaway effect and will not need a high laser power for scanning. This will help to process a higher melting point ceramic, which otherwise furnishes a cracked part [5]. Application of an electric field across a powder bed (PB) is also useful to decrease sintering time and temperature of ceramics [6]. Application of various forces (electric, magnetic, ultrasonic) across PB helps improve properties [7–9].

Application of microwaves in powder bed fusion as a primary heat source requires it to be concentrated at a small area so its effect can be realized at the desired zone of a processing area without damaging other parts getting built on PB. The application of a microwave localized heating applicator has demonstrated the possibility of the localized melting of metal powders. Using the applicator, microwaves of wavelength ~ 12 cm (2.45 GHz) can be concentrated to a diameter of ~1 mm, which is in contrast to lasers, for which spot size lower than the wavelength is limited by diffraction.

S. Kumar, *A Concise Encyclopedia of Additive Manufacturing*,
https://doi.org/10.1007/978-3-031-98394-8_105

References

1. Pinkerton, A. J. (2016). Lasers in additive manufacturing. *Optics and Laser Technology, 78A*, 25–32.
2. Leonelli, C., Veronesi, P., Denti, L., et al. (2008). Microwave assisted sintering of green metal parts. *Journal of Materials Processing Technology, 205*(1–3), 489–496.
3. Salehi, M., Maleksaeedi, S., Nai, M. L. S., & Gupta, M. (2019). Towards additive manufacturing of magnesium alloys through integration of binderless 3D printing and rapid microwave sintering. *Additive Manufacturing, 29*, 100790.
4. Curto, H., Thuault, A., Jean, F., et al. (2020). Coupling additive manufacturing and microwave sintering: A fast processing route of alumina ceramics. *Journal of the European Ceramic Society, 40*(7), 2548–2554.
5. Buls, S., Vleugels, J., & Hooreweder, B. V. (2018). Microwave assisted selective laser melting of technical ceramics. In *Solid freeform fabrication 2018: Proceedings of the 29th annual international* (pp. 2349–2357).
6. Hagen, D., Kovar, D., & Beaman, J. J. (2018). Effects of electric field on selective laser sintering of yttria-stabilized zirconia ceramic powder. In *Solid freeform fabrication 2018: Proceedings of the 29th annual international* (pp. 909–913).
7. Zeng, C., Huang, F., Xue, J., et al. (2024). Effect of combined direct current electric field and pulsed magnetic field on the transient melt Pool in laser additive manufacturing process. *3D Printing and Additive Manufacturing, 11*(5), e1887–e1898.
8. Du, D., Wang, L., Dong, A., et al. (2022). Promoting the densification and grain refinement with assistance of static magnetic field in laser powder bed fusion. *International Journal of Machine Tools and Manufacture, 183*, 103965.
9. Yan, Z., Trofimov, V., Song, C., et al. (2023). Microstructure and mechanical properties of GH5188 superalloy additively manufactured via ultrasonic-assisted laser powder bed fusion. *Journal of Alloys and Compounds, 939*, 168771.

Chapter 106
On-Demand Production and On-Location Production

On-demand production and on-location production are the advantages of AM. On-demand production means a product is made when it is required. On-location production means a product is made where (e.g., near the end-user) it is required [1]. Both advantages are dependent on each other. For example, on-demand production has no value if the product cannot reach the location because of logistics. Similarly, on-location production has less value if the order to make products cannot be at the top of the queue of orders.

Reference

1. Beiderbeck, D., Deradjat, D., & Minshall, T. (2018). *The impact of additive manufacturing technologies on industrial spare parts industries*. University of Cambridge repository.

S. Kumar, *A Concise Encyclopedia of Additive Manufacturing*,
https://doi.org/10.1007/978-3-031-98394-8_106

Chapter 107
On-Demand Production and Rapid Manufacturing

On-demand production and rapid manufacturing are named interchangeably, but there is a subtle difference between the two. On-demand production means a part can be printed fast when there is a need or demand. Rapid manufacturing means a part is printed fast because there is no longer a reason for which the part gets delayed. Thus, rapid manufacturing implies a single part may not be printed fast, but parts are printed fast, while on-demand production implies a single part can be printed fast, but many parts must not be demanded. While on-demand production indicates urgency, rapid manufacturing indicates a paradigm shift.

S. Kumar, *A Concise Encyclopedia of Additive Manufacturing*, https://doi.org/10.1007/978-3-031-98394-8_107

Chapter 108
Optimization of Parameters

Optimization of parameters means finding their right values that give the best product. When the application changes, the definition of the best product changes, which changes the right values. For example, if a polymer bearing is made, it must have a high surface finish. In this case, a product is best when it has a high surface finish. If a polymer door handle is made, the requirement is not to have a high but a low surface finish so it can provide adequate grip in use. In this case, the product is best when it has a low surface finish. Thus, the definition of a best product is not the same as that in the previous case, giving two sets of optimized parameters for two definitions.

This brings a question: which parameters to optimize? In most AM machines, machine parameters are process parameters, and they are optimized. But the best product obtained from machine parameters may not be the best from a wider perspective. To find the best product, parameters from the whole process chain need to be selected.

However, optimization of process parameters rather than all parameters gives information about the process. If the optimization is done with an aim to find the best product for a given application, it will inform whether a particular process for a particular application is best or not. Optimization of all parameters may furnish better products, but the optimization may not inform the deficiency of a process.

S. Kumar, *A Concise Encyclopedia of Additive Manufacturing*,
https://doi.org/10.1007/978-3-031-98394-8_108

Chapter 109
Origin of Overhangs

There are two types of overhangs in AM (Fig. 109.1):

1. First type: an angled surface—due to stair-stepping in reverse. Example: a cylinder of increasing diameter at one side (Fig. 109.2). The overhang makes an angle with a substrate. This type is due to the first type of layer arrangement.
2. Second type: a horizontal surface—due to a single layer. Example: a design made up of two upright cylinders, where one cylinder is not placed exactly on another cylinder. The horizontal displacement of the top cylinder with respect to another cylinder gives rise to an overhang that is a horizontal surface (Fig. 109.3a). The surface is parallel to the substrate, i.e., it does not make an angle. This type is due to the second type of layer arrangement.

Figure 109.2a shows an overhang that is made because layers are shifted outside a dotted line, i.e., outside the perimeter of each last layer. It is made because a number of layers contribute to forming it by shifting. Its profile depends on which layers shift how much. The shift has made a planar overhang, but it also makes other surfaces convex (Fig. 109.2b) or concave (Fig. 109.2c). At no point on these surfaces do the next layers shift inwards. If they shift, they will not make overhang but stair-steps.

Figure 109.3a shows that a cylinder is made on another cylinder by shifting the position of the former on the latter. When the next layer of the first cylinder shifts outside the perimeter of the last layer, a down-facing surface is made that consists of one layer that has just shifted. The rest of the second cylinder is made on this layer but does not contribute to the formation of this layer. Thus, this type is different from the first type. Because many layers contribute to the formation of the first type.

Figure 109.3b shows the next layer of the first cylinder shifts outside the red dotted line to create an overhang and a cylinder. The next layer of the second cylinder again shifts outside the green dotted line to create the second overhang and the start of the third cylinder.

S. Kumar, *A Concise Encyclopedia of Additive Manufacturing*,
https://doi.org/10.1007/978-3-031-98394-8_109

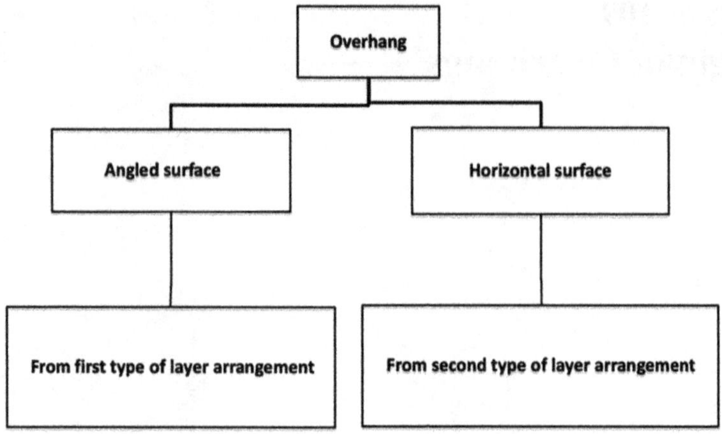

Fig. 109.1 Types of overhangs

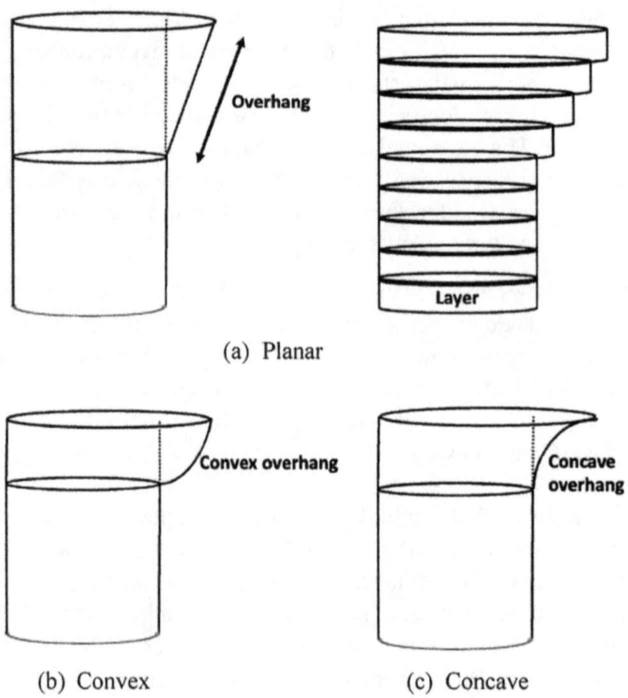

Fig. 109.2 First type of overhang: (**a**) planar, (**b**) convex, and (**c**) concave

When a layer changes once, one extra cylinder forms, creating one overhang (Fig. 109.3a). If changes occur twice, two extra cylinders form, creating two overhangs of the second type (Fig. 109.3b). Changes occur after an interval of many layers. Therefore, cylinders form. Because a cylinder consists of many layers. What

Fig. 109.3 Second type of overhang: (a) two cylinders, and (b) three cylinders

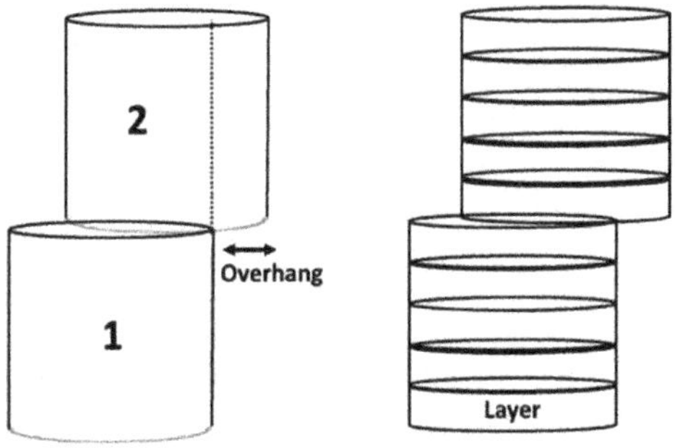

(a) Overhang

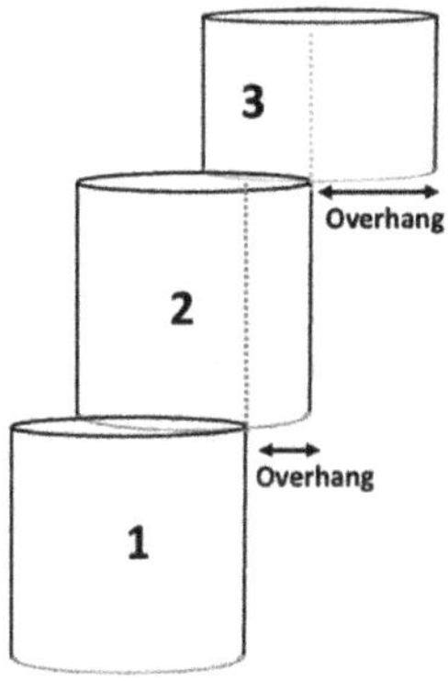

(b) Two overhangs

if the same change occurs with each layer? Then the cylinders will not form, but outward shifts will form. This is the same outward shift that is an overhang of the second type. These outward shifts make an angled surface, which is an overhang of the first type. Thus, the conversion of the overhang from the second type to the first type depends upon the interval between the two changes. If there is no interval between two changes, the first type forms. Otherwise, the second type forms. The frequency of changes determines which type of surface forms. If the overhang of the second type forms, it requires a support structure to manufacture. Manufacturing it may be a problem, but this is not an unsolvable problem. If the overhang of the first type forms, the overhang contains the gaps that cannot be filled up, making the problem unsolvable. Thus, there are two types of manufacturing problems (solvable and unsolvable) originating from two types of layer arrangements. One type of layer arrangement is where changes occur without any interval, i.e., changes occur consecutively; another type is where changes occur non-consecutively.

Figure 109.4 shows both types of overhangs. In the first type, the angle may be more than 0° and less than 90° (or more than 90° and less than 180°).

Figure 109.5a shows the design of a circular hole made on the vertical surface of a rectangular block. Which area of the hole is an overhang? The design can be

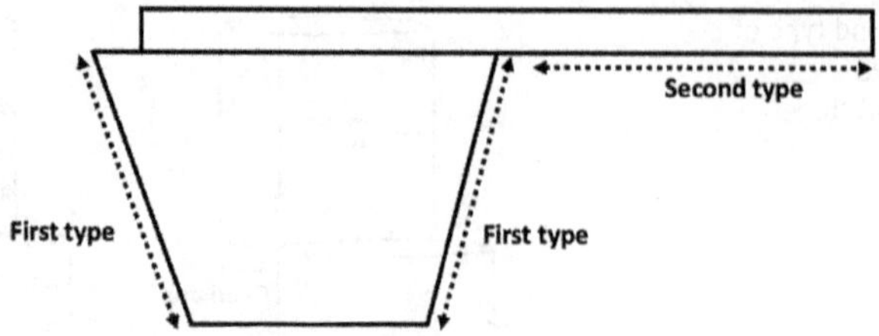

Fig. 109.4 First type and second type of overhang

Fig. 109.5 Overhang in circular hole: (**a**) circular hole and (**b**) overhang

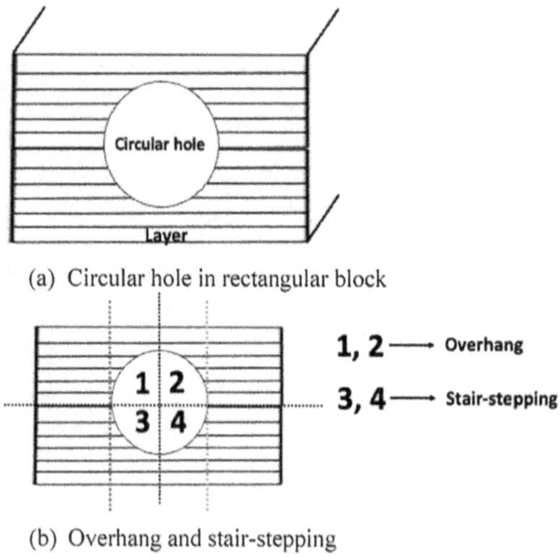

divided into four zones. The upper half of the hole, consisting of two zones, shows overhangs, while the lower half, consisting of two zones, shows stair-stepping (Fig. 109.5b).

In zone 1, the next layer shifts outside the perimeter of the last layer, as seen by the shifting outside the red dotted line; the shifting shows the first type of overhang (Fig. 109.6a). In zone 2, the next layer shifts outside the last layer, as seen by the shifting outside the green dotted line; the shifting again shows the first type (Fig. 109.6b). But in zones 3 and 4, the shifting does not take place outside. In zone 3, the next layer shifts inside the perimeter of the last layer, as seen by the shifting toward the dotted red line; the shifting shows the stair-stepping (Fig. 109.6c). In zone 4, the next layer shifts inside the last layer, as seen by the shifting toward the green dotted line; the shifting again shows the stair-stepping (Fig. 109.6d).

What if a rectangular hole is made on the vertical surface of a rectangular block? In this hole (Fig. 109.7a), there is no gradual shifting. The absence of the gradual shifting eliminates the possibility of either the first type of overhang or stair-stepping. When the hole is started to be made on a vertical surface, there is no connection between the two sides of the hole. At the end of the hole, when the next layer is deposited to end the state of no connection, the next layer does not shift with

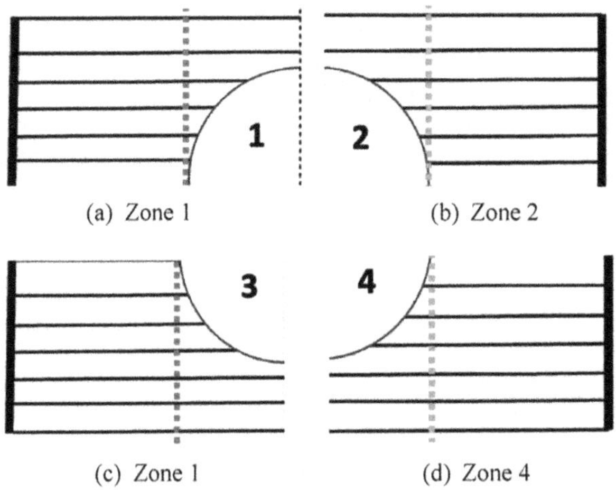

Fig. 109.6 Various zones of circular hole: (**a**) overhang, (**b**) overhang, (**c**) stair-stepping, and (**d**) stair-stepping

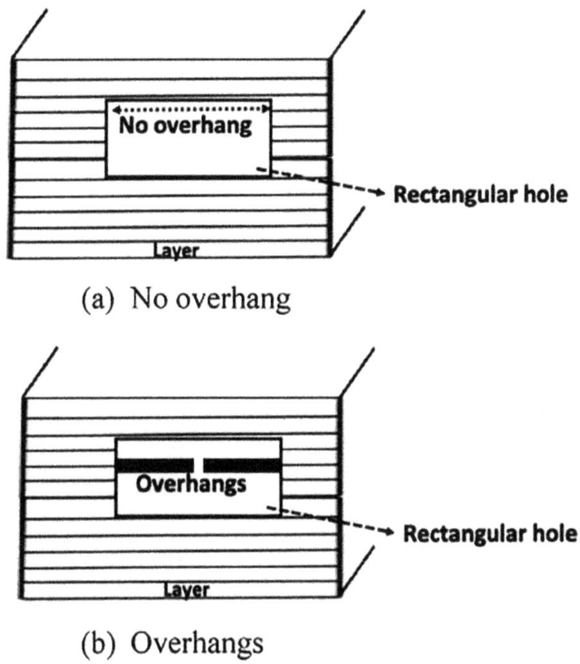

Fig. 109.7 Rectangular hole: (**a**) no overhang, (**b**) overhangs

respect to either the left or the right side of the hole. Because for the shift to take place, there must be a gap in the next layer, i.e., the next layer must be broken. In the absence of the gap, no overhang forms (Fig. 109.7a). While in the case of a circular hole, there always remain gaps for the next layer to shift until the layer reaches the end of the circular hole. However, if there is a gap in the next layer within a rectangular hole, overhangs of the second type form (Fig. 109.7b).

Chapter 110
Overhangs and Stair-Stepping Effect

Stair-stepping effect occurs when the vertical boundary of a layer does not match the contour of a design. Due to the mismatch, there remains a gap between the vertical boundary and the design. When the next layer is deposited, it, like the last layer, again does not match the contour, creating another gap. If these gaps, made by the consecutive layers, are seen, they look like the steps of a stair, giving the name "stair-stepping effect."

When the next layer shifts with respect to the last layer to match the contour, the shift creates a gap. Depending on the contour, the direction of the shift changes. When the contour of a design moves away from the next layer, the shift will be outwards to reach the contour. When the contour moves toward the next layer, the shift will be inwards so as not to reach across the contour.

There can be two possibilities for the next layer to shift (Fig. 110.1):

1. Inwards, so there will be a positive gap, and.
2. Outwards, so there will be a negative gap.

When the next layer shifts inwards, the gap is a right-angled triangle, which is one of the steps of a stair. When the next layer shifts outwards, the gap is still a right-angled triangle but reversed by 180°, which is similar to the step of a stair reversed by 180°. If the former shift gives a stair-stepping effect, the shift given by the latter must be called "stair-stepping effect in reverse," which is, though, called overhang.

Which type of contour gives only positive gaps? Or when the next layer shifts inwards, how far should it shift inwards to create only a positive gap? The outcome depends on its size and position with respect to the last layer. When it is smaller than the last layer, it does not need to shift either left or right to create a positive gap. The decrease in its area causes a shift in the position at either side of the last layer, e.g., a cylinder having a constantly decreasing diameter.

In a cylinder having a constantly decreasing diameter (Fig. 110.2a), the position of a layer does not shift with respect to the central axis, but the position of the

S. Kumar, *A Concise Encyclopedia of Additive Manufacturing*,
https://doi.org/10.1007/978-3-031-98394-8_110

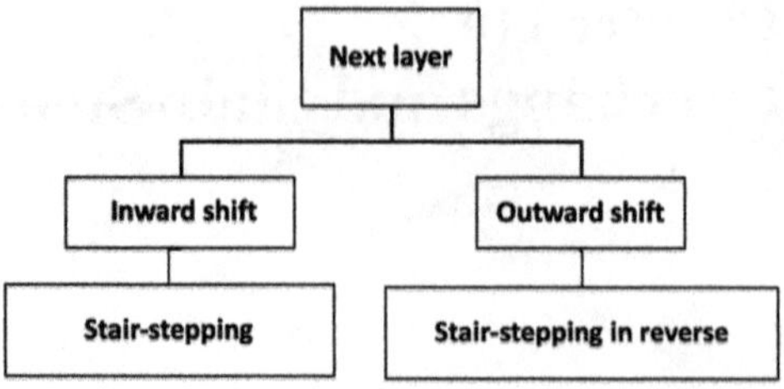

Fig. 110.1 Shifts of next layer

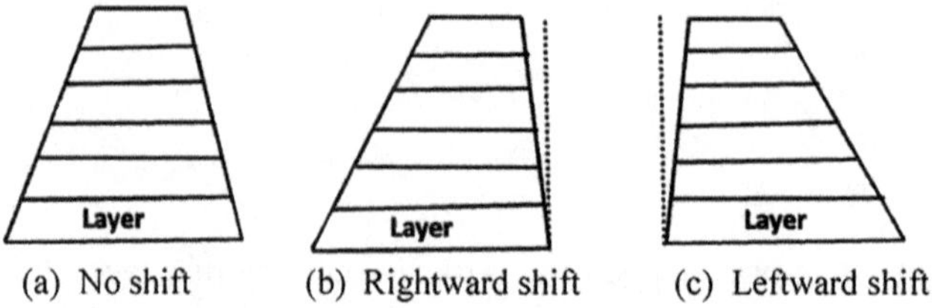

Fig. 110.2 Stair-stepping on both sides due to decreasing area of the next layer. (**a**) No shift. (**b**) Rightward shift. (**c**) Leftward shift

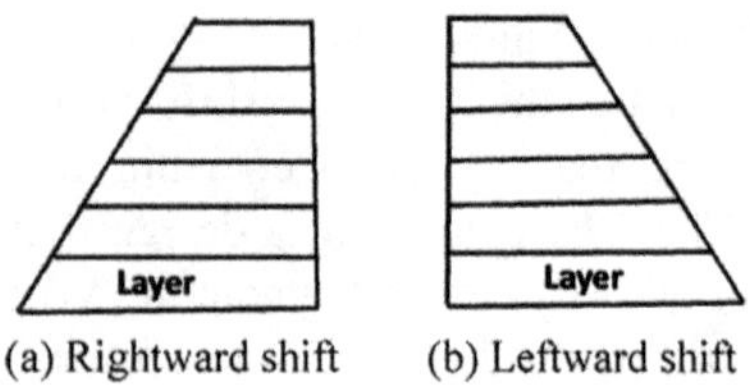

Fig. 110.3 Stair-stepping on one side due to decreasing area of next layer. (**a**) Rightward shift. (**b**) Leftward shift

vertical boundary of the next layer shifts with respect to that of the last layer due to the decreasing diameter. Since the shift happens only inwards at either side of the last layer, the shift creates a positive gap, a stair-stepping effect on either side. As long as the position of the next layer does not shift with respect to the central axis, there will be stair-stepping effects on either side.

The problem starts for a contour, where the next layer starts to shift with respect to the central axis. When it shifts rightwards, there will be more stair-stepping at the left side than at the right side (Fig. 110.2b), and vice versa (Fig. 110.2c). The cylinder then is no longer of a diameter that decreases uniformly from all sides but is of its left side more bent than the right side. Its shift has a limitation, i.e., the bending of the contour has a limitation. The shift must not cause it to overstep the boundary of the last layer at the right side. If it reaches the boundary of the last layer, the vertical boundaries of both the layers are aligned, giving no stair-stepping effect at the right side (Fig. 110.3a). If the same thing happens leftwards, there will be no stair-stepping effect at the left side (Fig. 110.3b). No stair-stepping effect means no inward shift, which means there is an inward shift at one side and no inward shift at the other side of the layer.

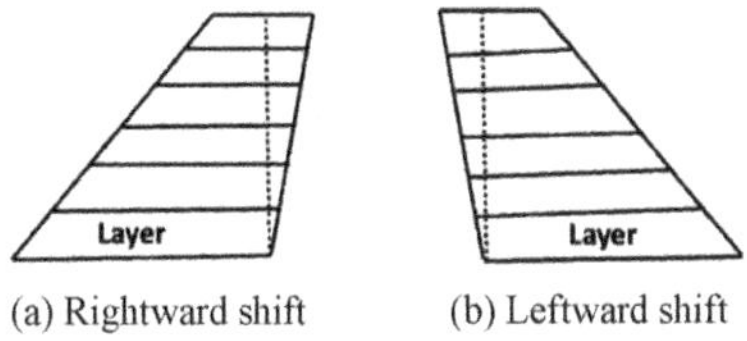

Fig. 110.4 Stair-stepping on one side and overhang on another side due to decreasing area of next layer. (**a**) Rightward shift. (**b**) Leftward shift

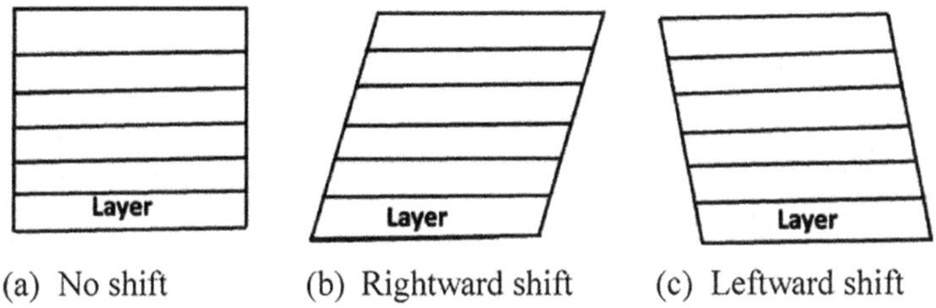

Fig. 110.5 Equal layers: (**a**) no stair-stepping and no overhang, (**b**) stair-stepping on the left side and overhang on the right side, and (**c**) overhang on the left side and stair-stepping on the right side

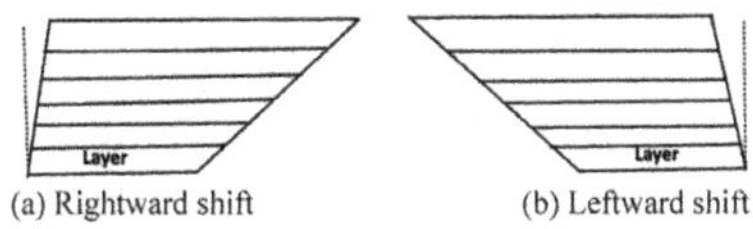

Fig. 110.6 Stair-stepping on one side and overhang on another side due to increasing area of next layer. (**a**) Rightward shift. (**b**) Leftward shift

What happens if the shift oversteps the boundary of the last layer? Then it is no longer an inward shift but an outward shift. The outward shift does not only mean a negative connotation of the inward shift. It means when the next layer oversteps, it brings the question of where to step. As long as there is an inward shift, the next layer has the support of the last layer. But with an outward shift, the next layer needs to be formed on the last layer that does not provide complete support. With only an inward shift, the formation of the next layer may not require a change in the formation parameter because the next layer does not lack the last layer as a support. But with an outward shift, the formation of the next layer may require a change in the formation parameter.

When the next layer is smaller than (Fig. 110.4), equal to (Fig. 110.5), or bigger than (Fig. 110.6) the last layer, an outward shift at one side means an inward shift at the other side. It means creating an overhang at one side means automatically creating a stair-step at another side. Creating an overhang at one side will not automatically mean creating a stair-step at another side only when the next layer is bigger than the last layer and the size difference between the two layers must be equal to the inward shift (Fig. 110.7).

When the next layer is bigger than the last layer, there will be overhang on both sides without the need to create overhang by shifting the next layer outwards, e.g.,

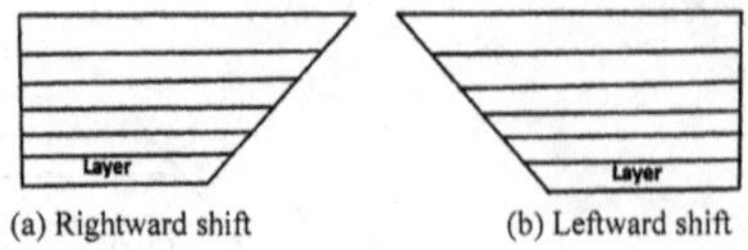

Fig. 110.7 Overhang on one side due to increasing area of next layer. (**a**) Rightward shift. (**b**) Leftward shift

Fig. 110.8 Overhang on both sides due to increasing area of next layer. (**a**) No shift. (**b**) Rightward shift. (**c**) Leftward shift

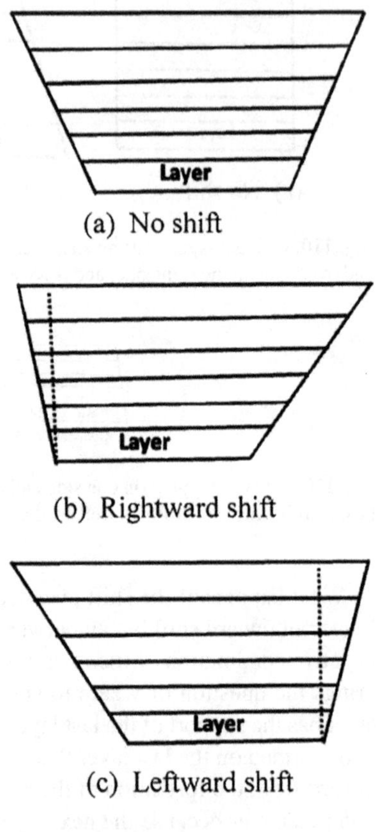

a cylinder with an increasing diameter (Fig. 110.8a). When the next layer shifts, there will be more overhang on one side than on the other side (Figs. 110.8b, c).

Stair-stepping effect requires having the steps of a stair. The steps can be made only when the curve is forward-bent rather than backward-bent. Without bending forward, it is not possible to climb using stairs. Therefore, all those curves that are forward-bent give stair-stepping effects, while those curves that are backward-bent give overhang problems.

Fig. 110.9 Reversing design by 180°: (**a**) design, (**b**) stair-stepping, (**c**) overhang, (**d**) reversing Fig. b, and (**e**) reversing Fig. c

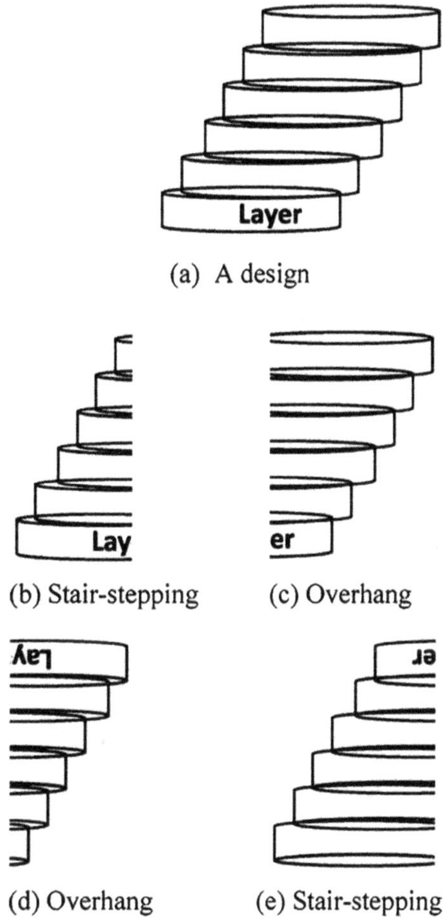

Stair-stepping is reversed to overhang, which can be checked by reversing a design (Fig. 110.9a) having both stair-stepping (Fig. 110.9b) and overhang (Fig. 110.9c). Figure 110.9b and c each shows one half of a design vertically divided into two halves. When the design is reversed by 180°, the stair-stepping becomes overhang (Fig. 110.9d), and vice versa (Fig. 110.9e) [1].

Reference

1. Kumar, S. (2025). A layer arrangement. In *A new theory of additive manufacturing* (pp. 11–60). Springer.

Chapter 111
Part Consolidation into Assembly

Instead of making many parts and joining them afterward, they can be made as one part. It will save the time it takes to find a suitable joining technique and the time it takes to join. It saves cost [1, 2]. Parts consolidation helps manage inventory [3] and makes the supply chain sustainable [4].

An example of part consolidation is when General Electric replaced 20 cast parts with a single AM part that was less expensive and more durable [5]. Replacing many conventionally designed parts into an assembled part by AM is an achievement, but it shows there is something missing. If AM can make an assembled design, it can also make better part designs for conversion. Therefore, if AM is to consolidate parts into an assembly, it needs to utilize the opportunity to redesign the parts before assembling [6, 7].

A consolidated part is not always good. If any section of it is damaged, the whole part needs to be either engaged in maintenance or replaced. It could be better if the consolidated part could be designed in such a way that it could be repaired well as well. If there is a choice between making one part from many parts without the convenience for easy repair afterward and making a few parts from many parts with the convenience for easy repair afterward, the latter choice is better [8].

References

1. Nie, Z., Jung, S., Kara, L. B., & Whitefoot, K. S. (2020). Optimization of part consolidation for minimum production costs and time using additive manufacturing. *Journal of Mechanical Design, 142*(7), 072001.
2. Nasabi, M. T. (2024). Assessing the economic implications of part consolidation in additive manufacturing through optimization analysis.
3. Knofius, N., van der Heijden, M. C., & Zijm, W. H. (2019). Consolidating spare parts for asset maintenance with additive manufacturing. *International Journal of Production Economics, 208*, 269–280.

S. Kumar, *A Concise Encyclopedia of Additive Manufacturing*,
https://doi.org/10.1007/978-3-031-98394-8_111

4. Son, D., Kim, S., & Jeong, B. (2021). Sustainable part consolidation model for customized products in closed-loop supply chain with additive manufacturing hub. *Additive Manufacturing, 37*, 101643.
5. Niaki, M. K., Torabi, S. A., & Nonino, F. (2019). Why manufacturers adopt additive manufacturing technologies: The role of sustainability. *Journal of Cleaner Production, 222*, 381–392.
6. Sossou, G., Demoly, F., Montavon, G., & Gomes, S. (2018). An additive manufacturing oriented design approach to mechanical assemblies. *Journal of Computational Design and Engineering, 5*(1), 3–18.
7. Jayapal, J., Kumaraguru, S., & Varadarajan, S. (2021). Part consolidation in design for additive manufacturing: A two-level approach using complexity metrics. In *Design for Tomorrow—Volume 2: Proceedings of ICoRD 2021* (pp. 881–892). Springer Singapore.
8. Kim, S., & Moon, S. K. (2020). A part consolidation design method for additive manufacturing based on product disassembly complexity. *Applied Sciences, 10*(3), 1100.

Chapter 112
Perspective of Complexity

In AM, a part is generally visualized as a set of horizontal layers. AM requires either having layers and a method to join them or a method to make and join them. To have layers means to bring them from somewhere if that is required. In almost all AM techniques, making a layer is accompanied by its automatic joining with an underlying layer (or a platform), implying that making a layer means both making and joining.

AM simplifies fabrication by reducing the fabrication to making layers and joining them. It is called simple on the assumption that making layers and joining them is simpler than making the whole part at once. If this is not the case, it is not simple. For example, if a cube is made from a bigger cube by machining, it is simple. If AM is used to make this cube layer by layer, AM does not simplify the fabrication but makes it complex. Because a number of layers need to be arranged and joined, while machining requires only trimming a bigger cube. Thus, only when a part is complex can AM be invoked to simplify the fabrication.

AM relies on a method to simplify the fabrication. The method is to fabricate layer by layer. The method relies on the fact that the actual happening of making layers and joining them is a simplification. It means a part being complex is an essential requirement to invoke AM but is not a sufficient requirement. The sufficient requirement is that making layers and joining them must not turn out to be more complex than the complexity of a part, i.e., an AM technique must be advanced enough that the technique does not find difficulty while making the complex feature of a part.

S. Kumar, *A Concise Encyclopedia of Additive Manufacturing*,
https://doi.org/10.1007/978-3-031-98394-8_112

Chapter 113
Post-Processing

There are three reasons to post-process:

1. The process is not good enough to make a part; therefore, post-processing is used to compensate for what the process cannot do [1–3]. The post-processing ranges from intensive cleaning to treatments lasting for several hours or days. If an AM process is developed well, post-processing is not required. For example, if a polymer part made from stereolithography is not cured, post-curing as a post-processing is applied to fully cure it. Since the process (stereolithography) did not give well-cured parts, the post-processing is done to compensate for the demerit of the process.
2. Process is not used for all purposes. Post-process is used when a part is required to have a property that is not provided by the process because the process is not meant to provide that.

That is, to achieve properties that are not achieved from AM not because AM does not work well but because the requirement is different [4, 5]. For example, post-processing as coating with hydroxyapatite to improve biocompatibility of a polymer part made from fused deposition modeling (FDM). In this case, post-processing is required not because FDM could not furnish a good coating but because it was not the job of FDM to do coating. Post-processing is not done to compensate for a substandard part (made from a process) but because the best part (made from the best process) could not fulfill the requirement.

3. Post-processing is used as a strategy where products are made by combining process and post-process [6, 7]. Post-processing is not problematic, as some AM materials are more post-processable than conventionally made materials [8].

S. Kumar, *A Concise Encyclopedia of Additive Manufacturing*, https://doi.org/10.1007/978-3-031-98394-8_113

References

1. Maleki, E., Bagherifard, S., Unal, O., & Guagliano, M. (2024). Hybrid intelligence approach to study post-processing impact on the mechanical performance of notched additively manufactured AlSi10Mg. *Materials & Design, 248*, 113462.
2. Bittner, F., Putta, A. K., Juerries, F., et al. (2024). The impact of post-processing heat treatment on the magnetic properties of additively manufactured Nd-Fe-B magnets. *Journal of Magnetism and Magnetic Materials, 603*, 172238.
3. Rauch, M., & Hascoet, J. Y. (2022). A comparison of post-processing techniques for additive manufacturing components. *Procedia CIRP, 108*, 442–447.
4. Villapún, V. M., Carter, L. N., & Cox, S. C. (2025). Plasma-electrolytic oxidation: A rapid single step post processing approach for additively manufactured biomedical implants. *Biomaterials Advances, 169*, 214186.
5. Martínez, R. Z., Bose-Bandyopadhyay, S., Burl, A., et al. (2025). Comparative analysis of machining and electropolishing for surface quality improvement of shape memory nitinol samples additively manufactured by laser powder bed fusion. *Additive Manufacturing Letters, 12*, 100261.
6. Dominguez, L. A., Xu, F., Shokrani, A., et al. (2020). Guidelines when considering pre & post processing of large metal additive manufactured parts. *Procedia Manufacturing, 51*, 684–691.
7. Billah, K. M. M., Kumar, V., Rathod, N., et al. (2025). Fast-rate joining of thermoplastic composites using integrated additive manufacturing and compression molding process. *Journal of Manufacturing Processes, 136*, 344–355.
8. de Oliveira Campos, F., Araujo, A. C., Munhoz, A. L. J., & Kapoor, S. G. (2020). The influence of additive manufacturing on the micromilling machinability of Ti6Al4V: A comparison of SLM and commercial workpieces. *Journal of Manufacturing Processes, 60*, 299–307.

Chapter 114
Post-Processing: Position in AM Chain

An AM process starts when shaping of material starts in a system. The process ends when the shaping ends. A process (or in-process) stage is preceded by a pre-process stage and followed by a post-process stage (Fig. 114.1).

A system can have material and information, but the process cannot start until the material starts to interact with process parameters. For example, an LPBF system can be filled up with powder, and a digital file can be transferred to it, but the process will not start until a powder layer is spread. The spreading of the layer is due to an interaction between the material and the parameter. In an extrusion deposition system, when a filament does not start to melt after being fed, the process does not start. The filament melts due to material–parameter interaction. The process stops when the interaction stops.

Stopping a process implies that shaping has stopped, but it does not imply that the processed material will no longer be processed. Though further processing will not come within an in-process stage. After the stopping of the process, if a product formed is inadequate, post-processing is required.

1 Parameter Types

Parameters used for doing experiments are experimental parameters. They can be pre-process, process, and post-process (Fig. 114.1). For example, if LPBF makes a steel product, parameters can be as follows: pre-process—powder type; process—laser power, scan speed; post-process—heat treatment. Thus, any parameters can be experimental, but all experimental parameters are not process parameters.

Process parameters are machine parameters, as a process requires a machine to execute. Thus, laser power and scan speed require an LPBF machine and are machine parameters as well. Another parameter, i.e., powder type, cannot be changed in the machine and, therefore, is not a machine parameter. Heat treatment

S. Kumar, *A Concise Encyclopedia of Additive Manufacturing*,
https://doi.org/10.1007/978-3-031-98394-8_114

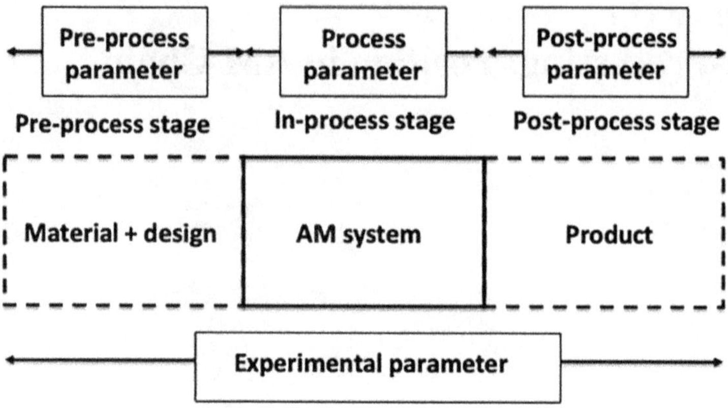

Fig. 114.1 Various stages of AM

requires another machine, i.e., a furnace, to execute and is not an LPBF machine parameter.

What if pre—and post-process parameters are executed on the same machine? If a machine stores many powder types, then, for example, after making a steel product, the machine uses another type to make another metal product. Though powder type is a pre-process parameter, in the presence of the facility to change it within a machine, it becomes a machine parameter. Thus, pre-process, process, and post-process parameters can be machine parameters, depending on the facility.

Inside an LPBF machine, if heat treatment is done with an existing heater or beam, it requires the part after fabrication to be left inside the machine. In this case, the heat treatment is a post-process parameter. Since it is done inside the machine, it is a machine parameter as well. But it is still different from other parameters (laser power, scan speed) that are executed in the same machine. For other parameters, the boundary of the process and that of the machine are the same; they are both process and machine parameters but not post-process parameters. Heat treatment is a process parameter if done after each layer or a couple of layers (for decreasing residual stress [1] and surface roughness or increasing accuracy, for example).

Reference

1. Roehling, J. D., Smith, W. L., Roehling, T. T., et al. (2019). Reducing residual stress by selective large-area diode surface heating during laser powder bed fusion additive manufacturing. *Additive Manufacturing, 28*, 228–235.

Chapter 115
Post-Processing: Pseudo Type

What if post-processing-type actions, such as milling, rolling, and heating, are used during the in-process stage? These improve the efficacy of a process. For example, in wire arc AM (WAAM), interlayer friction stir processing [1] or rolling [2] is used after some layers to improve grain refinement. As rolling is done before the completion of the shaping, it is part of a process. The rolling decreases the height of a build and influences the shape (i.e., proportion of features), but influencing a shape is not equal to creating a shape. Influencing is a negative development that is negated by changing the digital file.

Similarly, milling is part of a process when, after some layers, it is used to fine-finish a layer to improve surface roughness [3]. When machining periodically changes the shape so a new shape comes due to contributions from both AM and machining, influencing the shape by machining is not a negative development. As the development is preplanned to create the shape that is not possible by AM alone, it does not need to be negated by changing the digital file. Thus, the process responsible for creating the product is not AM but a new process due to both.

References

1. Wei, J., He, C., Dong, et al. (2024). Enhancing mechanical properties and defects elimination in 2024 aluminum alloy through interlayer friction stir processing in wire arc additive manufacturing. *Materials Science and Engineering A, 901*, 146582.
2. Huang, J., Fu, Y., Zhai, W., et al. (2024). Hybrid interlayer hot rolling and wire arc additive manufacturing of Al-Mg alloy: Microstructure, mechanical properties and strengthening mechanism. *Journal of Materials Research and Technology, 30*, 7037–7050.
3. Karunakaran, K. P., Suryakumar, S., Pushpa, V., & Akula, S. (2010). Low cost integration of additive and subtractive processes for hybrid layered manufacturing. *Robotics and Computer-Integrated Manufacturing, 26*(5), 490–499.

S. Kumar, *A Concise Encyclopedia of Additive Manufacturing*,
https://doi.org/10.1007/978-3-031-98394-8_115

Chapter 116
Post-Processing Does Not Slow AM Process

Post-processing is not the reason for a process to be slow because post-process comes after the process ends. Having an option for post-processing can and cannot expedite the process. It can expedite in some cases. For example, if a part when made in 10 h has a fine finish and when made in 2 h has a high surface roughness, the high roughness can change to a fine finish if there is an option for post-processing (e.g., fine machining). It is better to make a part in 2 h instead of 10 h and opt for post-processing. If post-processing takes 1 h, 7 h will be saved to make a part having a fine finish.

Post-processing does not help know how much the process was slowed. Its slowness was hidden because the process was not utilized to work for 8 h. AM is for fast production; the process time for two instead of 8 h is preferred. But the option for a 2-h process plus a 1-hour post-process shows the post-process was needed because the process was not good enough to give the required surface finish. It implies that the process has a demerit, which requires a post-process.

The demerit is that the process was not given time to work for 8 h. If the process had been given sufficient time, the process could have shown that it could furnish a fine finished product. The process was not given sufficient time because the product needed to be made fast. It is the demerit of the process that it cannot make a fine finished product fast. The requirement of the post-processing is the consequence of this demerit. Thus, post-processing as a disadvantage is not an independent disadvantage but a derivative of an independent disadvantage, i.e., process slowness. The slowness comes first; the requirement comes later.

What if the geometry of a part is complex so the process cannot give a fine finish? Then, it is also not expected from post-processing (fine machining) to provide a fine finish. If the geometry is complex, fine machining is not possible. If fine machining is possible, the geometry is not complex enough. If fine machining cannot be applied to a complex part, it cannot be considered post-processing. For that complex geometry, the requirement of post-processing is out of the picture and is therefore neither a disadvantage nor an advantage. Whatever AM provides should

S. Kumar, *A Concise Encyclopedia of Additive Manufacturing*,
https://doi.org/10.1007/978-3-031-98394-8_116

be accepted as it is. If AM provides that complex part quickly, AM has no demerit. If AM provides after a long time of processing, AM has a demerit as it is slow. The demerit of AM is not hidden. Because there is no post-processing to hide the demerit. There is no post-processing that can cause AM to rush and cut its long processing time for the reason that the post-processing will compensate for the lack of processing and provide a fine finish. The absence of the post-processing for this complex geometry shows the slowness of AM rather than the requirement of post-processing as a fundamental disadvantage.

It does not mean if a process is made to move slowly, post-processing will never be required. Post-processing is required because the performance of a process is inadequate. One of the reasons why it is inadequate is because the process is not given time to utilize its full potential, as the process was required to be expedited. If the process were given time, post-processing would not have been required. The process performance can be inadequate for other reasons as well, such as a lack of processable materials, lack of control, and lack of understanding of parameter–material interactions used for different applications. One of the reasons is lack of understanding of the process itself. For example, if a process gives a product having surface roughness, Ra 10 μm, the result is considered inadequate because Ra is not 5 μm. If the process gives Ra 5 μm, the result is still inadequate because the expectation has increased; Ra 1 μm is required.

A process can fail for many reasons, and these reasons cannot be removed by making the process slow. These reasons can or cannot be removed by post-processing. When post-processing removes, it is required. When it is required, AM gives required products; then the requirement of post-processing is essential for AM to give better products. Then, that requirement is considered a disadvantage of AM. This is erroneous. Because it deflects the cause from one place to another. It shifts the problem from process to post-process. It ignores the importance of process and emphasizes the solution by hiding the problem. It undermines the requirement of post-processing results due to other disadvantages of AM.

Chapter 117
PBF: Laser and Electron Beam

The difference between laser beam powder bed fusion (LPBF) and electron beam powder bed fusion (EPBF) is given below.

1 Powder

In EPBF, powder needs to be electrically conductive to facilitate dissipation of charges from it to surroundings. This limits the types of powders that are suitable to be processed; therefore, mostly metallic materials are used. Powders that form oxide or nitride layers on their surfaces due to contamination with gases diminish their suitability, as these layers have lower electrical conductivity.

In EPBF, powders are displaced due to charging, while LPBF does not have such a limitation. This implies big powder is suitable in EPBF, while smaller powder can be used in LPBF [1]—this leads to higher surface roughness in EPBF [2].

In LPBF, there does not exist a yardstick (such as electrical conductivity existing in EPBF), the lack of which makes the materials unprocessable.

2 Beam

In the case of conversion of electrical energy into a beam (electron or laser), known as wall-plug efficiency, the rate of conversion for e-beams is about 95%, while for l-beams used in LPBF, it is from 5% to 20% for CO_2 laser, up to 20% for Nd: YAG laser, from 30% to 40% for fiber laser, and up to 70% for diode laser [3, 4], showing the higher efficiency of e-beams over l-beams.

S. Kumar, *A Concise Encyclopedia of Additive Manufacturing*,
https://doi.org/10.1007/978-3-031-98394-8_117

3 Beam-Powder Interaction

E-beams are made up of electrons that, when irradiated on metal powders, get deflected by free electrons surrounding metal atoms. If the power of the e-beam is high, the e-beam will not be deflected by free electrons but will move deeper inside the metal until decelerated by a lattice of atoms and stopped. The interaction causes the lattice to vibrate and generate heat. Thus, the kinetic energy of moving electrons is transferred into heat energy, responsible for melting. If the size of the atom is big (higher atomic number), there is less chance for the e-beam to escape. Thus, a bigger atom size lets a higher percentage of kinetic energy be transferred, which increases the efficiency of the process. For an applied voltage of the order of kV, the size of e-beam impingement in a powder is of the order of microns.

An l-beam is made up of photons that are neutral. When it irradiates metal powders, photons interact with free electrons surrounding atoms. For a high number of free electrons, a photon has a chance to be reflected. For an l-beam of high intensity irradiating a metal having fewer numbers of free electrons, incoming photons interact with bound electrons, which causes the electrons to re-radiate or interact with lattice atoms, causing vibration. The l-beam is assumed to be high intensity because if it has low intensity, it will be less likely to reach the bound electrons. The metal is assumed to have a small number of free electrons, because if it has a high number of free electrons, the beam will be more likely to be deflected by these electrons without reaching the actual metal surface comprising bound electrons.

Photons deflected by bound electrons have less chance to escape (than the photons deflected by free electrons) and will reflect internally until they lose energy by interacting with the lattice. Thus, incoming laser energy is transferred to metal powders as heat energy. When an incoming photon reaches bound electrons without being deflected by free electrons, it is able to transfer its energy. This distance inside the powder (penetration depth) is about a few nm (Fig. 117.1).

This shows that l-beam-material interaction is a surface phenomenon, while e-beam–material interaction (penetration depth—few μm) is a relative bulk phenomenon.

While an l-beam is prone to be reflected from the metal powder, there is no such occurrence with an e-beam, making the e-beam more efficient in terms of transfer of energy to metals (as well as in terms of generating it from the electrical energy). The comparative inefficiency of the l-beam is somewhat compensated in PB processing when the reflected beam has a chance to be directed toward another powder and gets absorbed. This depends upon the surface roughness of powders, the gap between powders, and the angle of incidence.

Metals such as copper, aluminum, and silver that have high reflectivity are not suitable to be processed in LPBF but in EPBF. These reflective metals on developing an oxide layer, such as an alumina layer on aluminum, are no longer suitable for EPBF because of the charge accumulation, but they are suitable for LPBF because of their decreased reflectivity. This example demonstrates that what is good in one

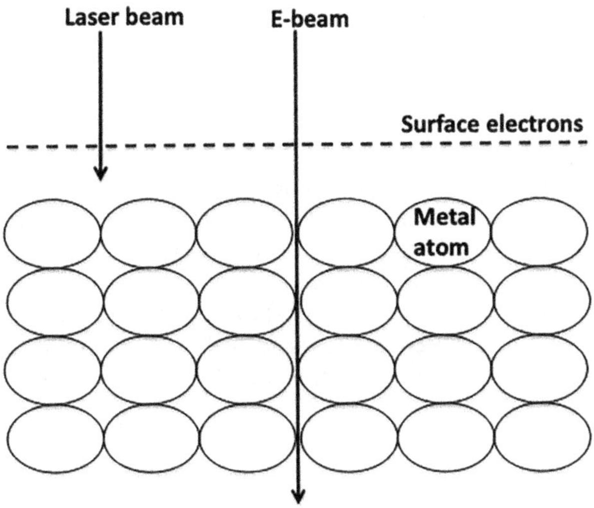

Fig. 117.1 Incidence of l—and e-beams at metal surface

process may not be good in another, and vice versa, implying that both processes can complement each other.

4 Parameter

The l-beam moves with a constant speed, i.e., the speed of light, while the e-beam has varied speeds depending upon the applied voltage. Applied energy density in EPBF can be changed by changing the speed of the e-beam, while there is no such option in LPBF, as the speed of light is a constant.

Laser power is an independent quantity. For a given setting of laser power in LPBF, it does not change with a change in other experimental parameters such as scan speed and scan strategy. In EPBF, power may change with a change in scan speed or scan strategy. This is the reason why keeping power constant and changing other experimental parameters gives useful information in LPBF, while the same gives misleading information in EPBF if all details are not taken into account.

LPBF can be carried out using an l-beam of different types of wavelengths, such as YAG laser and fiber laser. Different wavelengths have different absorptance for different materials. There is no such provision in EPBF that it can be carried out using an e-beam of a particular wavelength. Since an e-beam accelerated at an applied voltage has a particular wavelength (de Broglie wavelength), but during EPBF, the speed of the beam varies, bringing variation in wavelength, it implies that there is no EPBF with a fixed wavelength, though there is LPBF with a fixed wavelength.

5 Implication of Vacuum

Vacuum increases the evaporation rate of materials and changes the composition of a processed material, which implies that LPBF (without vacuum, with inert gas) will cause less composition change than EPBF does. Vacuum has an advantage, as there is no gas in a chamber, and therefore no gas will be absorbed by the liquid metal pool; absorption of the gas is a source of porosity.

LPBF has an edge over EPBF to process the material by increasing atmospheric pressure in the chamber. High pressure will decrease porosity, surface roughness [5], and sublimation. Besides, in LPBF, gas flow inside the processing chamber can be managed to change surface roughness and quality [6]. Flow parallel to the powder bed is used to remove particulates from the chamber and is used as a parameter to vary properties [7].

6 Scanning

EPBF works using an electromagnetic coil-based scanning system that enables scan speeds on the order of thousands of m/s, while LPBF works on a mirror-based scanning system that provides scan speeds on the order of 10 m/s. EPBF, due to its higher scan speed, furnishes distinct advantages over LPBF, e.g., potential for higher production rate, mitigation of thermal stress by use of simultaneous parallel beams, utilization of beam as a source of an additional heater, etc.

In EPBF, high scan speed combined with high beam power enables the transfer of high energy density at a faster rate, leading to fast melting of the whole layer; this fulfills one of the tasks for higher production rate. LPBF has a lower scan speed and does not match EPBF in this aspect. However, there is a limit by which scan speed in LPBF or EPBF can be increased. At higher scan speeds, there will be small beam-powder interaction time (of the order of nanoseconds for a beam size of 100 μm moving at 1000 m/s). This will cause ablation [8], i.e., non-melting of powders, generating evaporated materials interfering with vacuum and beam (particularly in EPBF), and pressing the molten material by recoil pressure.

Scan speed more than a certain speed cannot be employed. As a result, a beam power more than a certain power cannot be employed to scan. This is the reason why the laser power of several kW is available, but there is no LPBF system equipped with that. Nevertheless, higher power can be utilized by distributing it among several beams with a restriction that the power in a beam should not exceed the power limit. If the processing chamber is hot, as in EPBF, still lower power (or applied energy density) is enough to raise the temperature of the layer.

References

1. DebRoy, T., Wei, H. L., Zuback, J. S., et al. (2018). Additive manufacturing of metallic components–process, structure and properties. *Progress in Materials Science, 92*, 112–224.
2. Townsend, A., Senin, N., Blunt, L., et al. (2016). Surface texture metrology for metal additive manufacturing: A review. *Precision Engineering, 46*, 34–47.
3. Hecht, J. (2018). *Understanding lasers: An entry-level guide*. John Wiley & Sons.
4. Li, L. (2000). The advances and characteristics of high-power diode laser materials processing. *Optics and Lasers in Engineering, 34*(4–6), 231–253.
5. Bidare, P., Bitharas, I., Ward, R. M., et al. (2018). Laser powder fusion in high-pressure atmospheres. *The International Journal of Advanced Manufacturing Technology, 99*, 543–555.
6. Montgomery, C., Farnin, C., Mellos, G., et al. (2018). Effect of shield gas on surface finish of laser powder bed produced parts. In *Solid freeform fabrication 2018: Proceedings of the 29th annual international solid freeform fabrication symposium* (pp. 438–444).
7. Ferrar, B., Mullen, L., Jones, E., et al. (2012). Gas flow effects on selective laser melting (SLM) manufacturing performance. *Journal of Materials Processing Technology, 212*, 355–364.
8. Steen, W. M., & Majumder, J. (2010). *Laser material processing*. Springer-Verlag.

Chapter 118
PBF: Serial and Non-serial Production

An AM system generally makes parts serially, which has the potential to make large-volume production [1, 2]. In powder bed fusion (PBF), many parts can be made when the parts are small and the system is big [3].

Serial production happens through two types:

1. Fabrication of a second part starts after the completion of the first part fabrication (Fig. 118.1a).
2. Fabrication of the first layer of a second part starts after the fabrication of the first layer of the first part is completed (Fig. 118.1b).

The first type is akin to job production, while the second type is used for making several parts.

Figure 118.1a and b shows one part and four parts in a bed, respectively. The creation of a bed over a batch of parts for three consecutive times allows the fabrication of 16 parts in a system at once (Fig. 118.1c). Using more beds will further increase the number of parts.

The second type is serial production because it finishes some portion of a part and then starts fabricating some portion of another part (Fig. 118.2). The portion is one layer.

1 Fast Serial Production

Arcam multi-beam processing is an attempt in this direction where several beams process the bed at the same time, but this attempt is more to improve the efficiency for fabricating a part than to fabricate multiple parts. Digital micromirror device (DMD) is used for fabricating microparts [4]. Area scanning by DMD fabricates a part by exposing some area of the bed, which is similar to the concept of parallel scanning for fabricating multiple parts in a bed.

S. Kumar, *A Concise Encyclopedia of Additive Manufacturing*,
https://doi.org/10.1007/978-3-031-98394-8_118

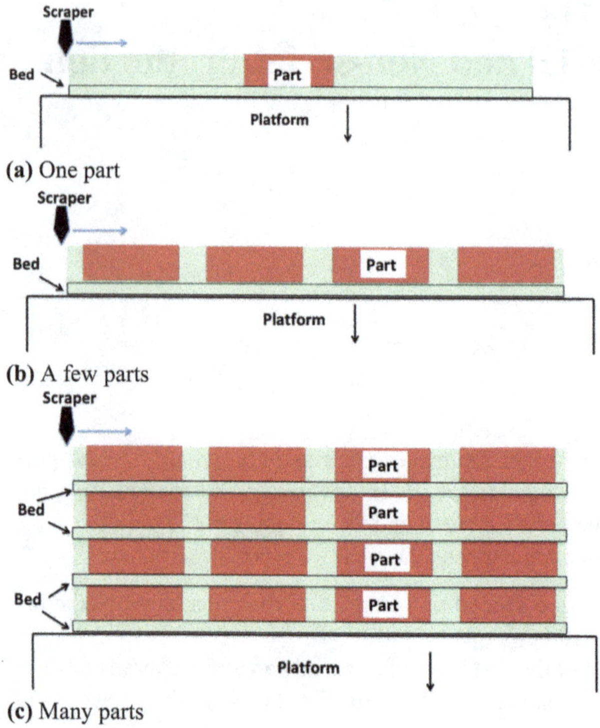

Fig. 118.1 Fabrication: (**a**) one part, (**b**) a few parts, (**c**) by creating successive beds

In laser PBF, using two beams to fabricate adjacent parts leads to the deterioration of quality due to plume [5] or melt pool [6] generated by the beams. Making many parts means dealing with an increased amount of vapor and vapor-induced part defects [7].

Thus, area scanning by DMD fabricates parts by exposing some portion of a bed and then exposing another portion of the bed. This type of scanning continues until the whole bed is covered. If there is a big part (to be fabricated) that covers the whole area of the bed, it is fabricated by multiple exposures of the same area (part) by moving from one portion to another. If there are many small parts (to be fabricated) that cover the whole area of the bed, they are fabricated by multiple exposures of the same bed by moving from one part to another (similar to Fig. 118.2b). Since there is a movement of processing from one part to another in the same bed, it is a serial production.

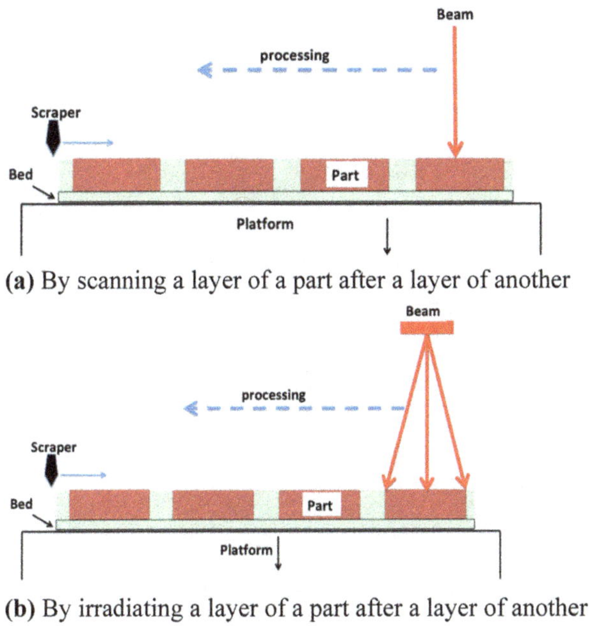

Fig. 118.2 Serial production. (**a**) By scanning a layer of a part after a layer of another. (**b**) By irradiating a layer of a part after a layer of another

2 Non-serial Production

Non-serial production in powder bed fusion (PBF) happens when the nth layer of all parts is scanned simultaneously, and not when scanning moves from the nth layer of one part to the nth layer of another. Scanning a layer is equal to fabricating a layer.

Non-serial production is possible in one of the following ways:

1. By arranging the number of beams equal to the number of parts (one beam for each nth layer) (Fig. 118.3a).
2. By exposing the nth layer of all parts instantly either by many beams (Fig. 118.3b).
3. By exposing the nth layer of all parts instantly either by one source (Fig. 118.3c).

If scanning moves from one part to another in a bed to fabricate the same ordinal number of layers, the same step (fabricating n^{th} layer) of production is done one by one. To speed up the production, fabricating the same ordinal number of layers (e.g., n^{th} layer) or the same step of production needs to be completed at the same time (Fig. 118.3). Thus, if a thousand parts are produced in an AM system (a serial production system), it can be mass production, implying the production is on a mass scale, but it is not mass production when the mass production implies only those productions that occur non-serially.

If the area exposed by a beam covers the whole bed, there is no need to move from one area to another of the bed for processing (Fig. 118.3c). Whatever number

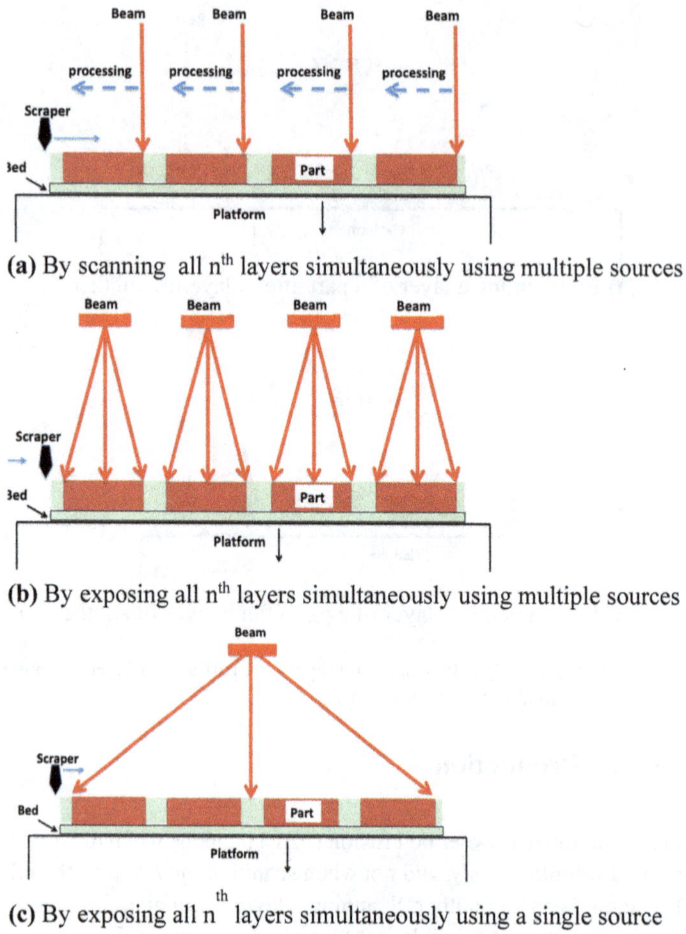

Fig. 118.3 Non-serial production. (**a**) By scanning all nth layers simultaneously using multiple sources. (**b**) By exposing all nth layers simultaneously using multiple sources. (**c**) By exposing all nth layers simultaneously using a single source

of parts need to be fabricated will have their designs accommodated in a bed and irradiated at the same time. This is a non-serial production.

3 Difference between Serial and Non-serial Production

The difference between serial and non-serial production in PBF is how a bed is exposed—either the whole bed is exposed so all parts grow simultaneously, or it is exposed partially so all parts do not grow simultaneously. Thus, the difference lies in what happens for one set of parts to be built in the bed.

After one set of parts is built in a bed, another set of parts needs to be built by creating an intermediate layer over these parts by moving in a build direction—this movement from one set to another is the same in both serial and non-serial production. This is a serial movement, which is a necessity if production on a mass scale is to be realized. Non-serial production cannot achieve mass production if production volume is not accumulated by producing serially with serial movement in the build direction.

This brings up the question of why non-serial production should not be considered serial production, as it is the serial component of the non-serial production that gives a bigger contribution to mass production. It is because mass production in all types of manufacturing is a result of serialwise production. For example, in injection molding (IM), it is realized by producing serially in a batch consisting of a few parts. Still, IM is mass production because these few parts in a batch are produced simultaneously.

References

1. Siebold, M. (2019). Additive manufacturing for serial production of high-performance metal parts. *Mechanical Engineering, 141*(05), 49–50.
2. Dröder, K., Heyn, J. K., Gerbers, R., & at al. (2016). Partial additive manufacturing: Experiments and prospects with regard to large series production. *Procedia CIRP, 55*, 122–127.
3. Wiese, M., Leiden, A., Rogall, C., et al. (2021). Modeling energy and resource use in additive manufacturing of automotive series parts with multi-jet fusion and selective laser sintering. *Procedia CIRP, 98*, 358–363.
4. Roy, N. K., Behera, D., Dibua, O. G., et al. (2019). A novel microscale selective laser sintering (μ-SLS) process for the fabrication of microelectronic parts. *Microsystems & Nanoengineering, 5*, 64.
5. Tenbrock, C., Kelliger, T., Praetzsch, N., et al. (2021). Effect of laser-plume interaction on part quality in multi-scanner laser powder bed fusion. *Additive Manufacturing, 38*, 101810.
6. Zhang, W., Hou, W., Deike, L., & Arnold, C. B. (2020). Using a dual-laser system to create periodic coalescence in laser powder bed fusion. *Acta Materialia, 201*, 14–22.
7. Greer, C., Nycz, A., Noakes, M., et al. (2019). Introduction to the design rules for metal big area additive manufacturing. *Additive Manufacturing, 27*, 159–166.

Chapter 119
PBF: Shape Limitation

Powder bed fusion (PBF) cannot make a hollow metallic sphere (Fig. 119.1b), though simpler than a hollow lattice cube (Fig. 119.1a) [1], as the closed sphere does not let powder drain out. Metal forming can make it, which implies the forming process is better than AM for making it. Although a hollow sphere can be made by incorporating drainage holes in the design [2].

PBF cannot economically make a solid metallic sphere (Fig. 119.1c), though it is again simpler than a hollow lattice cube. Machining or sintering does it, making conventional manufacturing (CM) better than AM [3].

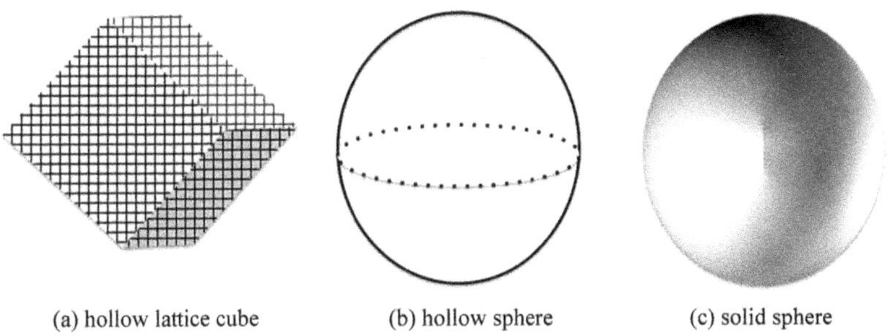

Fig. 119.1 Schematic diagram of (**a**) hollow cube having lattice surface, (**b**) hollow sphere, and (**c**) solid sphere

S. Kumar, *A Concise Encyclopedia of Additive Manufacturing*, https://doi.org/10.1007/978-3-031-98394-8_119

References

1. Korpela, M., Riikonen, N., Piili, H., et al. (2020). Additive manufacturing—Past, present, and the future. In M. Collan & K. E. Michelsen (Eds.), *Technical, economic and societal effects of manufacturing 4.0*. Palgrave Macmillan.
2. Noronha, J., Dash, J., Downing, D., et al. (2025). Thin-plate lattices in AlSi10Mg alloy via laser additive manufacturing: Highly enhanced specific strength and recovery. *Additive Manufacturing, 99*, 104664.
3. Hällgren, S., Pejryd, L., & Ekengren, J. (2016). Additive manufacturing and high speed machining -cost comparison of short Lead time manufacturing methods. *Procedia CIRP, 50*, 384–389.

Chapter 120
Process

The Oxford dictionary gives the definition of a process as "a series of actions or steps taken in order to achieve a particular end." A process is a set of tasks [1]. AM processes are a combination of various steps in sequence. Such steps, for example, are what the materials are, how they are brought, and how they are converted. For example, three steps in powder bed fusion (PBF), which defines the process, are that materials are in the form of a powder, they are placed in the form of a bed, and they are converted by fusion.

Processes that come under PBF follow the same steps but differ in the sources responsible for fusing them. For example, electron beam melting (EBM) and selective laser melting (SLM) are two PBFs that differ because they are fused by e-beam and l-beam, respectively. Selective laser sintering (SLS) and SLM are two PBFs that are fused by the same type of applied energy source, i.e., l-beam, but differ because both use the l-beam for different purposes. SLS uses it for partial melting, while SLM uses it for full melting.

All these three PBFs, i.e., SLS, SLM, and EBM, differ in the last step of the process. There is no example of PBF that differs in the first or second step. A process is executed in a machine or system (machine and system are used interchangeably in AM). Some machines execute more than one process, while some processes can be executed in more than one machine. For example, an SLM machine can execute both processes, SLM and SLS. By decreasing laser power and preventing full melting, SLS can be achieved. Similarly, an SLS machine can execute both processes, SLS and SLM. By choosing low melting point materials such as tin or bronze and facilitating full melting, SLM can be achieved. Thus, a process (SLS or SLM) can be executed in two machines (SLS & SLM).

Most AM machines are process-specific and therefore execute a single AM process each. However, if other non-AM processes can be included to understand the difference between a process and a machine, these AM machines can execute more than one process. Most of the AM processes have similar beginnings, and respective AM machines can execute two processes—one their initial process and the other

S. Kumar, *A Concise Encyclopedia of Additive Manufacturing*,
https://doi.org/10.1007/978-3-031-98394-8_120

their existing process. For example, wire arc AM (WAAM) is developed from arc welding; a WAAM machine can execute two processes—arc welding (initial process) for repair [2] and WAAM (existing process) for making complex channels [3].

References

1. Alexopoulos, K., & Chryssolouris, G. (2018). Process. In S. Chatti & T. Tolio (Eds.), *CIRP encyclopedia of production engineering*. International Academy for Production Engineering.
2. Lee, J. H., Lee, C. M., & Kim, D. H. (2022). Repair of damaged parts using wire arc additive manufacturing in machine tools. *Journal of Materials Research and Technology, 16*, 13–24.
3. Marqués, A., Dieste, J. A., Monzón, I., et al. (2024). Improvements in injection Moulds cooling and manufacturing efficiency achieved by wire arc additive manufacturing using conformal cooling concept. *Polymers, 16*(21), 3057.

Chapter 121
Processable Materials

AM has approximately 29 metallic powders [1, 2], 15 metallic wires, 5 polymer powders, 5 polymer filaments, and 5 photopolymer resins. These numbers are constantly growing. Even though they are minuscule compared to thousands of materials in conventional manufacturing (CM) [3, 4]. A small number of polymer materials does not mean their use is less than that of metals. The industrial sale of plastics is 80%, while that of metals is only 6% [5].

There are commercial metallic materials having different trade names [6]. A total of 129 polymer materials, including 66 for powder bed fusion (PBF), are listed [7]. But these are either the same basic AM materials or developed from them for a specific application or machine. These customized materials thus do not increase the number of basic AM materials.

Reasons for a few processable materials are the following:

- Most AM research exploits the design advantage of AM. For this, creating a material is not required as much as testing new designs using existing materials. To make a new part using a new design does not get setback due to the lack of materials, e.g., polyamide is used to make chairs as well as implants.
- Material has to fulfill higher requirements in AM than in CM [8]. In sintering, a mold shapes a material. In AM, coalescing of a material is required to make a shape [9]. Thus, whatever the requirement in CM to be a successful material is not sufficient in AM.
- The material developed for one AM process is not transferable to another, e.g., powder that works in laser PBF has no guarantee to work in electron beam PBF.

What if the number of processable materials increases in the future? Will this disadvantage still exist? The disadvantage has more to do with the reason for its existence than to do with the number of processable materials. Due to extensive research, if the number of processable materials increases and becomes on par with CM, it will no longer be a disadvantage. But its absence will not help the reason for

S. Kumar, *A Concise Encyclopedia of Additive Manufacturing*,
https://doi.org/10.1007/978-3-031-98394-8_121

its existence to go away. The reasons are the lack of confining space compared to mold-based CM, the need to process each bit of material compared to machining, etc. The disadvantage shows more of a characteristic of AM than the relation of AM with prevalent research and development efforts.

References

1. Qian, M. (2015). Metal powder for additive manufacturing. *JOM, 67*, 536–537.
2. Anderson, I. E., White, E. M. H., & Dehoff, R. (2018). Feedstock powder processing research needs for additive manufacturing development. *Current Opinion in Solid State and Materials Science, 22*(1), 8–15.
3. Martin, J., Yahata, B., Hundley, J., et al. (2017). 3D printing of high-strength aluminium alloys. *Nature, 549*, 365–369.
4. Ashby, M. F., Brechet, Y. J. M., Cebon, D., & Salvo, L. (2004). Selection strategies for materials and processes. *Materials & Design, 25*(1), 51–67.
5. Cotteleer, M., & Joyce, J. (2014). 3D opportunity: Additive manufacturing paths to performance, innovation, and growth. *Deloitte review, 14*(1), 3–19.
6. Korpela, M., Riikonen, N., Piili, H., et al. (2020). Additive manufacturing—Past, present, and the future. In M. Collan & K. E. Michelsen (Eds.), *Technical, economic and societal effects of manufacturing 4.0*. Palgrave Macmillan.
7. Wiese, M., Thiede, S., & Herrmann, C. (2020). Rapid manufacturing of automotive polymer series parts: A systematic review of processes, materials and challenges. *Additive Manufacturing, 36*, 101582.
8. Chen, Z., Han, C., Gao, M., et al. (2022). A review on qualification and certification for metal additive manufacturing. *Virtual and Physical Prototyping, 17*(2), 382–405.
9. Johnson, N. S., Vulimiri, P. S., To AC, et al. (2020). Invited review: Machine learning for materials developments in metals additive manufacturing. *Additive Manufacturing, 36*, 101641.

Chapter 122
Production Volume Flexibility

Since there is no mold that comes as an obstacle between a design and parts in AM, the conversion of a design into parts does not depend on the mold. The maximum production volume is no longer limited by the quality of a mold. In AM, producing a few parts is not the underutilization of an AM system, while in a mold-based system, producing a few parts can mean the underutilization of the system. The conversion in AM has the flexibility to choose serial production for a low volume and either serial or non-serial production for a high volume, which is not available in mold-based processes. Machining has the advantage of selecting either serial or non-serial type to change the volume, but machining depends on tools and fixtures, which do not give the same flexibility as AM does [1, 2].

References

1. Scholz, S., Mueller, T., Plasch, M., et al. (2016). A modular flexible scalable and reconfigurable system for manufacturing of microsystems based on additive manufacturing and e-printing. *Robotics and Computer-Integrated Manufacturing, 40*, 14–23.
2. Cleeman, J., Bogut, A., Mangrolia, B., et al. (2022). Scalable, flexible and resilient parallelization of fused filament fabrication: Breaking endemic tradeoffs in material extrusion additive manufacturing. *Additive Manufacturing, 56*, 102926.

S. Kumar, *A Concise Encyclopedia of Additive Manufacturing*,
https://doi.org/10.1007/978-3-031-98394-8_122

Chapter 123
Rapid Manufacturing

Making a new product or verifying a new product design frequently is time-consuming through injection molding (IM). Since with a change in a digital file, a new mold needs to be fabricated first to make a new product.

There is no requirement for a mold (an intermediate step) in AM, which allows AM to reach the product stage from the digital stage without going through the intermediate step. There is thus one step less in AM compared to IM. This makes AM fast—it is rapid manufacturing (RM). RM does not imply that AM is always faster than conventional manufacturing (CM) but is faster than only those CM that get delayed due to the intermediate step.

AM is RM because it can rapidly convert a new digital file into a product, which can help in an emergency situation if required [1]. But AM cannot be RM for making a product that is made rapidly by other processes, e.g., forging to make a big part or constant-thickness part, or turning to make an axisymmetric part [2]. In this case, the absence of the intermediate step does not help AM enough to be faster than CM. In another example, if investment casting (IC) makes a complex part, it requires several steps. While in AM, the same is made in a single step. AM is therefore faster. But for a big part, e.g., an aircraft wing, AM may not be faster than casting even if additional steps are needed in other processes.

References

1. Tareq, M. S., Rahman, T., Hossain, M., & Dorrington, P. (2021). Additive manufacturing and the COVID-19 challenges: An in-depth study. *Journal of Manufacturing Systems, 60*, 787–798.
2. Oliveira, J. P., & Santos, T. G. (2024). Demystifying "absolute truths" of additive manufacturing. *CIRP Journal of Manufacturing Science and Technology, 54*, 57–62.

S. Kumar, *A Concise Encyclopedia of Additive Manufacturing*, https://doi.org/10.1007/978-3-031-98394-8_123

Chapter 124
Rapid Prototyping

AM is rapid prototyping when AM makes a part, not meant as an end-use product, to verify a design fast [1]. Rapid prototyping is one of the oldest and current applications of AM that makes both visual and functional prototypes. A visual prototype must look like its digital image, while it does not need to be as strong as a final product. For example, if it does not disintegrate on checking its appearance, it is strong.

The functional prototype must be able to go through testing [2]. For example, if the prototype of a nozzle is made to test the flow of air through it, it should have adequate properties to test. These properties are (1) strength, so it must not break due to the flow of air when it is fitted with an air source, and (2) a smooth inner surface like that of a final product to test the flow.

If a functional prototype is complex, while the aim is to check only mechanical properties, a simple instead of a complex shape will work. If the aim of the prototype is to check everything that a final product furnishes, such as design, complexity, functionality, and property, there is no difference between a prototype and a final product. This prototype to check mechanical properties is known as a mechanical benchmark [3].

References

1. Vido, M., de Oliveira Neto, G. C., Lourenço, S. R., et al. (2024). Computer-aided design and additive manufacturing for automotive prototypes: A review. *Applied Sciences, 14*(16), 7155.
2. Fagundes, A. P., de Brito, L. J. O., Padoin, N., et al. (2022). Additive manufacturing of functional devices for environmental applications: A review. *Journal of Environmental Chemical Engineering, 10*(3), 108049.
3. Rebaioli, L., & Fassi, I. (2017). A review on benchmark artifacts for evaluating the geometrical performance of additive manufacturing processes. *The International Journal of Advanced Manufacturing Technology, 93*, 2571–2598.

S. Kumar, *A Concise Encyclopedia of Additive Manufacturing*,
https://doi.org/10.1007/978-3-031-98394-8_124

Chapter 125
Repair and Refurbishment

The aim of repair is to bring a damaged part to its original condition [1], while the aim of refurbishment is to improve a damaged part and make it better than its original condition in some or all respects [2]. Fig. 125.1 shows schematic diagrams of a part before and after damage; one of the features is lost in a damaged part. Figure 125.2 shows the repaired part, the same as the original part, consisting of a repaired feature (red color).

Repair intends to match the material, property, and design of a damaged part to those of an original part after repair, so a damaged part regains its properties and shape that existed before damage. Therefore, a new material is not sought to make up the material lost due to damage. There is no search for a repairing technique that can furnish better property than that of the original part. Besides, a new design is not sought to improve functions.

Refurbishment brings a change, which means there must be a change in design, property, or material, causing a change in the characteristics of an original part. Therefore, it is considered an opportunity to make the best out of a damaged part; conveying damage is not the start of its devaluation. If repair aims at maintaining the status quo, refurbishment breaks it.

Figure 125.3a shows the refurbishment of a damaged part. AM adds an extra feature so the refurbished part becomes better. If the damaged part cannot be brought back to the original condition, some new features can be added to compensate for the lack of repair (Fig. 125.3b). During refabrication, AM can improve material properties, e.g., facilitating the growth of a single crystal during the turbine blade repair [3].

The damaged surface first needs to be prepared by machining to check the fitability of a part for refabrication and to maintain the accuracy of addition by AM [4]. If accuracy after AM is not right, it can be machined. Machining can be planned to occur several times—this will lead to mixing additive repair with subtractive repair [5].

S. Kumar, *A Concise Encyclopedia of Additive Manufacturing*, https://doi.org/10.1007/978-3-031-98394-8_125

(a) Original condition (b) Damaged part

Fig. 125.1 A part: (**a**) original, (**b**) damage

Fig. 125.2 Repaired part

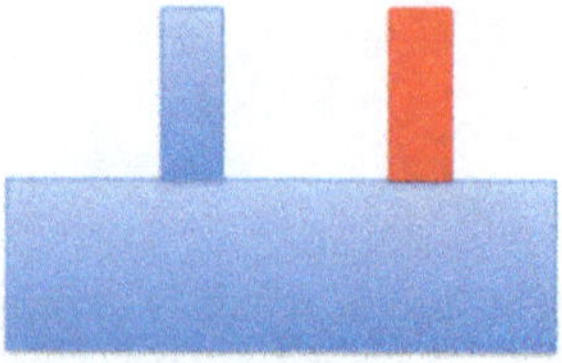

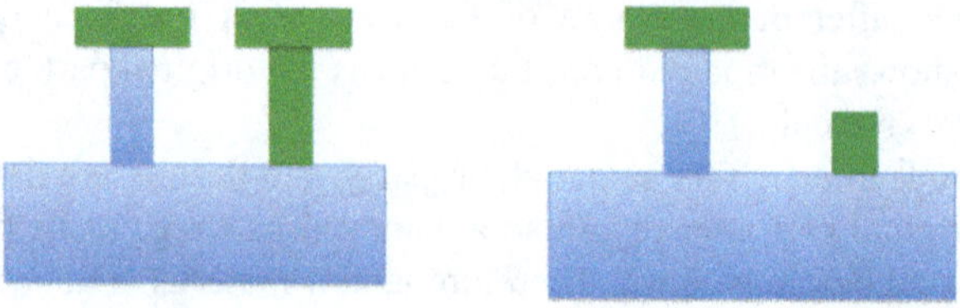

(a) better than original in all respects (b) better than original in some respects

Fig. 125.3 Refurbished parts. (**a**) Better than original in all respects. (**b**) Better than original in some respects

AM is used to repair turbine blades [6], crankshaft surfaces, gears, molds, compressors, etc. [7]. Considering the repair of a turbine blade tip, if its strength increases after repair, it will not be repair but refurbishment. In this case, there should be repair, not refurbishment, because a tip having higher strength will bring a strength gradient across the blade, which will decrease its fatigue property.

References

1. Wilson, J. M., Piya, C., Shin, Y. C., et al. (2014). Remanufacturing of turbine blades by laser direct deposition with its energy and environmental impact analysis. *Journal of Cleaner Production, 80*, 170–178.
2. Brinker, J., Gembarski, P. C., Hagen, S., & Thomas, O. (2020). Anwendungspotenziale von Additive Repair und Refurbishment für Service-orientierte Geschäftsmodelle. In R. Lachmayer, K. Rettschlag, & S. Kaierle (Eds.), *Konstruktion für die Additive Fertigung* (pp. 43–54).
3. Chandra, S., Tan, X., Wang, C., et al. (2018). Additive manufacturing of a single crystal nickel-based superalloy using selective electron beam melting. In *Proceedings of the 3rd international conference on Progress in additive manufacturing (pro-AM)* (pp. 427–432).
4. Petrat, T., Graf, B., Gumenyuk, A., & Rethmeier, M. (2016). Laser metal deposition as repair Technology for a gas Turbine Burner Made of Inconel 718. *Physics Procedia, 83*, 761–768.
5. Li, L., Li, C., Tang, Y., & Du, Y. (2017). An integrated approach of reverse engineering aided remanufacturing process for worn components. *Robotics and Computer-Integrated Manufacturing, 48*, 39–50.
6. Wang, D., Liu, Z., Deng, G., et al. (2024). A laser powder bed fusion–based methodology for repairing damaged nickel-based turbine blades: Investigation of interfacial characteristics and hot isostatic pressing treatment. *Materials Characterization, 212*, 113948.
7. Chen, Y., Yang, C., Fan, C., & Wang, M. (2022). Microstructure evolution and mechanical properties of a nickel-based superalloy repaired using wire and arc additive manufacturing. *Materials Characterization, 193*, 112315.

Chapter 126
Repair in PBF

There are not many cases of repairs in powder bed fusion (PBF); though, cases recently increased [1, 2]. Repair can be employed if the following two conditions are met:

1. Damage should not be an obstacle for the damaged surface to be kept parallel to the platform or build plane, which will ensure no collision with a roller during powder spreading [3].

For example, if all features marked as 1, 2, 3, 4, and 5 (Fig. 126.1a) get snapped from a rectangular block ABCD, it is possible to keep the damaged surface AB parallel to the build plane and then repair the part. The front view of the block ABCD and surface AB is shown by rectangle ABCD and line AB, respectively.

If these features are on an oblique surface (AB) of a triangular part (Fig. 126.1c), it is not possible to repair when any of the features on AB get damaged, as AB cannot be kept parallel to the build platform. If its orientation is changed to align it parallel, and fixtures are used to fix it parallel, it can compromise the accuracy of the repair.

2. The shape of a part should be such that when the part is fixed into a platform, the damaged surface must remain parallel to the platform [4]. For example, the lower surface CD (Fig. 126.1a) of the block ABCD is plane without having any complex feature on it, which is suitable to be fixed on the platform, keeping the surface AB parallel to it (Fig. 126.2a). If the lower surface CD is tilted (Fig. 126.3a) or has a long, thin feature (Fig. 126.3b), the part may not be fitted well to enable working of the PBF system.

Figure 126.1a shows an original part having five features, while Fig. 126.1b shows the same part having one feature numbered 3 (feature-3) missing. Features are blocks of equal height having the same cross-section. To repair the damaged part (Fig. 126.1b), the missing feature needs to be regrown at the same location of the part.

S. Kumar, *A Concise Encyclopedia of Additive Manufacturing*, https://doi.org/10.1007/978-3-031-98394-8_126

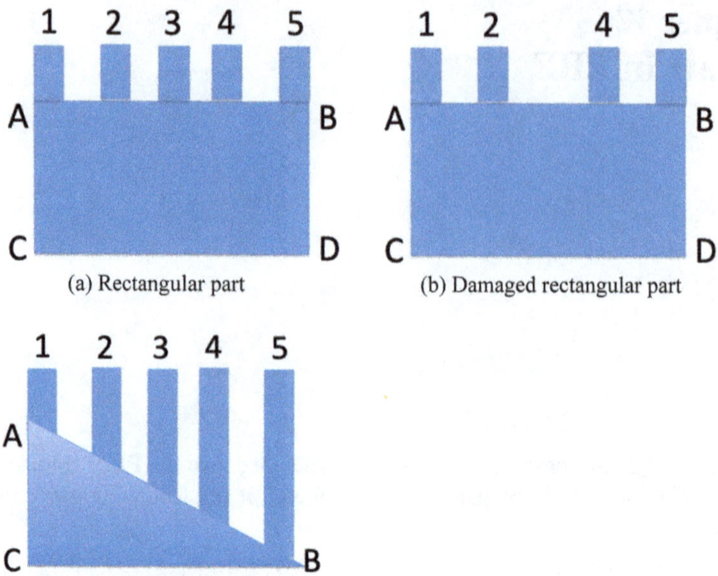

(a) Rectangular part (b) Damaged rectangular part (c) Triangular part

Fig. 126.1 Schematic diagrams of original and damaged parts. (**a**) Rectangular part. (**b**) Damaged rectangular part. (**c**) Triangular part

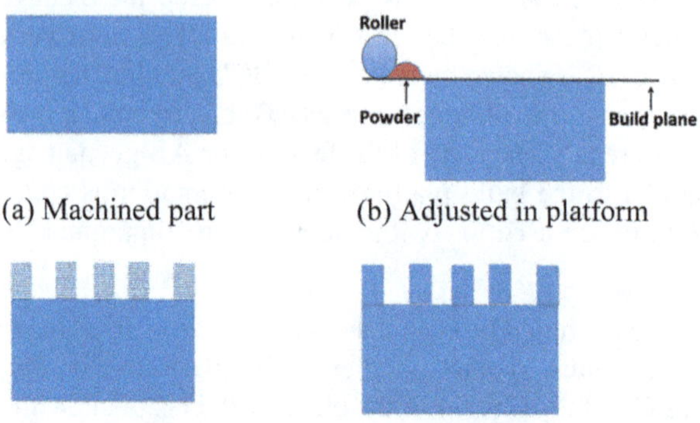

(a) Machined part (b) Adjusted in platform (c) CAD file for repair (d) Repaired part

Fig. 126.2 Four stages of a repair. (**a**) Machined part. (**b**) Adjusted in platform. (**c**) CAD file for repair. (**d**) Repaired part

It is not possible to regrow a feature selectively unless in an exceptional case when the height of the feature is approximately equal to the layer thickness—when the height of feature-3 is equal to the layer thickness, the part can be repaired selectively.

For repair, what is required is the powder in a vacant space between feature-2 and feature-4, and selective consolidation in the vacant space as per the contour of

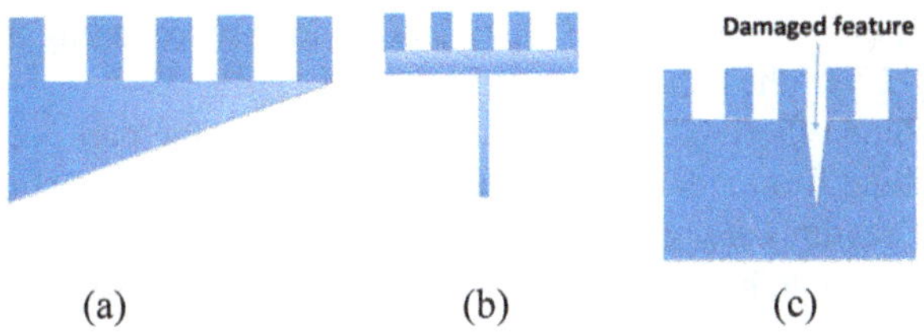

Fig. 126.3 Non-repairable parts (**a**) oblique lower surface, (**b**) thin feature in lower surface, (**c**) deep damaged feature

feature-3. The powder is brought by spreading it using a roller, which fills up the vacant space. As the upper surface of features-1,2,4,5 will come in contact with the roller, it is assumed that these features are strong enough not to break due to friction with the moving roller. This again brings limitations to the type of damage a part can have to make it repairable.

As scanning is optimized for one layer thickness, the formation of the feature will not face the problem of inadequate bonding between the base of the feature and a rectangular block. If the height of the feature is bigger, scanning will not furnish adequate bonding.

This example of a missing feature having height equal to a layer thickness is unlikely to occur, but it, nevertheless, shows the problems while repairing. The problem is not the presence of a missing feature (feature-3), but the presence of other features (features-1,2,4,5). The other features block the access of a coating device (roller) to the site of the missing feature. In this case, these features have not blocked the access because all these features are of equal height, which does not prevent the roller from rolling. Besides, it is assumed that there is a need for deposition of only one layer, as all features are of one layer height—it means the roller has a role to play. If the features were of more than one layer height, the rolling had no utility, as the consolidation after rolling could not have been achieved—the consolidation needs to happen layerwise.

If features are required to be made from several layers, all features must be removed by machining to build all new features instead of just one new feature. Figure 126.2a shows a machined part without any features, which can be fitted well into the platform (Fig. 126.2b) because surface AB is parallel to the build plane. Lower parallel surface CD will help fix it. A CAD file having all features sliced (Fig. 126.2c) is used to build features to make a final part (Fig. 126.2d).

Need to remove four features for repairing just one feature brings a question mark on the efficiency of powder bed fusion for repair, but economically and resource-wise it is better than both—not using the damaged part and opting for a new part.

If an original part is damaged severely (Fig. 126.3c), where there is a triangular cut reaching inside the part, repair may not be an option. If it is machined parallel to AB as done earlier (Fig. 126.2a), approximately 3/4 of the part is machined out. If it is machined parallel to BC so the repair can be done by changing the orientation, approximately 2/5 of the part is removed.

If it is machined from side AC, approximately 3/5 of the part goes. There is no option to machine from side CD, as approximately 4/5 of the part is removed, as well as all five features are separated. In all cases, remaining parts after machining are not large enough to bring significant difference between repair and new fabrication, questioning the reason for repair [5].

Since a damaged part needs to meet some conditions before it can be repaired, it could be better if the design of a part could be changed, keeping in view the future need for repair [6].

References

1. Wang, D., Liu, Z., Deng, G., et al. (2024). A laser powder bed fusion–based methodology for repairing damaged nickel-based turbine blades: Investigation of interfacial characteristics and hot isostatic pressing treatment. *Materials Characterization, 212*, 113948.
2. Chen, Y., Yin, Y., & Yi, M. (2024). Powder bed fusion repair of titanium with surface damage: Molecular dynamics study on microstructure and mechanical properties. *Additive Manufacturing, 84*, 104096.
3. Sawczuk, P., Kluczyński, J., Sarzyński, B., et al. (2023). Regeneration of the damaged parts with the use of metal additive manufacturing—Case study. *Materials, 16*(10), 3772.
4. Andersson, O., Graichen, A., Brodin, H., & Navrotsky, V. (2017). Developing additive manufacturing technology for burner repair. *Journal of Engineering for Gas Turbines and Power, 139*(3), 031506.
5. Kumar, S. (2022). Application. In *Additive manufacturing solutions* (pp. 93–110). Springer.
6. Aziz, N. A., Adnan, N. A. M., Wahab, D. A., & Azman, A. H. (2021). Component design optimisation based on artificial intelligence in support of additive manufacturing repair and restoration: Current status and future outlook for remanufacturing. *Journal of Cleaner Production, 296*, 126401.

Chapter 127
Repair in PBF: Customized Bed Concept

The platform in PBF is planar, and therefore, for repair, parts having a nonplanar base are not suitable [1]. Besides, the size of the base needs to be small enough to fit within the platform. Although the size of the platform is enlarged to accommodate a bigger base [2, 3]. What if the shape of a platform or bed can be changed by a nut, screw, and bolt? It can repair various types of shapes.

For example, if a bed is bent inward, making a half-circle, such that the bed is not a rectangular planar surface but a curved surface (Fig. 127.1a), and if the radius of the half-circle is slightly more than the radius of a cylinder, the cylinder can fit in the bed. A cylinder machined into half along its central axis will conform to the half-circular space of the bed (Fig. 127.1b). The machined surface will not be different from the substrate surface used in PBF. This type of bed can work well to repair cylindrical objects, as fitting the objects will not require machining to make their circumference partially planar (Fig. 127.1c). If the inner space of the bed is adjustable, it will allow objects of various sizes to fit. Since most cylindrical objects are formed by a lathe, this PBF system will be the lathe equivalent of a general PBF system.

This example demonstrates why a part compatible with PBF may not be compatible with all PBF systems. In the state-of-the-art PBF system, bed architectures in all systems are the same. There is no PBF system whose bed is compatible with nonplanar objects. PBF systems of different models and those using different energy beams are different in overall architecture, but for the compatibility of different part shapes with the bed, all are the same. This similarity is the limitation of the state-of-the-art PBF system, which has led to the conjecture that what is compatible with PBF is also compatible with a PBF system.

S. Kumar, *A Concise Encyclopedia of Additive Manufacturing*, https://doi.org/10.1007/978-3-031-98394-8_127

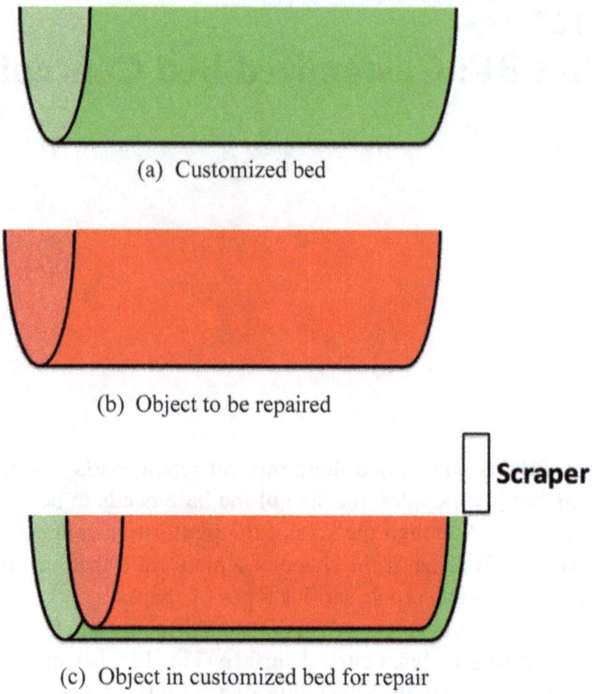

Fig. 127.1 Repair in customized bed. (**a**) Customized bed. (**b**) Object to be repaired. (**c**) Object in customized bed for repair

1 Problem with Planar Bed

Even if a customized bed is developed, it will not localize the repair. It will help fit a part that matches the bed shape, which will remove the need to shape a part. And it will eliminate the need to refabricate a machined surface to regain the lost shape. For example, if a cylinder is repaired on its side surface, its side surface and the other side surface are machined; after repair, the effect of the machining on the other side surface needs to be undone.

To refabricate, powder or wire welding may be used. If it does not give an adequate finish, secondary machining is required. What if instead of welding, PBF is used? Then, the repair done by PBF on the side surface needs to be removed to make the cylinder again conform to the planar bed. It means if the machined surface is refabricated by PBF, there will be no repair left anymore on the side surface. Though PBF will not be used to refabricate, this shows the limitation of a planar bed. This is why there are a few repairs done in the PBF system. The number can increase if a bed is customized. Though the customized bed is not feasible if the majority of damaged parts are not of the same shape.

References

1. Sato, N., Matsumoto, M., Ogiso, H., & Sato, H. (2022). Challenges of remanufacturing using powder bed fusion based additive manufacturing. *International Journal of Automation Technology, 16*(6), 773–782.
2. Walachowicz, F., Bernsdorf, I., Papenfuss, U., et al. (2017). Comparative energy, resource and recycling lifecycle analysis of the industrial repair process of gas turbine burners using conventional machining and additive manufacturing. *Journal of Industrial Ecology, 21*(S1), S203–S215.
3. https://www.eos.info/industries/customer-success-stories/siemens-3d-printing-turbine-repair-burner

Chapter 128
Selective Laser Melting

It is a type of AM in which a part is formed by melting a powder bed [1–3].

References

1. Song, B., Wen, S., Yan, C., et al. (2020). *Selective laser melting for metal and metal matrix composites*. Academic Press.
2. Gao, B., Zhao, H., Peng, L., & Sun, Z. (2022). A review of research progress in selective laser melting (SLM). *Micromachines, 14*(1), 57.
3. Aufa, A. N., Hassan, M. Z., Ismail, Z., et al. (2024). Current trends in additive manufacturing of selective laser melting for biomedical implant applications. *Journal of Materials Research and Technology, 31*, 213–243.

S. Kumar, *A Concise Encyclopedia of Additive Manufacturing*,
https://doi.org/10.1007/978-3-031-98394-8_128

Chapter 129
Selective Laser Sintering

It is a type of AM in which a part is formed by not melting a powder bed [1–3].

References

1. Song, Y., Ghafari, Y., Asefnejad, A., & Toghraie, D. (2024). An overview of selective laser sintering 3D printing technology for biomedical and sports device applications: Processes, materials, and applications. *Optics & Laser Technology, 171*, 110459.
2. Yehia, H. M., Hamada, A., Sebaey, T. A., & Abd-Elaziem, W. (2024). Selective laser sintering of polymers: Process parameters, machine learning approaches, and future directions. *Journal Of Manufacturing And Materials Processing, 8*(5), 197.
3. Yan, C., Shi, Y., Zhaoqing, L., et al. (2020). *Selective laser sintering additive manufacturing technology*. Academic Press.

S. Kumar, *A Concise Encyclopedia of Additive Manufacturing*,
https://doi.org/10.1007/978-3-031-98394-8_129

Chapter 130
Scanning Type

1 Linewise

In high-speed sintering (HSS), when a thermal lamp starts to scan from one end of a layer, in the first few moments, it irradiates the whole breadth but a fraction of the length of the layer. Thus, the lamp does not scan any area but just a line on the layer. After a few more moments, it covers some more length, and then it has covered some area. It starts from scanning a line, and the line moves in a direction normal to it. The line becomes wider with the progress of scanning, which is an area. This is a linewise scanning.

Linewise scanning is fast because it starts from a line. It would not be fast if it had to create a line from scratch by adding many fragments or points. It does not constitute a line by controlling many points, which is reserved for another scanning (pointwise scanning).

Examples of linewise scanning are stacking of l-beam emitters in a bar across a powder bed (PB) and processing it without deflecting beams but switching on/off some emitters and moving the bar [1, 2], or movement of an infrared lamp across PB in multi-jet fusion [3]. HSS is an example of double linewise scanning—the first linewise scanning takes place when an array of nozzles deposits material on PB, while the second takes place when an infrared lamp moves.

2 Pointwise

If scanning starts from a point, it is a pointwise scanning that makes many overlapping points to make a single line. It will not be as fast as a linewise scanning, but it controls points and has the ability to add points. Thus, it can make a line of any size in any direction, a combination of many lines of various sizes. If the linewise

S. Kumar, *A Concise Encyclopedia of Additive Manufacturing*,
https://doi.org/10.1007/978-3-031-98394-8_130

scanning is for fast scanning, it is for fine scanning. If the linewise is to provide a high production rate, it is to provide high complexity. If the linewise can make both a line and an area, it can make all three: a point, a line, and an area. The aim of the pointwise scanning is to make a layer to fulfill layer-by-layer addition. If it is not used to make a layer, as in two-photon polymerization [4], it will be layerless AM.

Its examples are scanning using a point source such as an e-beam, an l-beam, or a jet from a single nozzle. In electron beam powder bed fusion (EPBF), the diameter of an e-beam is the size of a point source; in laser powder bed fusion (LPBF), the laser spot size of the l-beam is the size of a point source; and in binder jet three-dimensional printing (BJ3DP), the size of the drop on the surface from a nozzle is the size of a point source.

3 Areawise

Linewise scanning starts from a line and ends up making an area, while pointwise scanning starts from a point and ends up making an area. If they do not make an area, a part will not be made. How they start is important, but whether they are able to end up in an area is more important. If the area is important, why not have the whole area be scanned at a time?

Consider a thermal lamp that does not follow linewise scanning by starting from one end of a layer and reaching the other end but faces the whole area and then scans instantly. It means there is a lamp situated over PB, which is switched on and then off. This will cause the whole area to be scanned without any physical movement of the lamp. This instant scanning is faster than linewise scanning. But it has no power of judgment, as it cannot identify the area that needs to be consolidated. If all the scanned area is solidified, there will always be a rectangular block and not a part as per design.

Areawise scanning needs help for making a useful part. It can get help if thermal radiation coming from the lamp is prevented from reaching every part of an area. This can be done by using a physical mask between the lamp and PB so the mask will prevent the radiation from reaching PB. Therefore, though the whole area is scanned, the effect of the radiation will not be realized below the masked area [5].

It can also get help if, instead of creating a physical mask, a mask-type mechanism is created within the lamp so when the lamp is switched on, the radiation does not emanate from the whole area of the lamp but only from its unmasked area [6]. If the radiation comes from the selected area, it will affect only the selected area of PB. Thus, the selected area will be solidified, leaving the remaining area unsolidified, and a part as per design will be formed.

It can again get help if, instead of using any type of masks, powders of the selected area of PB are changed so the selected area acts differently when the whole area is irradiated. If a selected area is mixed with a radiation-absorbing material (the same in high-speed sintering), that area will be solidified. If mixing is done as per a design, solidification will happen as per the design, which will help make parts as per the design [7]. If the selected area has higher melting point material, that area will not be solidified, but other areas will be solidified, which will again help make parts as per design [8].

An example of areawise scanning is microheater array powder sintering, where a microheater array is set above PB. Areawise scanning has been called layerwise scanning [9]. But the layerwise scanning implies layer-by-layer scanning and has been used as such for long [10]. It will thus be difficult to distinguish between two meanings of the layerwise scanning: (1) scanning a layer, completing it, going to the next layer, and again scanning, and (2) scanning every point of a layer at a time. The name areawise scanning, unlike the name layerwise scanning, has no such precedence and is free from ambiguity.

References

1. Arredondo, M. Z., Boone, N., Willmott, J., et al. (2017). Laser diode area melting for high speed additive manufacturing of metallic components. *Materials and Design, 117*, 305–315.
2. Dallarosa, J., O'neill. W., Sparkes, M., & Payne, A. (2016). Multiple beam additive manufacturing. Patent WO2016201309A1.
3. Cai, C., Tey, W. S., Chen, J., et al. (2021). Comparative study on 3D printing of polyamide 12 by selective laser sintering and multi jet fusion. *Journal of Materials Processing Technology, 288*, 116882.
4. Geng, Q., Wang, D., Chen, P., & Chen, S. C. (2019). Ultrafast multi-focus 3-D nano-fabrication based on two-photon polymerization. *Nature Communications, 10*(1), 2179.
5. Hermann, D. S., & Larson, R. (2008). Selective mask sintering for rapid production of parts, implemented by digital printing of optical toner masks. In *NIP & digital fabrication conference*.
6. Farsari, M., Claret-Tournier, F., Huang, S., et al. (2000). A novel high-accuracy microstereolithography method employing an adaptive electro-optic mask. *Journal of Materials Processing Technology, 107*(1–3), 167–172.
7. Ellis, A., Noble, C. J., & Hopkinson, N. (2014). High speed sintering: Assessing the influence of print density on microstructure and mechanical properties of nylon parts. *Additive Manufacturing, 1–4*, 48–51.
8. Khoshnevis, B., Zhang, J., Fateri, M., & Xiao, Z. (2014). Ceramics 3D printing by selective inhibition sintering. In *SFF proceedings* (pp. 163–169).
9. Holt, N., Horn, A. V., Montazeri, M., & Zhou, W. (2018). Microheater array powder sintering: A novel additive manufacturing process. *Journal of Manufacturing Processes, 31*, 536–551.
10. Deckard, C. R., Beaman, J. J., & Darrah J F (1992). Method for selective laser sintering with layerwise cross-scanning. US5155324A.

Chapter 131
Scanning Type Differences

If the area of a powder bed (PB) is big and the exposure area of a lamp is small, the whole area of PB cannot be scanned by just switching the lamp on and off. The lamp needs to move to cover the whole area. If the area of the PB is big and the exposure area of the lamp is very small, the lamp needs to move not only in one direction but also in other directions to cover the whole area. When the lamp is still and is still being used for scanning, it is areawise scanning. But if it moves as a point source in different directions, it gives rise to pointwise scanning.

This brings up questions. How small is very small for an area to become a point? Is there anything more than a physical dimension (point, line, or area) that sets these three types of scanning apart?

A symbolic point has no dimension; therefore, pointwise scanning is 0D scanning [1]; although it is also called 1D scanning. A line has one dimension; therefore, linewise scanning is 1D scanning. An area has two dimensions; therefore, areawise scanning is 2D scanning. What could be 3D scanning, i.e., volumewise scanning? The extension of areawise scanning in the third dimension is a scanning that will cover 3D volume. Therefore, as per the name, it will be 3D scanning. But it will still be 2D scanning if it makes an object by additive layer manufacturing. Because additive layer manufacturing happens layer by layer, even if a layer has a measurable volume. Since all layers have measurable volumes, additive layer manufacturing happens through volume by volume when all the volumes have the same build direction. What if the volumes do not have the same build direction? Then it will not come under additive layer manufacturing but volumetric additive manufacturing, a layerless manufacturing [2, 3].

What if a product is made by a single scan, i.e., scanning in one go? Whether this scanning is 2D or 3D scanning. Whether the product is made by additive layer manufacturing or by volumetric additive manufacturing?

If a product is made by a single scan, the geometry must be simple enough so that the product does not need to be made by dividing its design into layers. Besides, there must be a capable machine that can convert the material into a product in one

S. Kumar, *A Concise Encyclopedia of Additive Manufacturing*,
https://doi.org/10.1007/978-3-031-98394-8_131

scan. If the product was initially conceptualized to be made layerwise but was not made by making many layers because many layers were not required, then the method is still additive layer manufacturing. It is a special case of additive layer manufacturing in which the number of layers is only one. Though additive layer manufacturing is a layerwise manufacturing and is not meant to be the manufacturing of only a layer, this special case can be understood as facilitated by the scanning of the top layer of the design, in which the extent of the scanning reaches the bottom layer. This is a 2D scanning that has a layer thickness equal to the height of the design. The presence of a layer thickness does not convert a 2D scanning into a 3D scanning, because every 2D scanning in additive layer manufacturing is supposed to have a 3D effect, without which a product will not be made. Changing the thickness of a layer does not change the type of scanning.

However, the method cannot be additive layer manufacturing if the design of the product was not conceptualized in the form of layers. Because if there is no layer, there will be no scanning on the basis of a layer. The scanning will not be done perpendicular to a build direction. In the absence of layers, there will be no build direction to guide the scanning. In additive layer manufacturing, there is a relation between the positioning of layers and the build direction. It is possible to have multiple build directions, which happens in multi-axis AM. But build directions maintain a relation to the positioning of layers. The moment the build direction changes, the positioning of the layer changes, so the direction will always be normal to the layer plane. Therefore, if a product is conceptualized in the form of layers, the scanning that will follow will not happen in a random direction. The scanning will be called 2D scanning not based on its 3D effect on materials, but only on the basis of its 2D effect on materials. The purpose of all three scanning methods (pointwise, linewise, and areawise) is to make a product through the formation of a layer, even if not forming the layer will give faster and better products.

Thus, the purpose of these scans is to make an area. In a pointwise scanning, the area is made when a point moves in both x- and y-directions in an xy plane. If the point moves only in one direction, it will end up making a line instead of an area. If the point does not move at all, it will not even make a line; there will be just a point. In a linewise scanning, the line needs to move in only one direction (normal to the line) [4] to make an area. If a line does not move, it will not make an area, but there will be just a line at the end of the movement. If the line moves in a plane just like a point, i.e., moves in both x- and y-directions, it will make a bigger area. But for making just an area, movement in one direction is sufficient. In an areawise scanning, without any movement of a scanner or lamp, an area can be made.

Hence, in a pointwise scanning (Fig. 131.1a), there is a minimum requirement of movement in two dimensions to scan an area. While in a linewise scanning (Fig. 131.1b), there is a minimum requirement of movement in one direction to scan an area. In an areawise scanning (Fig. 131.1c), there is no requirement of any movement. Though it can be stated that in an areawise scanning, there is a minimum requirement of movement in zero direction or dimension to scan an area. These are the basic differences among three types of scanning.

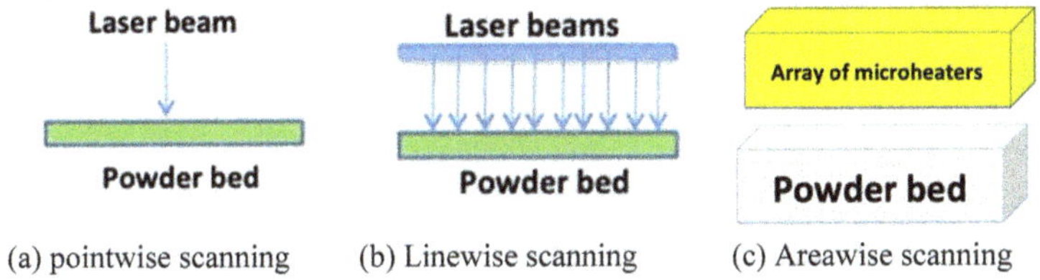

Fig. 131.1 Schematic diagrams of pointwise, linewise, and areawise scanning. (**a**) pointwise scanning. (**b**) Linewise scanning. (**c**) Areawise scanning

References

1. Yeung, H., & Neira, J. (2023). Toward voxel level control for laser powder bed fusion additive manufacturing processes.
2. Whyte, D. J., Doeven, E. H., Sutti, A., et al. (2024). Volumetric additive manufacturing: A new frontier in layer-less 3D printing. *Additive Manufacturing, 84*, 104094.
3. Lin, W., Chen, D., & Chen, S. C. (2020). Emerging micro-additive manufacturing technologies enabled by novel optical methods. *Photonics Research, 8*(12), 1827–1842.
4. Zhu, L., Cheng, J., & Zhou, H. (2000). Research of rapid prototyping process using linear array of high power laser diodes. In *In high power lasers in manufacturing, Proceeding of SPIE* (p. 3888).

Chapter 132
Shaping in Manufacturing

Shaping is a change obtained in the external shape of a raw material when the raw material is processed. Raw material is a material or a part that is not yet processed. There are many shaping techniques. To categorize them, a simple algebra of addition and subtraction can help.

There are the following three shaping categories based on a change in the amount of a raw material:

1. Subtraction: the product weighs less than the raw material; the technique is subtractive.
2. No change: the product weighs similar to the raw material; the technique is formative.
3. Addition: the product weighs more than the raw material; the technique is additive.

Subtractive techniques are machining, drilling, honing, etc. For some shaping techniques, their category is not straightforward, whether they are formative or additive. Let us see which technique belongs to which category.

- **Casting:** Casting has dual loyalty. It comes under both additive and formative techniques.
- In casting, a product is formed when liquid from a container is poured into a mold. The solidified liquid is a product. The liquid contains a number of liquid drops. The product is made not due to one drop but because of the aggregation of many drops. The product has more material than a drop has. This satisfies the condition of an additive process where the product must have more weight or material than an individual raw material has. Here, the individual raw material is a drop of liquid. Therefore, it is categorized as an additive technique [1].
- The casting would not have been additive if the liquid in the container could not have been visualized as broken into many drops. Then, the container of the liquid

S. Kumar, *A Concise Encyclopedia of Additive Manufacturing*,
https://doi.org/10.1007/978-3-031-98394-8_132

could be considered as a unit. And the product could have been considered as a transformation of the unit that is liquid into another form that is solid, which could have made the technique formative and not additive.

- Thus, casting can be considered a formative technique when it is considered that liquid as a drop plays a less significant role than the liquid as a whole unit. Because the addition of liquid is not independent of the total space provided by a mold. The mold either allows making a product utilizing the total space or does not allow making any product at all. That is, the mold either allows utilizing the whole liquid as a unit or does not allow utilizing any of its drops. The mold does not recognize drops as a building block. If it could have recognized, the shape and size of the product would not have been a constant outcome but would have varied by varying the number of drops. In the absence of the drop as a building block, what is left is the liquid as a unit as a building block. Therefore, the raw material is not a drop but the total liquid. Therefore, the comparison should not be made with the drop but with the total liquid. When a comparison is made between the product and the unit, there is no change in the amount of the material, and therefore, the technique is formative. Therefore, casting is categorized as a formative technique [2–5].
- What if casting happens differently? Let us consider a mold that is vertically long, has many contours, and is open. When liquid is poured into it to fill the fraction of it, the liquid is allowed to solidify. It makes a product. The product can be removed from the mold and can be used. If it is not removed from the mold and more liquid is poured onto it, the liquid rises to reach another contour of the mold, making another product that can be removed and used. Thus, the shape of a product (or the product) depends upon the number of times the liquid is poured. This making of the product happens by addition, because with the addition of more materials, new contours are added, giving rise to a new product. A new product is the result of how many fractions of liquid are added. The product always has more materials than a fraction of liquid that was added after some interval—this is a comparison to verify the technique type. This type of casting will fit more in additive techniques than in formative techniques.
- Therefore, whether casting fits in a particular technique type depends not on what casting is but on what casting is used for. This gives an opportunity to fit the casting where it is more convenient to be fitted in the presence of other techniques. It means it needs to be known what other techniques there are and whether moving this (casting) technique from one category to another category gives inconvenience to other techniques.
- Let us categorize casting as an additive technique. Then, a question comes up as to whether casting adds layer-by-layer or not. The question is asked because additive techniques are defined by those techniques that add layer-by-layer [6, 7]. If the question were asked whether it adds, it could have been proved that it adds, citing the above-mentioned example. But if the question is asked whether it adds layerwise, proving that it certainly adds is not sufficient.

- It is not time to ask why a definition was created for a particular action (layer-by-layer) and the name was given for a general action (additive one). Is this not an appropriation? If the name were, for example, 3D printing instead of additive manufacturing, there would not have been such appropriation. What will happen if subtractive manufacturing is defined as a technique that subtracts only layer-by-layer, or if formative manufacturing is defined as a technique that fills a mold only layer-by-layer or fills with a high or low pressure, etc.?

- Before the definition came, casting could have been fitted in additive techniques. If it were not fitted, it was because casting as a formative technique was more logical, not because the addition was an addition only when it happened layer-wise. After the definition came, casting is not an additive technique; therefore, it has only one choice left: to be a formative technique.

- Why not beat the definition and prove that the type of casting that adds fraction-by-fraction is additive manufacturing? When liquid is poured into the mold after each interval, the amount of liquid makes a layer, and therefore, the addition goes layer-by-layer. But a computer model of a product is needed, and the layer formation should be controlled by a computer as per the model as well. This can happen by taking a CAD model of the internal cavity of the mold, and the amount of liquid that is added to fulfill one layer can be computer-controlled using this model. But the definition says this should not be a formative technique, as the liquid must not be confined with the help of the mold wall. To beat this condition, the mold will be considered as a substrate or a platform. And the liquid will be deposited to fill the substrate as per the model. This will prove that this casting fulfills all conditions of AM, and, therefore, casting is AM. Mold can be taken as a substrate, as the definition does not define a substrate, and therefore, the definition does not exclude a mold as a substrate.

- Use of a substrate or platform to make a part on it is an essential requirement of AM. If AM uses a substrate, AM is not completely free from being a formative technique. If casting is a formative technique because it confines the material from all sides, AM is also partially formative because it confines the material from one side. AM, though being additive, does not have the freedom to add material through the substrate, in the same way casting does not have the freedom to add material through the walls of the mold. AM could be really additive if materials were thrown in the air layer-by-layer, resulting in a product hanging in the air. If casting is excluded from AM techniques because it is not AM, AM should also be excluded from AM techniques because it is also not completely AM. However, some AM techniques are coming that make parts without a substrate, hanging in a medium (of liquid, not air) [8, 9].
- **Welding:** Welding adds [10], but it needs to be known whether the addition leads to shaping. If a wire is broken and the wire is welded to join, this addition does not change the shape. Therefore, it cannot be considered a shaping technique.

- In another case, if welding is used to join two rectangular sheets, where one sheet is normal to another sheet, making a structure "T," it is a shaping technique.

Because welding changed the shape of the raw material (sheet) to "T." It is an additive technique because the weight of the structure is more than an individual sheet. Again, it is not an additive technique because it is not a layer-by-layer technique. If there is a categorization of shaping techniques in which there are only three categories, it will not come under any of the three categories. Because it is not a subtractive technique, it is again not a formative technique, and it must not come under additive technique because its addition does not follow either layerwise or voxelwise.

- Since the majority of times, welding is used for joining and not for making a shape that does not follow layerwise, welding as a technique should be excluded from being a shaping technique. It will make the categorization an approximate one and not a perfect one but will let retain the simplicity of the categorization. The categorization will be approximate because it is not meant to include a technique depending on what it is able to do but on what it usually does. For example, welding is not fit to be included depending on its ability to make a "T" structure, which it usually does not do. This means any technique that cannot be subtractive or formative is not part of the categorization if the technique does not fall under the definition. If welding is used in such a way that it falls under the definition, it is AM [11].
- Welding, like other techniques such as soldering, brazing, etc., is classified as a joining technique in manufacturing, but this classification does not answer whether they are shaping techniques.
- **Cladding**: Cladding is used to make a thin layer on a part surface, e.g., to protect it [12–14]. The addition of a thin layer does not change the part shape, and therefore, it is not a shaping technique. It can be a shaping technique if the layer is so thick that it is not able to follow the up and down of the contour of the part surface, or if the layer does not conform to the part surface, bringing a change in the surface contour. It could be part of a categorization that was there to show how any technique can sometimes play different roles. However, it is not part of an approximate categorization.
- **Heat treatment**: it does not change the external structure of the raw material, and therefore, it is not a shaping technique. If it changes the shape, it is called heat-induced forming, e.g., hot forming, forming [15], forging, incremental sheet forming [16], etc., which are formative techniques.

The shaping [17] and non-shaping techniques are given in Table 132.1.

Table 132.1 Shaping and non-shaping techniques in manufacturing

Materials	Technique name	Examples
Shaping techniques		
Added	Additive	3D printing, layerwise, voxelwise
Subtracted	Subtractive	Machining, drilling, honing, sawing, milling, turning, cutting, grinding, finishing for shape change
No change	Formative	Casting, sintering, forming, stamping, forging, injection molding, extrusion, bending, sand casting, die casting, vacuum molding, hot isostatic pressing, rolling, drawing
Non-shaping techniques		
Joining, welding, soldering, brazing, coating, chemical vapor deposition, cladding, heat treatment, spraying, annealing, bonding, finishing, etc., for no shape change		

References

1. Kosky, P., Balmer, R., Keat, W., & Wise, G. (2013). Manufacturing engineering. In *Exploring engineering* (pp. 205–235). Elsevier.
2. Luo, H., Zhang, Y., Yu, J., et al. (2023). Additive, subtractive and formative manufacturing of glass-based functional micro/nanostructures: A comprehensive review. *Materials & Design, 233*, 112285.
3. DeBoer, B., Nguyen, N., Diba, F., & Hosseini, A. (2021). Additive, subtractive, and formative manufacturing of metal components: A life cycle assessment comparison. *The International Journal of Advanced Manufacturing Technology, 115*(1), 413–432.
4. Zivanovic, S. T., Popovic, M. D., Vorkapic, N. M., et al. (2020). An overview of rapid prototyping technologies using subtractive, additive and formative processes. *FME Transactions, 48*(1), 246–253.
5. Yoon, H. S., Lee, J. Y., Kim, H. S., et al. (2014). A comparison of energy consumption in bulk forming, subtractive, and additive processes: Review and case study. *International Journal of Precision Engineering and Manufacturing-Green Technology, 1*, 261–279.
6. Pérez, M., Carou, D., Rubio, E. M., & Teti, R. (2020). Current advances in additive manufacturing. *Procedia CIRP, 88*, 439–444.
7. Rashid, A. (2019). Additive manufacturing technologies. In S. Chatti & T. Tolio (Eds.), *CIRP encyclopedia of production engineering*. Springer, Berlin.
8. Kelly, B. E., Bhattacharya, I., Heidari, H., et al. (2019). Volumetric additive manufacturing via tomographic reconstruction. *Science, 363*(6431), 1075–1079.
9. Loterie, D., Delrot, P., & Moser, C. (2020). High-resolution tomographic volumetric additive manufacturing. *Nature Communications, 11*(1), 852.
10. Chaturvedi, M., & Vendan, S. A. (2022). *Advanced welding techniques*. Springer.
11. Singhal, T. S., Jain, J. K., Kumar, M., et al. (2024). A comprehensive comparative review: Welding and additive manufacturing. *International Journal on Interactive Design and Manufacturing (IJIDeM), 18*(3), 1829–1843.
12. Saha, M. K., & Das, S. (2016). A review on different cladding techniques employed to resist corrosion. *Journal of the Association of Engineers, 86*(1–2), 51–63.
13. Ko, J., Kim, J. W., Min, H. W., et al. (2022). Review of manufacturing technologies for coated accident tolerant fuel cladding. *Journal of Nuclear Materials, 561*, 153562.

14. Mohamed, G. H., Karuppasamy, K., Alrwashdeh, M., et al. (2024). Unveiling the latest progresses in chromium-coated Zircaloy cladding ATF materials: Fabrication techniques, performance metrics, and beyond. *Alexandria Engineering Journal, 108*, 13–39.
15. Das, B., & Biswas, P. (2018). A review of plate forming by line heating. *Journal of Ship Production and Design, 34*(02), 155–167.
16. Liu, Z. (2018). Heat-assisted incremental sheet forming: A state-of-the-art review. *The International Journal of Advanced Manufacturing Technology, 98*(9), 2987–3003.
17. Jaksch, A., Cholewa, S., Bartels, D., et al. (2025). Introduction to powder and beam based additive manufacturing. In D. Drummer & M. Schmidt (Eds.), *Progress in powder based additive manufacturing* (Springer tracts in additive manufacturing). Springer.

Chapter 133
Slurry Bed Process

Slurry is a mixture of solid and liquid in which the solid is suspended in the liquid. To suspend solid particles, a dispersant is required. The dispersant gets adsorbed on the surface of solid particles and does not let them come in contact with each other to agglomerate. Increasing more dispersant is counterproductive as it decreases dispersion and increases agglomeration by bonding solid particles with dispersant polymer chains.

The liquid medium is called a solvent that may be water or other polymeric solutions (ethanol, acetic acid, etc.). But water is not suitable as a solvent for some materials that get oxidized or solidified when mixed with it [1]. Slurry requires a plasticizer to increase its fluidity. Besides, it requires a binder to hold particles together when dried.

The slurry bed process has the following advantages over powder bed fusion (PBF):

1. It is not convenient to make a layer using sub-micron-sized powders, as they repel each other due to electrostatic forces. Using slurry made from them helps make a thin layer not achievable in PBF.
2. PBF does not work well with powders having broad size distribution, non-spherical shape, or high differences in density, as spreading these powders is affected by their properties, while spreading a slurry made by them is not affected by their properties.
3. A powder layer possesses lower density. Using slurry instead will increase the layer density. An initial higher density of the layer may help achieve higher part density.

In this process, with the help of a doctor blade, slurry is laid on a bed, which is then dried so the deposition of the next layer should not deform it (Fig. 133.1). For an aqueous-based slurry, drying can happen by raising the substrate temperature to more than the boiling point of water. Instead of taking a slurry and making the bed,

S. Kumar, *A Concise Encyclopedia of Additive Manufacturing*, https://doi.org/10.1007/978-3-031-98394-8_133

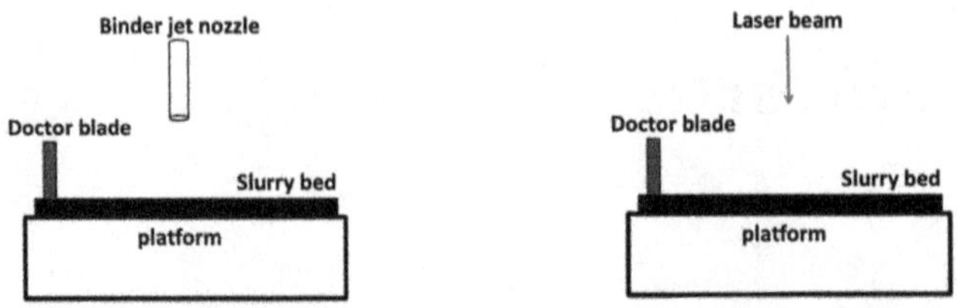

(a) Slurry bed laid by a doctor blade and shaped by jetted binder (b) Slurry bed laid by a doctor blade and shaped by an l-beam

Fig. 133.1 Shaping in slurry bed process due to: (**a**) binder jetting, (**b**) scanning by an l-beam

the slurry can also be made in-situ by depositing powder and liquid using separate nozzles on the platform, followed by spreading using a doctor blade [2].

Without increasing the temperature, a layer can also dry by capillary action when the solvent is absorbed by previous layers. The dried layer is selectively treated by an l-beam [3] or a binder jet [4] to create a pattern on it. The processed part, known as a green part, goes through post-processing to improve its properties [5].

For smooth spreading of a layer, viscosity needs to be small. But if it is small, there will be a higher liquid content in the slurry, which needs to be removed during drying—this may give rise to cracks in the slurry. Low viscosity implies there is no high content of solid particles, which gives rise to low density of the layer and then of the part. Low density means low strength which may not be desired unless a porous part is needed for some applications.

A dried slurry layer, unlike a powder layer, has particles bonded with each other. This bonding is not meant to stay permanently but to get dissolved during post-processing. The layer needs to be additionally treated in the same way as treated in PBF to make a pattern on it by creating additional bonds among slurry particles at selected areas. These areas, being the constituent of a desired part, are not meant to be dissolved during post-processing. The additional bond is created by an l-beam, the same as in LPBF, when the beam partially melts slurry and vaporizes polymers. This process is known as laser slurry deposition [4]. The bond is also created by jetted binder, the same as in binder jet three-dimensional printing, when the binder fills in pores, locks separate particles, and permeates through the slurry layer to bond with an underlying layer—this process is known as laser slurry deposition-print [6].

Slurry mixed with photopolymer can be spread in the form of a bed to make a layer. Since the solid content (about 40 vol. %) in photopolymer is low [7] in comparison to the high-content process [8], the low-content process is nearer to the photopolymer bed process rather than the slurry bed process. This type of low-content slurry is conveniently processed akin to photopolymers. The low solid content means the low deflection of the l-beam, allowing an adequate curing thickness more than a layer thickness. This adequate thickness needs to be obtained to ensure sufficient bonding between two layers. The method of bonding is curing by an l-beam, similar to what happens in stereolithography [7].

References

1. Wu, W., Liu, W., Jiang, J., et al. (2019). Preparation and performance evaluation of silica gel/tricalcium silicate composite slurry for 3D printing. *Journal of Non-Crystalline Solids, 503–504*, 334–339.
2. Nguyen, H. X., Suen, H., Poudel, B., et al. (2020). Development of an innovative, high speed, large-scaled, and affordable metal additive manufacturing process. *CIRP Annals, 69*(1), 177–180.
3. Muhler, T., Gomes, C., Ascheri, M., et al. (2015). Slurry-based powder beds for selective laser sintering of silicate ceramics. *Journal of Ceramic Science and Technology, 06*(02), 113–118.
4. Zocca, A., Lima, P., Diener, S., et al. (2019). Additive manufacturing of SiSiC by layerwise slurry deposition and binder jetting (LSD -print). *Journal of the European Ceramic Society, 39*(13), 3527–3533.
5. Wang, H. R., Cima, M. J., Kernan, B. D., & Sachs, E. M. (2004). Alumina-doped silica gradient-index (GRIN) lenses by slurry-based three-dimensional printing (S-3DP™). *Journal of Non-Crystalline Solids, 349*, 360–367.
6. Hoffmann, M., Schubert, N. H., Günster, J., et al. (2025). Additive manufacturing of glass-ceramic dental restorations by layerwise slurry deposition (LSD-print). *Journal of the European Ceramic Society, 45*, 117235.
7. Harrer, W., Schwentenwein, M., Lube, T., & Danzer, R. (2017). Fractography of zirconia-specimens made using additive manufacturing (LCM) technology. *Journal of the European Ceramic Society, 37*, 4331–4338.
8. Binobaid, A., De Lisi, M., Camilleri, J., et al. (2025). Zirconia-calcium silicate bioactive composites for dental applications using DLP additive manufacturing. *Bioprinting, 45*, e00377.

References

Chapter 134
Slurry Deposition Process

In a slurry deposition process, where a bed is not formed by the slurry, the slurry is deposited from a nozzle (Fig. 134.1). Use of slurry helps make ceramic or metal components. If slurry is not used, there are no other ways but melting to make big metal components. Melting a high-melting-point material requires extra effort. For example, it requires control of microstructure development and crack mitigation, devices for melting, and dedicated systems for heat management—these all increase cost as well [1].

There are expensive systems based on powder or wire, requiring l-beams or e-beams. Melting low-melting-point materials such as a tin alloy [2] or an aluminum alloy [3] is not so difficult. There are processes available for depositing these materials [4], but parts made are weak and cannot be a substitute for high-strength parts. Hence, these processes are incapable, as they cannot be used for processing high-melting-point materials. Slurry deposition is a solution to these problems—it is neither as difficult as a deposition process related to high-melting-point materials nor as incapable as a process related to low-melting-point materials.

Slurry deposition can be a substitute for a deposition process related to high-melting-point materials. Slurry extruded on a platform should not be deformed by its own weight and should sustain its form when another layer is deposited on it. Higher solid loading provides strength to an extruded material but requires high force to extrude. The high force may break the continuity of the extruded material, which will cause defects in part fabrication [5].

Due to the presence of binder or gel in a slurry, extruded materials are bonded with other extruded materials or a previously deposited layer before they solidify. Gel is preferred in some materials (tricalcium silicate) because it prolongs solidification of extruded filaments, which gives filaments sufficient time to bond. Besides, gel increases the flow property of the slurry, which helps it to be extruded uniformly [6].

S. Kumar, *A Concise Encyclopedia of Additive Manufacturing*,
https://doi.org/10.1007/978-3-031-98394-8_134

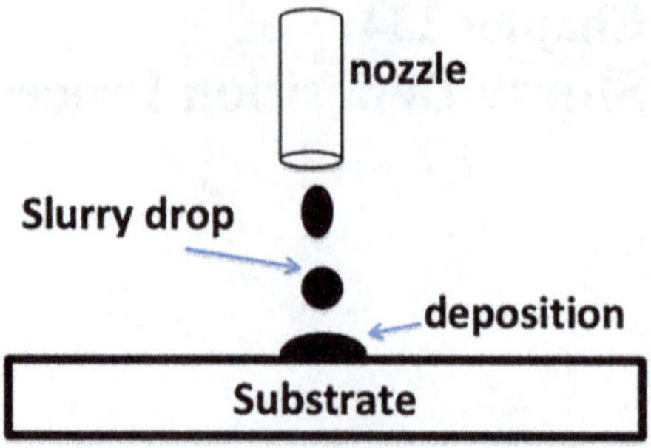

Fig. 134.1 Schematic diagram of slurry deposition

References

1. Sames, W. J., List, F. A., Pannala, S., et al. (2016). The metallurgy and processing science of additive manufacturing. *International Materials Reviews, 61*(5), 1–46.
2. Vega, E. J., Cabeza, M. G., Muñoz-Sánchez, B. N., et al. (2014). A novel technique to produce metallic microdrops for additive manufacturing. *International Journal of Advanced Manufacturing Technology, 70*, 1395–1402.
3. Zuo, H., Li, H., Qi, L., & Zhong, S. (2016). Influence of interfacial bonding between metal droplets on tensile properties of 7075 aluminum billets by additive manufacturing technique. *Journal of Materials Science and Technology, 32*(5), 485–488.
4. Fang, X., Wei, Z., Du, J., et al. (2017). Forming metal components through a fused-coating based additive manufacturing. *Rapid Prototyping Journal, 23*(5), 893–903.
5. Tang, S., Yang, L., Li, G., et al. (2019). 3D printing of highly-loaded slurries via layered extrusion forming: Parameters optimization and control. *Additive Manufacturing, 28*, 546–553.
6. Wu, W., Liu, W., Jiang, J., et al. (2019). Preparation and performance evaluation of silica gel/tricalcium silicate composite slurry for 3D printing. *Journal of Non-Crystalline Solids, 503–504*, 334–339.

Chapter 135
Solid Freeform Fabrication

AM is called solid freeform fabrication [1] or freeform fabrication [2] because it can fabricate a solid object having a form (shape) that is free from any limitation. Other manufacturing processes offer limited shapes and are therefore not called solid freeform fabrication.

For example, in injection molding (IM), the shape of the cavity of the mold restricts the form of a solid object fabricated; the object therefore has no freedom to take any form other than that given by the mold.

1 Disadvantage of Mold-Based Process

Complexity in IM depends upon how complex a mold can be formed, how well the mold can be filled up during injection, and how well the mold allows the injected material to be ejected [3]. If there is incompetency at any stage of the process, it will affect the maximum complexity of the solid form that can be achieved.

Mold has disadvantages—if a thin wall or a small pin needs to be made, a mold having a thin cavity is required that is not possible to fabricate with ease. Similarly, if a narrow hole needs to be made, it may not be possible to make a mold having a narrow protrusion for such a requirement. Even in that type of mold, it is not possible to make objects easily [4, 5]. Even if it is possible to make objects from such molds, the molds will not last long enough to justify cost investment. Thus, if there is a design that consists of such features, IM is not the right choice to make such objects.

S. Kumar, *A Concise Encyclopedia of Additive Manufacturing*, https://doi.org/10.1007/978-3-031-98394-8_135

2 Advantage of Mold-Based Process

If the shape of an object in IM were not limited by the shape of the mold, it could have the freedom to make any shape, but the freedom would not come free. The confining space of the mold has the advantage that it does not let material go astray, which gives a better surface finish of a product than AM does [6]. In the absence of the confining space, the right parameters need to be searched for the material to remain confined within a designated space. But due to the confining space offered by the mold, materials can be stacked rather fast. Otherwise, every bit of the material requires to be checked whether the material is at the right place before solidification or whether it has not shifted due to shrinkage after the solidification. When every bit of material takes attention, the process will not be as fast as in the presence of a confining space where such attention is not required. This is one of the reasons for the slowness of AM, where every bit of material addition needs attention.

If there is something (e.g., mold) that gives a confining space, then after the fabrication of one object, another object can be fabricated, and so on—this can go fast. This advantage is absent when there is no mold (or no confining space as an outcome). In AM, each object needs to be grown, even if the object to be fabricated is always of the same shape—there is no certainty that the two objects fabricated will have no difference due to the growth. Thus, a mold gives an advantage of repeatability. If the same object always needs to be fabricated, IM is a preferred choice [7]. This is the advantage gained with mold-based process.

Thus, mold has given advantages. It has made IM faster than AM. It has given repeatability [8], which has caused IM rather than AM to be a suitable process for mass production.

But if a customized product rather than mass production is a priority, if fabrication of a product is more important than the speed of the fabrication, and if an old product having guaranteed property due to the confining space is less preferred than an innovative product, AM is better than IM.

References

1. Alimardani, M., Fallah, V., Iravani-Tabrizipour, M., & Khajepour, A. (2012). Surface finish in laser solid freeform fabrication of an AISI 303L stainless steel thin wall. *Journal of Materials Processing Technology, 212*(1), 113–119.
2. Shu, X., Chen, G., Wang, C., et al. (2025). Mechanism, microstructure and performance of novel electron beam freeform fabrication compared with conventional method. *Vacuum, 233*, 113912.
3. Masato, D., Sorgato, M., & Lucchetta, G. (2021). A new approach to the evaluation of ejection friction in micro injection molding. *Journal of Manufacturing Processes, 62*, 28–36.
4. Song, M. C., Liu, Z., Wang, M. J., Yu, et al. (2007). Research on effects of injection process parameters on the molding process for ultra-thin wall plastic parts. *Journal of Materials Processing Technology, 187*, 668–671.

5. Glojek, A., Milfelner, M., & Tavčar, J. (2025). Mold design and injection molding simulations for polymer gear. In *Polymer gears* (pp. 301–331). Elsevier.
6. Matetich, C., & Vold, J. (2024). A comparison of additive manufactured to injection molded Martian regolith based polymer matrix composites. *Advances in Space Research, 74*(8), 3997–4008.
7. Ituarte, I. F., Coatanea, E., Salmi, M., et al. (2015). Additive manufacturing in production: A study case applying technical requirements. *Physics Procedia, 78*, 357–366.
8. Wesemann, C., Spies, B. C., Schaefer, D., et al. (2021). Accuracy and its impact on fit of injection molded, milled and additively manufactured occlusal splints. *Journal of the Mechanical Behavior of Biomedical Materials, 114*, 104179.

Chapter 136
Speed: Methods to Increase

In the deposition process, a high fabrication rate is achieved by increasing the number of nozzles [1], by increasing material deposition [2–5], by developing the material so the material should solidify early [6], etc. In powder bed fusion (PBF), there are a number of methods by which a high rate can be achieved, which are given.

1 Increasing Layer Thickness

Increasing layer thickness decreases the number of layers required to make a part [7]. Therefore, the number of times powder is deposited to make all layers decreases, which in turn decreases the fabrication time and energy consumption [8]. However, this method has a limitation—increasing it more than a certain thickness does not let the effect of melting reach the base of the layer [9]—powders either bind by solid-state sintering if build continues for long or remain unaffected. Thus, there is a limitation to which the rate can be enhanced by this method.

2 Dividing Processing Area

Dividing a processing area into a number of zones and processing all zones simultaneously increases the rate [10]. The layer is processed by a number of beams in parallel (Fig. 136.1a). Fabrication time is proportional to the total area divided by the number of beams when each beam scans an equal area. Therefore, the time decreases with an increase in the number of beams—this method can lead to mass production [11].

The reason for an increase in the rate is the addition of efforts. The effort is to join materials, which is limited by the size of a machine, as a small machine can

S. Kumar, *A Concise Encyclopedia of Additive Manufacturing*,
https://doi.org/10.1007/978-3-031-98394-8_136

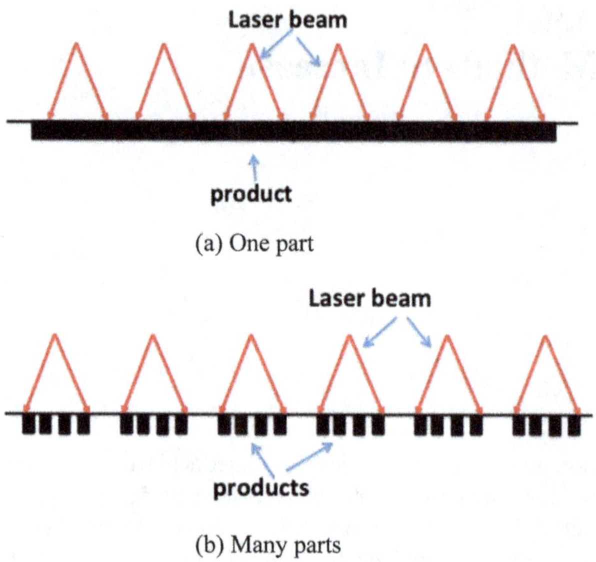

Fig. 136.1 Application of a number of beams for making (**a**) one part, (**b**) many parts

accommodate only a small number of beams. Besides, this method is limited by managing a sequence of operations involving many beams. It is further limited by sustainability—if one beam fails, the beam needs to be fixed before all beams can be used together.

The method is again limited by cost-effectiveness and logic—why not make many machines each equipped with a single laser rather than a single machine having many lasers? As the percentage of the cost of the laser is high compared to the total cost of a simple laser-based machine, it would not be much expensive to have many machines than a single big multi-laser machine. The presence of many machines will allow working with different materials in separate machines and making different products. However, big multi-laser machines are made [12–14].

The method has limitations for making a small part where the size of the boundary between two zones is not significantly smaller than the area of the zones when two zones are processed by separate beams. Different sizes of melt pools will be formed by different beams that have small variations or noise. For a small part, unlike a big part, this variation will be significant.

For making a number of small parts in PBF that is equipped with a number of beams to employ parallel scanning (Fig. 136.1b), the task of processing will be shared equally among each beam [15]. It will increase the overall fabrication rate but will neither increase the rate for a single part nor require synchronization of beams to process the part (as a part is processed by only one beam). The single part thus fabricated in a multi-beam setting will not be different from a part if it is fabricated in a similar setting but using a single beam.

Figure 136.1a shows beams from six laser sources contribute to make one part, while Fig. 136.1b shows beams from six laser sources; they each scan equal space (1/6 of the space) and make four parts each.

For scanning a layer by parallel beams, if their heat-affected zones are overlapping, the development of thermal stress across the layer will be minimal. Thus, parallel scanning increases the production rate while minimizing the thermal stress.

3 Optimizing Process

If high strength is not required for all sections of a part, the section that does not require the high strength will be scanned with a fast scan speed, which will increase the fabrication rate [16]. Alternatively, the section can also be scanned with less overlap between two adjacent scans, which will decrease the number of scans required to cover the section, causing a decrease in scanning time. Besides, if instead of high strength an optimum strength is required, there can be several build rates [17].

Hence, optimizing process parameters as per the requirement of different sections will not only pave the way for increasing the rate but also decrease the energy required. The type of gas used in a processing chamber can be one of the process parameters, as some gas allows obtaining high density at a high rate and thus help increase the fabrication rate [18].

If various beam diameters are available, using the large diameter will decrease the time to scan the whole layer and increase the fabrication rate [19]. Increasing the roller speed can be another option [20].

4 Changing Orientation

A part can be made in many ways by changing its orientation (changing the angle between its features and build direction) [21]. If changing the orientation results in a decrease in the number of layers required for the completion of the build, fabrication time will decrease. However, this method has limitations: a decrease in mechanical properties (heat accumulated on a top layer will not be the same in all orientations), need for support structures [22], change in the grain size (length of a columnar grain will not be the same in all orientations), property changes [23], and a decrease in surface roughness and geometrical accuracy.

The effect of the orientation change is shown in Fig. 136.2 for making a hollow rectangular part, which is open at one side surface (to let powder drain out after fabrication). There will be an increase in the fabrication rate with a decrease in the height of the build, but it will result in a big overhang (Fig. 136.2a). Changing the orientation by 90° (Fig. 136.2b) will increase the height of the build, which will decrease the rate. But it will decrease the size of an overhang required to be built and, therefore, be free from difficulties in making a big overhang. Hence, the

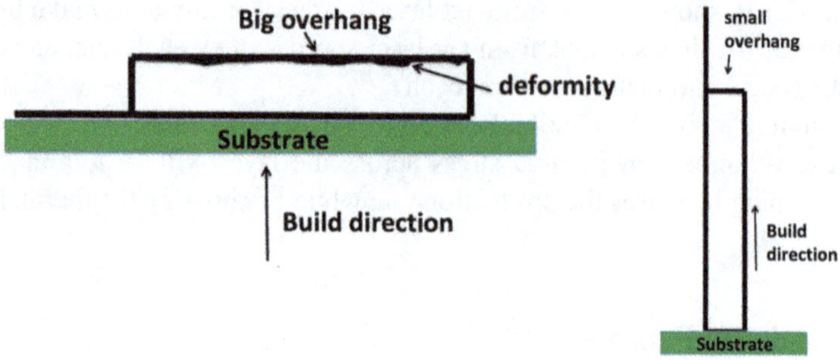

(a) Big overhang (b) Small overhang

Fig. 136.2 Different orientations: (**a**) big overhang, (**b**) small overhang

orientation (Fig. 136.2a) furnishes a lower surface finish for a big area (underneath the big overhang), while the orientation (Fig. 136.2b) furnishes a lower surface finish for a small area (underneath the small overhang). Besides, making a big overhang without a support structure means losing the planarity of the overhang surface, causing a decrease in the dimensional accuracy.

5 Increasing Scan Speed

A scan speed can be termed low or high depending upon materials and experiments. A low scan speed of 689 mm/s is considered high when it gives porous parts [24].

Increasing scan speed increases the fabrication rate. Though the highest speed used is limited by the following:

1. Evaporation of powders rather than melting—at high speed of a high-power l-beam, the top section of a powder accumulates a high amount of heat, leading to its evaporation.
2. Formation of a longer melt pool that breaks into shorter pools, giving rise to discontinuity, which is due to Rayleigh instability states that if the length-to-width ratio is more than π (a critical value depending upon interface), the pool will break. Longer pool will no longer be long if it solidifies on the way, i.e., if the solidification rate is faster than or equal to the scan speed. For an aluminum alloy scanned at 1 m/s, the solidification rate is almost the same [25]. The cooling rate remains high in an experimental setup, giving a high solidification rate—this is one of the reasons why Rayleigh instability is not the main problem for consolidation.

Which one of the above two reasons can be the main reason for failure at high scan speeds (>100 m/s)? At a certain high speed, only evaporation takes place, and there will be no melt pool. In the absence of any melt pool, the first reason is the cause of failure. Increasing the pressure of a processing chamber will let increase the speed, as the pressure suppresses the evaporation—more pressure means higher speed. There is a limit beyond which application of the pressure does not work—either due to its effect on the melt pool, such as depression, or due to the safety aspect involved with working at high pressure.

Increasing the speed further without increasing the pressure will bring forth a result no more different than evaporation unless laser power is not increased proportionally. In the absence of increased laser power, there will be neither evaporation nor melting, but just heating. With a decrease in speed, there will be a certain speed at which evaporation will stop and melting will start. If the solidification rate is similar to the speed, this mechanism will be able to furnish a part. If the solidification rate is low, two cases will happen: (a) part formation will fail due to Rayleigh instability, or (b) part formation will not fail if some measures are taken.

One measure is increasing cooling rate by decreasing the temperature of a substrate, which will increase solidification rate, but a part may still fail due to the formation of cracks. Another measure is to employ laser pulses (~ ms in order to prevent evaporation) instead of a continuous l-beam, which will create smaller pools. Applying the mechanism at high speed will leave few options to exercise for making a defect-free part, decreasing the freedom to vary microstructures and properties.

3. High speed requires high laser power to maintain the required laser energy density for melting [26]. High laser power means a high cost for a machine. Working with high laser power means managing high heat generated, which requires upgrading a device (laser, mirror, etc.) and incorporating a new cooling device to safeguard the machine.

   In EPBF, there will not be any damage to scanning devices at high speed, but the lifetime of electrodes is diminished by the constant generation of high power.

6 Adopting Areawise Scanning

Pointwise scanning means scanning using a point source (such as laser spot size), which takes time for processing a layer—point source has to process one point on PB then it has to move to another point. In fast pointwise scanning, the processing moves fast but has to cover many points, which is the limitation.

Changing from pointwise to linewise or areawise scanning is another method to increase the rate [27]. In the linewise scanning, a number of l-beam sources (diode emitters) are fixed together on a horizontal line, and the line moves to process layers. In the areawise scanning, all heat point sources are fixed in an area over PB and process the bed simultaneously without either movement of the heat source or deflection of heat radiation [28]. In the case of photopolymer, the whole area can be cured simultaneously [29].

7 Material

Using a material that can be consolidated without requiring a high-power heat source allows speeding up the processing. For example, a polymer, i.e., a low melting point material, requires low energy to be melted. This low requirement allows developing systems for linewise or areawise scanning, which speeds up the processing. Besides, this low requirement can be fulfilled even if a laser source is not used, e.g., using infrared rays to fuse polymers in multi-jet fusion [30, 31].

References

1. Chauvette, J. F., Brzeski, D., Hia, I. L., et al. (2021). High-speed multinozzle additive manufacturing and extrusion modeling of large-scale microscaffold networks. *Additive Manufacturing, 47*, 102294.
2. Büßenschütt, K., Köhnen, P., Kies, F., et al. (2024). High-speed direct energy deposition as a high-throughput design tool for laser-based additive manufacturing. *Additive Manufacturing Letters, 8*, 100188.
3. Schaible, J., Sayk, L., Schopphoven, T., et al. (2021). Development of a high-speed laser material deposition process for additive manufacturing. *Journal of Laser Applications, 33*(1), 012021.
4. Wang, K., Du, D., Liu, G., et al. (2021). A study on the additive manufacturing of a high chromium nickel-based superalloy by extreme high-speed laser metal deposition. *Optics & Laser Technology, 133*, 106504.
5. Huang, W., Zhong, H., Li, X., et al. (2025). Formation mechanism of hump defects in high-speed laser wire additive manufacturing. *Optics & Laser Technology, 183*, 112239.
6. Brzeski, D., Hia, I. L., Chauvette, J. F., et al. (2021). Design of thermoset composites for high-speed additive manufacturing of lightweight sound absorbing micro-scaffolds. *Additive Manufacturing, 47*, 102245.
7. Shi, X., Ma, S., Liu, C., et al. (2016). Performance of high layer thickness in selective laser melting of Ti6Al4V. *Materials, 9*, 975.
8. Shamsdini, S., Shakerin, S., Hadadzadeh, A., et al. (2020). A trade-off between powder layer thickness and mechanical properties in additively manufactured maraging steels. *Materials Science and Engineering A, 776*, 139041.
9. Leicht, A., Fischer, M., Klement, U., et al. (2021). Increasing the productivity of laser powder bed fusion for stainless steel 316L through increased layer thickness. *Journal of Materials Engineering and Performance, 30*, 575–584.
10. Yu, K. S., Cheng, C. W., Lee, A. C., et al. (2022). Additive manufacturing of NdFeB magnets by synchronized three-beam laser powder bed fusion. *Optics & Laser Technology, 146*, 107604.
11. Kumar, S. (2022). Mass production. In *Additive manufacturing solutions* (pp. 145–167). Springer.
12. https://nikon-slm-solutions.com/slm-systems/nxg-xii-600/
13. https://velo3d.com/sapphire-xc/
14. https://www.amproinnovations.com/sp1200
15. Wong, H., Dawson, K., Ravi, G. A., et al. (2019). Multi-laser powder bed fusion benchmarking—Initial trials with Inconel 625. *International Journal of Advanced Manufacturing Technology, 105*, 2891–2906.

16. Kniepkamp, M., Harbig, J., Seyfert, C., & Abele, E. (2018). Towards high build rates: Combining different layer thicknesses within one part is selective laser melting. In *2018 international solid freeform fabrication symposium* (pp. 2286–2296).
17. Schwerz, C., Schulz, F., Natesan, E., & Nyborg, L. (2022). Increasing productivity of laser powder bed fusion manufactured Hastelloy X through modification of process parameters. *Journal of Manufacturing Processes, 78*, 231–241.
18. Pauzon, C., Forêt, P., Hryha, E., et al. (2022). Argon-helium mixtures as laser-powder bed fusion atmospheres: Towards increased build rate of Ti-6Al-4V. *Journal of Materials Processing Technology, 279*, 116555.
19. Sow, M. C., De Terris, T., Castelnau, O., et al. (2020). Influence of beam diameter on laser powder bed fusion (L-PBF) process. *Additive Manufacturing, 36*, 101532.
20. Chen, H., Cheng, T., Li, Z., et al. (2022). Is high-speed powder spreading really unfavourable for the part quality of laser powder bed fusion additive manufacturing? *Acta Materialia, 231*, 117901.
21. Zhang, Y., Bernard, A., Harik, R., et al. (2017). Build orientation optimization for multi-part production in additive manufacturing. *Journal of Intelligent Manufacturing, 28*(6), 1393–1407.
22. Calignano, F. (2014). Design optimization of supports for overhanging structures in aluminum and titanium alloys by selective laser melting. *Materials and Design, 64*, 203–213.
23. Barba, D., Alabort, C., Tang, Y. T., et al. (2020). On the size and orientation effect in additive manufactured Ti-6Al-4V. *Materials and Design, 186*, 108235.
24. Gong, X., Lydon, J., Cooper, K., & Chou, K. (2014). Beam speed effects on Ti–6Al–4V microstructures in electron beam additive manufacturing. *Journal of Materials Research, 29*(17), 1951–1959.
25. Tang, M. (2017). *Inclusions, porosity and fatigue of AlSi10Mg parts produced by selective laser melting*. PhD thesis,. Carnegie Mellon University.
26. Sun, Z., Tan, X., Tor, S. B., & Yeong, W. Y. (2016). Selective laser melting of stainless steel 316L with low porosity and high build rates. *Materials and Design, 104*, 197–204.
27. Gibson, I., Rosen, D., Stucker, B., & Khorasani, M. (2021). *Additive manufacturing technologies*. Springer.
28. Holt, N., Horn, A. V., Montazeri, M., & Zhou, W. (2018). Microheater array powder sintering: A novel additive manufacturing process. *Journal of Manufacturing Processes, 31*, 536–551.
29. Nguyen, H. X., Suen, H., Poudel, B., et al. (2020). Development of an innovative, high speed, large-scaled, and affordable metal additive manufacturing process. *CIRP Annals, 69*(1), 177–180.
30. Khorasani, M., MacDonald, E., Downing, D., et al. (2024). Multi jet fusion (MJF) of polymeric components: A review of process, properties and opportunities. *Additive Manufacturing, 91*, 104331.
31. Nazir, A., & Jeng, J. Y. (2020). A high-speed additive manufacturing approach for achieving high printing speed and accuracy. *Proceedings of the Institution of Mechanical Engineers, Part C: Journal of Mechanical Engineering Science, 234*(14), 2741–2749.

Chapter 137
Speed: Paradox

One of the advantages of AM is that it is fast. As a consequence, AM is termed rapid prototyping [1–2], rapid manufacturing [3–4], and rapid tooling [5–8]. One of the disadvantages of AM is that it is slow [9–11]. How can it be fast and slow at the same time? If it is fast, it is not slow, and vice versa. What is the paradox?

When AM makes a part, it takes time. This time can be long or short, depending on the time this time is compared with. If, after comparison, it is long, AM is slow. If, after comparison, it is small, AM is fast.

If AM makes a part, it takes time because it has to fulfill steps required to make a part: having a file, starting a machine, optimization of parameters, fabrication after some trials, post-processing if required, etc. If AM makes two parts instead of one, the time taken to make two parts will be twice or less than twice that of one part. If more parts are fabricated, time will increase more. But if the fabrication of these parts is compared with the fabrication due to an injection molding (IM) machine that produces them in thousands, AM takes a long time to fabricate. AM becomes slow because there is a comparison. If there were no comparison, AM would not have been dwarfed. But what is the comparison about? The comparison is about numbers, about quantity. In a given time, there are only one or two parts with AM, while there are thousands with IM. AM cannot compete with IM when the comparison is with only the number. AM can compete when there is something other than the number to compare with, such as variety of parts or the variety in quality.

When AM makes a part, it takes time. This time-taking is termed as slow because this is compared with IM. When AM makes another part with a change in design, it takes time. This time-taking needs to be compared with the time IM takes when design changes. The resulting comparison will enlighten whether this time-taking should be termed slow or not. When the design changes, IM cannot start making a part. IM waits for a new mold to be made. AM takes time for making a part, but this time-taking is still shorter when IM takes a long time for a mold to be ready. AM is slow but is faster when IM becomes slow. IM can make thousands of parts but could not make a single part when the competition was there to make a single part with a

S. Kumar, *A Concise Encyclopedia of Additive Manufacturing*,
https://doi.org/10.1007/978-3-031-98394-8_137

design change. IM could have made a thousand parts, but before it started, the competition was over.

AM is fast when a comparison is made with a design change. AM is slow when a comparison is made without a design change. AM can thus be made slower or faster by changing the reference frames of comparison.

References

1. Huang, Y., Lin, G., Huang, C., Li, L., et al. (2024). Application of laser 3D printing technology in rapid prototyping manufacturing. In *International conference on optics, electronics, and communication engineering (OECE 2024)* (Vol. 13395, pp. 22–27). SPIE.
2. Khoo, H., Allen, W. S., Arroyo-Currás, N., & Hur, S. C. (2024). Rapid prototyping of thermoplastic microfluidic devices via SLA 3D printing. *Scientific Reports, 14*(1), 17646.
3. Al Rashid, A., & Koç, M. (2023). Additive manufacturing for sustainability and circular economy: Needs, challenges, and opportunities for 3D printing of recycled polymeric waste. *Materials Today Sustainability, 24*, 100529.
4. Agrawaal, H., & Thompson, J. E. (2021). Additive manufacturing (3D printing) for analytical chemistry. *Talanta Open, 3*, 100036.
5. Saxena, P., Pagone, E., Salonitis, K., & Jolly, M. R. (2021). Sustainability metrics for rapid manufacturing of the sand casting moulds: A multi-criteria decision-making algorithm-based approach. *Journal of Cleaner Production, 311*, 127506.
6. Guerrero-Vacas, G., Gomez-Castillo, J., & Rodríguez-Alabanda, O. (2024). Manufacture of thermoplastic molds by fused filament fabrication 3D printing for rapid prototyping of polyurethane foam molded products. *Rapid Prototyping Journal, 30*(11), 32–49.
7. Nguyen, L., Buhl, J., Israr, R., & Bambach, M. (2021). Analysis and compensation of shrinkage and distortion in wire-arc additive manufacturing of thin-walled curved hollow sections. *Additive Manufacturing, 47*, 102365.
8. Gebhardt, A., & Hoetter, J. S. (2019). Rapid Tooling. In S. Chatti & T. Tolio (Eds.), *CIRP encyclopedia of production engineering*. Springer, Berlin.
9. Dejene, N. D., & Lemu, H. G. (2023). Current status and challenges of powder bed fusion-based metal additive manufacturing: Literature review. *Meta, 13*(2), 424.
10. Gutowski, T., Jiang, S., Cooper, D., et al. (2017). Note on the rate and energy efficiency limits for additive manufacturing. *Journal of Industrial Ecology, 21*(S1), S69–S79.
11. Saha, S. K., Wang, D., Nguyen, V. H., et al. (2019). Scalable submicrometer additive manufacturing. *Science, 366*(6461), 105–109.

Chapter 138
Stair-Stepping

During fabrication of a curved part, layers do not exactly coincide with the periphery of a curve, which causes gaps between layers and the curve. These gaps are inherent deficiencies and cannot be eliminated because a straight line will never coincide with a curve.

If a planar layer is replaced with a curved layer, gaps can be eliminated in some geometry [1]. But there is no industrial AM for curved layers, though there have been attempts with extrusion-based deposition by changing the build direction [2–4].

The cause of the staircase effect is the progress of a feature making an angle with a vertical direction. For example, if a right cylinder is made with its base on a platform, there will not be any staircase effect because the build direction does not make an angle with the central axis of the cylinder [5] (Fig. 138.1a). The cylinder is made up of horizontal planar layers with circular peripheries (discs). These discs are vertically aligned, constituting a cylinder. There is no gap and, therefore, no staircase effect.

But if the same cylinder is fabricated in a tilted position, horizontal planar layers or discs are required to fill up the tilted cylinder. Since discs grow in a vertical direction, they have vertical straight boundaries. These boundaries do not conform to the oblique boundary of the tilted cylinder. The gap thus generated between the vertical and oblique boundaries is a staircase effect (Fig. 138.1b, c). If the discs had oblique boundaries, there would not be any staircase effect.

If the build direction is oblique, the disc will have oblique boundaries. If the build direction is not fixed but tilts with a tilt in a feature with an aim to match the contour of the feature, there will not be any staircase effect. Therefore, if a process has variable build directions, this will be an antidote to the staircase effect.

This effect arises because of the position of one layer over another layer (Fig. 138.1c, showing gap). This implies if a layer is perfectly made, geometries within the layer are perfectly demarcated and filled up, and even then, these will have no bearing on the staircase effect. However, if the build direction tilts with a tilt in the cylinder, i.e., if there are multiple build directions to make a part, there will

S. Kumar, *A Concise Encyclopedia of Additive Manufacturing*,
https://doi.org/10.1007/978-3-031-98394-8_138

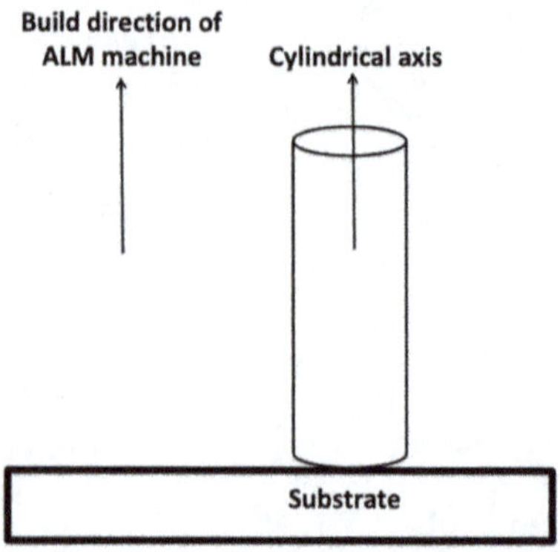

(a) Cylindrical axis and build direction are same

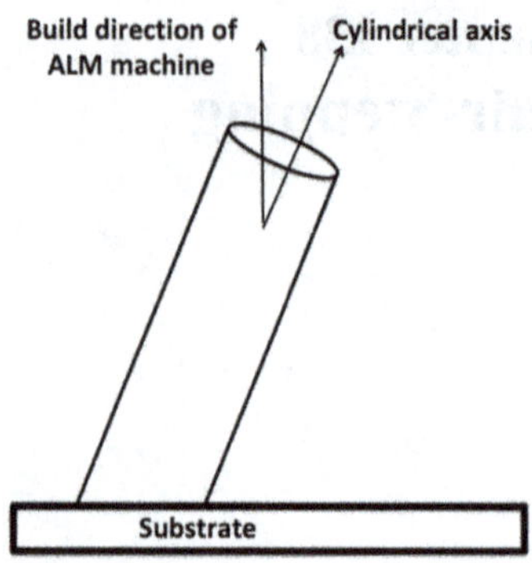

(b) Cylindrical axis and build direction are not same

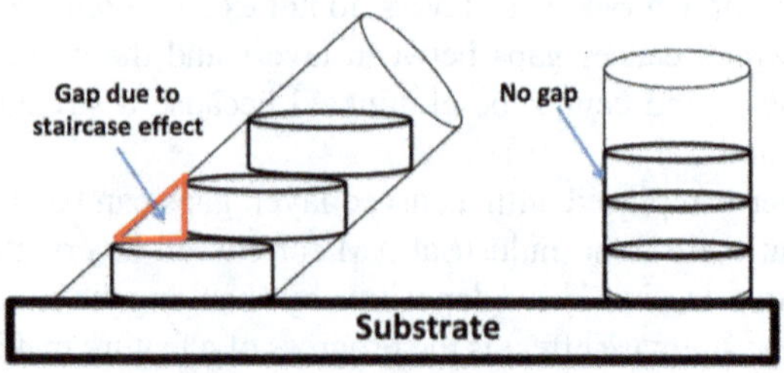

(c) Gap due to staircase effect in oblique cylinder, no gap in vertical cylinder

Fig. 138.1 Staircase effects: (**a**) an example of no staircase effect, (**b**) an example of staircase effect, (**c**) gap and no gap

be no staircase effect [6]. Having multiple build directions means having multiple slicing directions [7].

1 Removal

To remove the staircase effect, the gap needs to be removed, which can be done in two ways:

1. It can be filled up. But filling up is not possible because, in most applications, the gap is too small to be filled up. Even if it can be filled up, this may not be a convenient option, as this will take time, which will decrease the fabrication speed. In binder jetting, a small binder is used to fill the gaps created by big powders [8]. In material extrusion, thinner filaments will fill the gap created by thick filaments [9].
2. The corner, or overfill, or negative gap can be removed by not letting it form. In stereolithography, if the beam is defocused at the boundary, the corner will not

be formed. Because the defocused beam will decrease the curing depth [10]. Thus, optimizing the curing depth helps improve the process [11].

2 Removal by Simple Machining and Consequences

The corner encompassing the gap is machined away by conventional [12] or non-conventional machining [13]. But this makes the part smaller. In order to have the same part size, the part needs to be made bigger. The increase in the size to get the original part size is called offset.

In Fig. 138.1c, to remove a gap, a machining tool needs to move parallel to the side surface of the cylinder so the red triangle is removed. This means, after machining, the side surface of the cylinder passes through the tip of the red triangle, which decreases the size of the cross-section of the cylinder. The maximum decrease in the size of the cross-section is equal to the distance from the tip to the base of the triangle or to the side surface. While the minimum decrease in the size is zero.

The cylinder is tilted in an x-direction in an x-y plane. Therefore, when the disc is visualized to be fitted into the cylinder, the maximum gap is in the x-direction. The amount of the gap depends upon the angle of the tilt. If there is more tilt or the angle is big, the gap is bigger or the staircase effect is more. But, in any case, there is no gap in the y-direction. Therefore, when machining is done to remove the gap, maximum material is removed in the x-direction while no material is removed in the y-direction.

Since the cross-section of the disc is circular, it means the diameter of the disc in both the x- and y-directions is the same. When the material is removed from the x-direction, the diameter of the disc (or cylinder) in the x-direction becomes smaller, while no material is removed from the y-direction, so the diameter of the cylinder in the y-direction remains unchanged. This causes the circular cross-section of the cylinder to change into an elliptical cross-section. The big diameter of the ellipse is the same as the circular cross-section of the disc, while the small diameter is equal to the big diameter minus the separation due to the maximum gap caused by the tilt.

To remove the staircase effect, when the machining is employed, the milling tool has to remove materials from the tilted cylinder. The milling tool has to remove no material in the y-direction and maximum material in the x-direction when it is moving around the disc of the tilted cylinder. The spinning tool has just to touch the periphery of the disc in the y-direction without removing any material and has to move toward the periphery in the x-direction with an aim to remove maximum material. The job of the tool is to move from the point of no material removal to the point of the maximum material removal, gradually increasing its removal amount.

When the tool starts to remove the material, it needs to remove the minimum material. How much minimum it can remove depends upon the precision of the tool. How much minimum removal can be accepted depends upon the requirement of an application. If a tool is not able to furnish that requirement, a finer tool is needed. When the tool reaches another point at the periphery where it needs to remove the maximum material, how much maximum it can remove depends upon the robustness of the tool. If the tool is not robust, a more robust tool is required.

Thus, the requirement of a tool is to be fine and robust. This requirement may be conflicting if the height, diameter, and tilt angle of a cylinder are high. If the diameter and tilt angle are not uniform, i.e., they increase and decrease frequently, the unavailability of the best tool is not a problem, but the reach of the tool to all peripheral points of the cylinder is the problem. Thus, machining as a means to remove the staircase effect is ridden with shortcomings.

The transformation of a circular cylinder, after machining, into an elliptical cylinder can be avoided if original layers are changed. If, instead of original circular layers or discs, bigger-sized elliptical layers or discs are made, then an elliptical cylinder will be transformed into a circular cylinder after the machining. This can be a solution, but this is not an ideal solution, as making a bigger part only to have a smaller part by machining does not fit with the goal of less carbon emission. Besides, this solution does not help have an AM that will be free from machining, though it can help machining for AM improve.

References

1. Yigit, I. E., & Lazoglu, I. (2020). Spherical slicing method and its application on robotic additive manufacturing. *Progress in Additive Manufacturing, 5*(4), 387–394.
2. Xu, S., Liu, J., He, D., et al. (2025). Self-support structure topology optimization for multi-axis additive manufacturing incorporated with curved layer slicing. *Computer Methods in Applied Mechanics and Engineering, 438*, 117841.
3. Tang, P., Zhao, X., Shi, H., et al. (2024). A review of multi-axis additive manufacturing: Potential, opportunity and challenge. *Additive Manufacturing, 83*, 104075.
4. Pierre, J., Verville, M., Chenier, G., et al. (2024). Non-planar material-extrusion additive manufacturing of multifunctional sandwich structures using carbon-reinforced polyetheretherketone (PEEK). *Additive Manufacturing, 84*, 104124.
5. Pereira, S., Vaz, A. I. F., & Vicente, L. N. (2018). On the optimal object orientation in additive manufacturing. *The International Journal of Advanced Manufacturing Technology, 98*, 1685–1694.
6. Schuh, G., Bergweiler, G., Lukas, G., et al. (2020). Feature-based print method for multi-axis material extrusion in additive manufacturing. *Procedia CIRP, 93*, 85–89.
7. Basnet, B., Huang, R., Tabatabaei, B. T., & Choi, J. W. (2025). Additive manufacturing of binder-coated zirconia on a freeform surface. *Journal of the European Ceramic Society, 45*(5), 117127.
8. Hartmann, C., van den Bosch, L., Spiegel, J., et al. (2022). Removal of stair-step effects in binder jetting additive manufacturing using grayscale and dithering-based droplet distribution. *Materials, 15*(11), 3798.
9. Bhatt, P. M., Kulkarni, A., Malhan, R. K., et al. (2022). Automated planning for robotic multi-resolution additive manufacturing. *Journal of Computing and Information Science in Engineering, 22*(2), 021006.
10. Shan, Y., Krishnakumar, A., Qin, Z., & Mao, H. (2022). Reducing lateral stair-stepping defects in liquid crystal display-based vat photopolymerization by defocusing the image pattern. *Additive Manufacturing, 52*, 102653.

11. Fan, X., Zhang, M., Hu, L., et al. (2025). Modeling and spatio-temporal optimization of grayscale digital light processing 3D-printed structures with photobleaching resins. *Additive Manufacturing, 99*, 104659.
12. Arora, P., Dehgahi, S., Butt, S. U., et al. (2025). Fabrication of sacrificial wax pattern through large-scale fused granulated fabrication (FGF-AM) hybrid manufacturing system. *Progress in Additive Manufacturing, 10*(2), 1325–1339.
13. Boban, J., & Ahmed, A. (2024). Electric discharge aided surface post-treatment of laser powder bed fused non-planar metallic components for enhanced form accuracy. *Journal of Manufacturing Processes, 109*, 564–582.

Chapter 139
Stair-Stepping: An Unsolvable Problem

When an upright cylinder is deformed by tilting it sideways while keeping its base fixed, the shape of its layers changes from a circular disk (Fig. 139.1a) to an elliptical disk (Fig. 139.1b). The elliptical disk does not have an upright but a tilted boundary. This is a problem. Because all layers in AM have an upright boundary, they do not allow their boundaries to match the required tilted boundary (Fig. 139.2).

Why cannot an elliptical disk with a tilted boundary be created? Because layer formation is not completely an artificial process. For example, it can be checked with powder bed fusion (PBF) how a layer forms.

In PBF, when material is pushed sideways to create a bed, the creation does not entirely depend on how the bed is pushed but also on how the material is settled afterwards. Pushing the material is an artificial component of the process, while the material being settled is a natural component of the process. How the material is pushed can be controlled, but how the material settles cannot be controlled. Because if material settling is controlled, it cannot be said that the material is allowed to settle. Different devices used to push the material sideways give different densities [1, 2]. Depending upon the device, the material can be pushed downward as well. Then the material is deprived of some chances to go downward on its own and settle, which decreases the amount of material that settles and becomes part of the bed. If the bed is pressed afterwards, which is not a general practice, it will decrease the influence of settling on bed density. But it will not be able to overcome the influence of material settling on the contour of the bed. Though the contour of a bed is neither of any use nor important, the contour is mentioned because when material flows to make a bed, the flow has a pattern that can be noticed when the contour of the flow, i.e., of the bed, is checked. Therefore, whether material is pressed or not, the formation of the bed is not free from the natural component of the process.

A bed is not formed to have a shape that will contribute to the shape of a product that is formed on the bed. This simplifies the formation of the bed. This is also the reason why its formation is not completely controlled by removing the natural component of its formation. This distinguishes the processing of material to form a bed

S. Kumar, *A Concise Encyclopedia of Additive Manufacturing*,
https://doi.org/10.1007/978-3-031-98394-8_139

Fig. 139.1 Layer shape: (**a**) of cylinder, and (**b**) of deformed cylinder

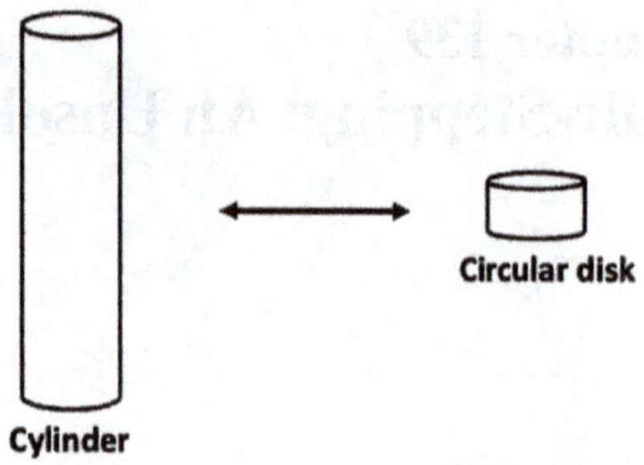

(a) Circular disk

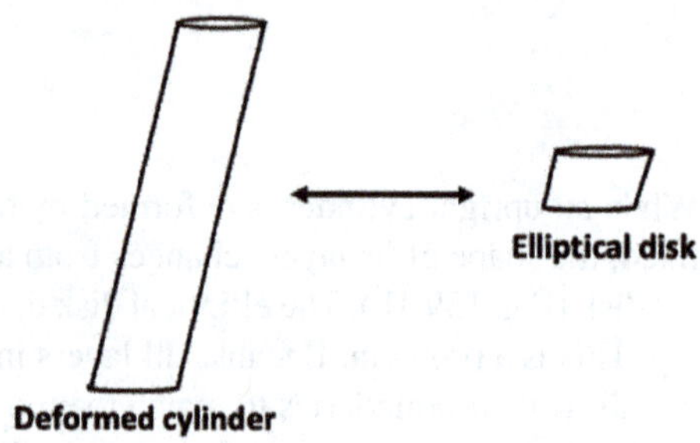

(b) Elliptical disk

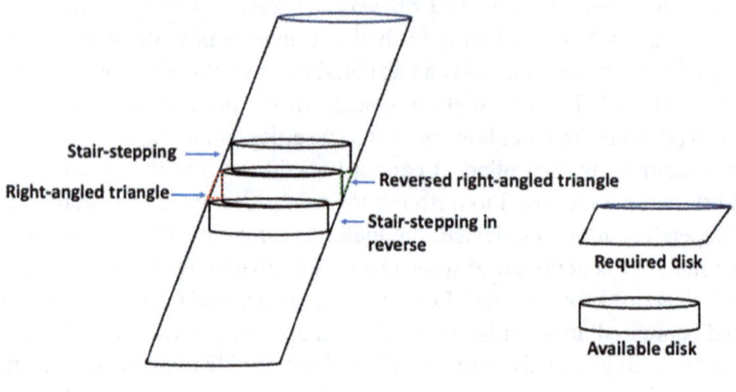

Fig. 139.2 Fitting of circular disks in deformed cylinder

in AM from the processing of materials to form other shapes in non-conventional techniques, such as sintering.

If the formation of a bed cannot be completely controlled, whatever formed on the bed cannot be completely controlled. For example, when a liquid binder is dropped on a porous powder bed, the binder will mostly flow downwards through porous channels. It will result in consolidation normal to the bed, i.e., if the binder is dropped along a line on the bed, it will result in a vertical wall. Suppose the formation of the bed is controlled so that it is densified. If the binder is dropped on a dense bed, the binder will not go downwards as far as it went in a porous bed; the binder will also have time to flow sideways. If the binder is dropped along a line, it will not result in a vertical wall but a wedge-shaped wall having a wider upper end.

The shape that is created on a bed depends on the properties of the bed. The property of the bed depends upon how much the formation of the bed is controlled. If the formation of the bed is not controlled, resulting in a porous bed, it is not possible to be free from the consequences of the lack of control on the wall that is formed on the bed. It is not possible to make a wedge-shaped wall instead of a vertical wall only by controlling the jetted binder. Lack of control means reliance on the natural component of the process. When the binder is jetted on the bed, the flow of the binder within the bed relies on gravity; that is again the natural component. When the binder is jetted, the purpose of the jetting is to enable the binder to reach the bed; the purpose is not to control the flow of binder within the bed so the binder will travel within the bed as per the contour of the design. If this were the case, there was no need to accept a vertical wall but to make a tilted wall that would have matched the contour, giving no stair-stepping effect.

Controlling the flow of the binder within the bed is difficult; besides, controlling the density of the bed is difficult, while not controlling them is easier. In AM, the easier options are chosen. This simplification, which hides the absence of control, gives rise to the stair-stepping effect.

What if the density of the powder bed is controlled? It will require a device. A device used to control the density of powder is called a mold as well. It means going backward in time to re-adopt those devices that are left behind to get freedom from them. AM will then lack the benefits that come with the absence of the mold. If jetting and binder are controlled so the binder will flow according to the contour, it will require time for the binding to take place. AM will then not be a fast technique. Thus, to solve the stair-stepping problem, AM must be practiced differently than how it is practiced presently without the advantages it provides. It implies that AM will no longer be AM, which makes stair-stepping an unsolvable problem.

However, if there comes a technique that densifies a bed without using any physical device or mold-type device, or if there comes a smart powder that gives the required density, then AM will not lose its advantage that comes with the absence of a mold. If there comes some smart binder and technique that lets the binder move within the bed as per the contour and the binding does not take much time (equivalent to the layer-spreading time), AM solves the stair-stepping problem without becoming slow.

References

1. Wang, L., Yu, A., Li, E., et al. (2021). Effects of spreader geometry on powder spreading process in powder bed additive manufacturing. *Powder Technology, 384*, 211–222.
2. Miao, G., Du, W., Pei, Z., & Ma, C. (2022). A literature review on powder spreading in additive manufacturing. *Additive Manufacturing, 58*, 103029.

Chapter 140
Stereolithography

Stereolithography [1] is a generic process, also known as vat photopolymerization [2]. Its variants are microstereolithography, digital light processing [3], large area maskless photopolymerization, etc.

In this process, photosensitive polymers are irradiated with light (UV or other lasers), which causes photons to interact with electrons and polymer molecules. The interaction creates either free radicals by breaking bonds or ions (cations) by removing electrons. The free radicals or cations bond with other polymer molecules, which enables them to bond again with other polymers, resulting in an increase in chain length, viscosity, gelling, and molecular weight.

Polymer chain formation is known as polymerization. Since a low-energy photon is sufficient to trigger polymerization, a laser power of a few mW is sufficient, which is in contrast to the high laser power of hundreds of watts used in metal powder-based AM, as high-energy photons are required to break metallic bonds and melt them.

With photopolymer, the process can work in many orientations. This could not be possible if the photopolymer were solid particles. In the majority of cases, the process works in a general orientation (Fig. 140.1a) where the l-beam is irradiated on the photopolymer from the top and the solidified part moves down inside the liquid. This orientation allows making big parts or many parts.

But what if the orientation is turned upside down? The approach to the substrate will also be turned upside down. The l-beam needs to approach the substrate by being present not above but below the substrate. The photopolymer will be required to be constrained from spilling. The substrate will have no room to go down but to go up, and, therefore, the solidified part has to move up layer after layer. In this orientation (Fig. 140.1b), the l-beam passes through a glass window and solidifies the liquid present between the substrate and the glass. To prevent the solidified material from being attached to the glass, a film is sealed on the glass. Successive layers are formed by moving the substrate up and filling up the gap between the solidified layer and the glass by recoating.

S. Kumar, *A Concise Encyclopedia of Additive Manufacturing*,
https://doi.org/10.1007/978-3-031-98394-8_140

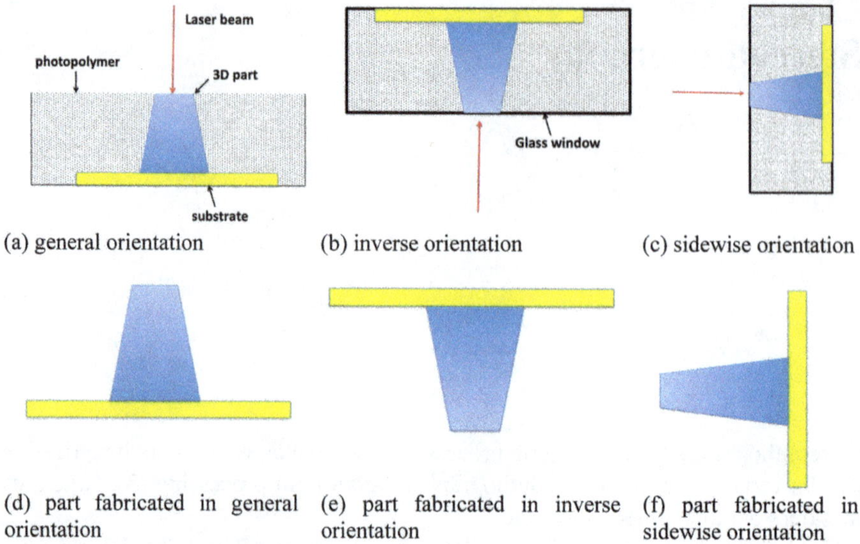

Fig. 140.1 Three orientations (from Fig. a to c) to make parts (from Fig. d to f). (**a**) General orientation. (**b**) Inverse orientation. (**c**) Sidewise orientation. (**d**) Part fabricated in general orientation. (**e**) Part fabricated in inverse orientation. (**f**) Part fabricated in sidewise orientation

This orientation may not allow making a big part as made in the general orientation but facilitates making a part with a minimum supply of liquid enough to fill the gap between the window and the ongoing solidified part [4]. Other orientations (e.g., Fig. 140.1c) are also possible, which allow parts being made in other orientations (Fig. 140.1d–f).

References

1. Husna, A., Ashrafi, S., Tomal, A. A., et al. (2024). Recent advancements in stereolithography (SLA) and their optimization of process parameters for sustainable manufacturing. *Hybrid Advances, 7*, 100307.
2. Dileep, C., Jacob, L., Umer, R., & Butt, H. (2024). Review of vat photopolymerization 3D printing of photonic devices. *Additive Manufacturing, 86*, 104189.
3. Cheng, J., Yu, S., Wang, R., & Ge, Q. (2024). Digital light processing based multimaterial 3D printing: Challenges, solutions and perspectives. *International Journal of Extreme Manufacturing, 6*(4), 042006.
4. Chi, Z., Yong, C., Zhigang, Y., & Behrokh, K. (2013). Digital material fabrication using mask-image-projection-based stereolithography. *Rapid Prototyping Journal, 19*(3), 153–165.

Chapter 141
Support Structure

Support structure (SS) is required, for example, when a design consisting of two parallel walls connected by a plate on top of them is made. It is made, e.g., by fabricating walls and horizontal sections between the walls (Fig. 141.1).

There are methods to avoid SS, which are as follows: (1) changing orientation, (2) self-support structure, and (3) reusable support.

1 Changing Design Orientation

If the orientation of the design is changed so that a plate is made first and walls are constructed on it thereafter, the same part (Fig. 141.1) can be constructed without SS. But changing the orientation is not always possible. For example, design (Fig. 141.2) always needs SS (shown by red lines), whether it is oriented at 90°, 180°, or 270° from the build direction (z-axis). Besides, if changing the orientation means increasing the number of layers to complete the design, it will not decrease the fabrication time.

2 Changing Build Orientation

If it is possible to change the build direction by changing the orientation of a substrate or build platform [1], some geometries without SS can be made. This can also work if, instead of the platform, the orientation of a nozzle [2] or a tool [3] is changed.

S. Kumar, *A Concise Encyclopedia of Additive Manufacturing*,
https://doi.org/10.1007/978-3-031-98394-8_141

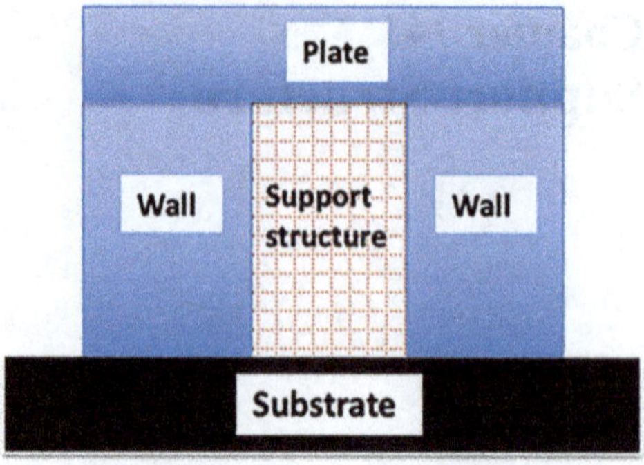

Fig. 141.1 Design consisting of SS

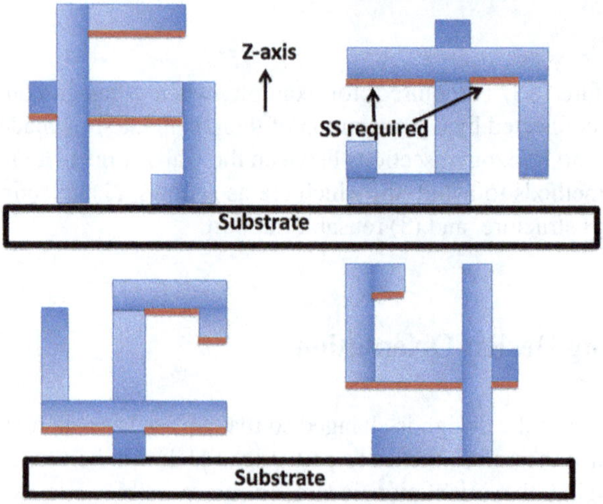

Fig. 141.2 Changing the orientation to avoid SS is not possible

3 Self-Support Structure

If a design has a curve structure, there is no need for either SS or to change the orientation (Fig. 141.3). The deposition will not be detached due to gravitational force. This partial attachment, if it continues, leads to an arch-type structure. Depending upon the thickness, length, and angle, all types of arch structures are not possible.

In the extrusion process, if there is a continuous fiber reinforcement [4], the deposited bead does not fall through a gap. Thus, no SS is required (Fig. 141.4a). But for thermoplastic filaments without fiber reinforcement, the deposited bead may cave in the gap (Fig. 141.4b). Although lattice structures can be made without SS [5].

Fig. 141.3 Self-support structure —arch similar to opposite 'V'

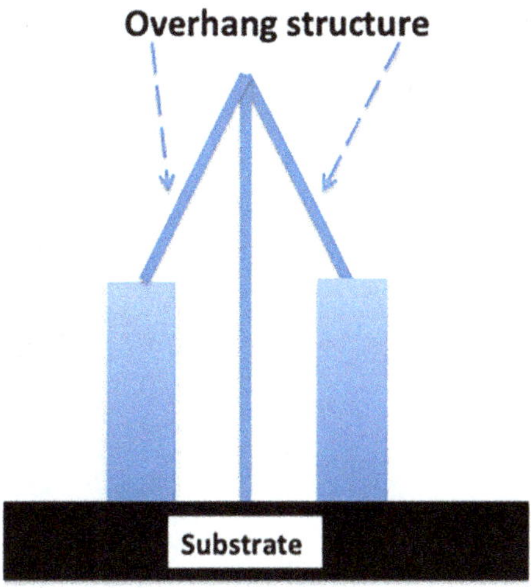

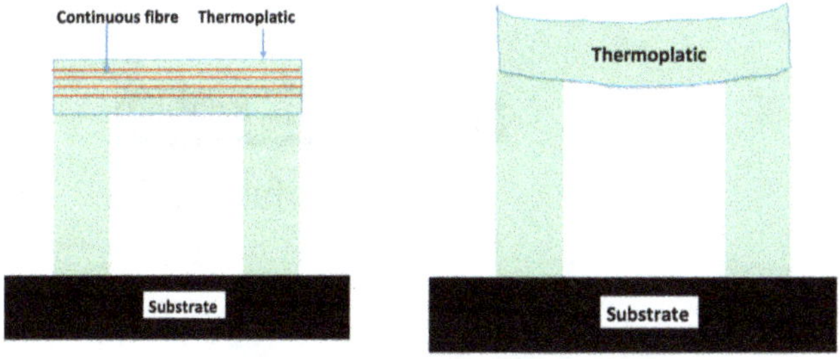

(a) Non-bending (b) Bending

Fig. 141.4 Deposited beads: (**a**) No SS required, (**b**) Bending in absence of SS

4 Reusable Support

If a reusable SS is used, making SS for recurring fabrication is not needed. SS is elevated from the platform to support the fabrication of overhangs. After the fabrication, SS is detached from the overhang. Reusable support requires the modification of a platform to include the provision (nut, bolt, fixture, jack) for using SS [6] (Fig. 141.5).

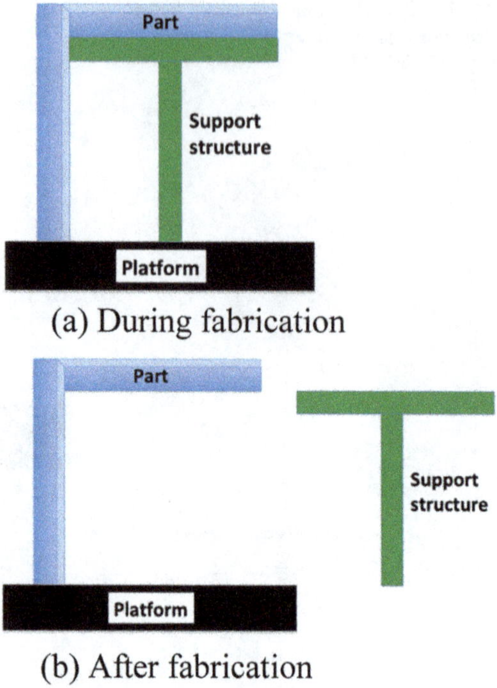

Fig. 141.5 Support: (**a**) used during fabrication, and (**b**) retained after fabrication

References

1. Bhatt, P. M., Malhan, R. K., Rajendran, P., & Gupta, S. K. (2020). Building free-form thin shell parts using supportless extrusion-based additive manufacturing. *Additive Manufacturing, 32*, 101003.
2. Li, X., Liu, W., Hu, Z., et al. (2024). Supportless 3D-printing of non-planar thin-walled structures with the multi-axis screw-extrusion additive manufacturing system. *Materials & Design, 240*, 112860.
3. Kaji, F., Jinoop, A. N., Zardoshtian, A., et al. (2023). Robotic laser directed energy deposition-based additive manufacturing of tubular components with variable overhang angles: Adaptive trajectory planning and characterization. *Additive Manufacturing, 61*, 103366.
4. Blok, L. G., Longana, M. L., Yu, H., & Woods, B. K. S. (2018). An investigation into 3D printing of fiber reinforced thermoplastic composites. *Additive Manufacturing, 22*, 176–186.
5. Parsompech, N., Suwanpreecha, C., Noraphaiphipaksa, N., et al. (2024). Supportless lattice structure of 316L stainless steel fabricated by material extrusion additive manufacturing: Effect of relative density on physical, microstructural and mechanical behaviour. *Materials Science and Engineering A, 915*, 147270.
6. Xu, Y., Wang, Z., Gong, S., & Chen, Y. (2021). Reusable support for additive manufacturing. *Additive Manufacturing, 39*, 101840.

Chapter 142
Support Structure: Difficulty in PBF

A bed acts as a support structure (SS) in powder bed fusion (PBF), which allows polymers more than metals or ceramics to avoid SS because polymers are lighter than metals and ceramics [1]. Due to the presence of bed as SS, PBF, instead of the deposition process, gives more opportunity to avoid SS. However, there are difficulties that arise when robust support is required.

Powder needs to provide the support but caves in when a heavy section is built. Besides, it cannot form a smooth underneath surface (Fig. 142.1). A thin section does not require a support irrespective of the orientation, but the support can be used in the horizontal orientation [2].

A powder layer has different thermal properties than a solidified layer has. Therefore, a section made on the powder layer acquires different properties [3]. Besides, the section warps upward, making a curl surface [4], while the section made on a solidified layer is free from warping [5]. Therefore, fabrication on a powder layer that acts as SS requires extra optimization for making a downfacing surface (Fig. 142.1c). SS is fabricated to avoid it. If an overhang is built at a certain orientation from a build direction, SS is not required while the quality of downfacing surfaces does not deteriorate [4].

In comparison, making a section on liquid is different from that of making one on a cured photopolymer in stereolithography. During solidifying, the curing depth needs to be maintained so that the curing does not exceed and solidify more liquid. The exceeding deteriorates the surface and the dimension. If a section is made on a cured layer, exceeding is accommodated as overcuring of the cured layer. Thus, making SS will facilitate maintaining the quality of that type of part whose qualities deteriorate due to reliance on liquid as a support [6].

S. Kumar, *A Concise Encyclopedia of Additive Manufacturing*,
https://doi.org/10.1007/978-3-031-98394-8_142

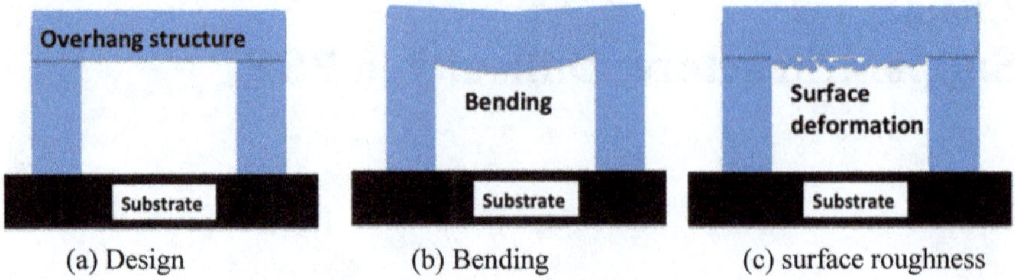

Fig. 142.1 Overhang: (**a**) design, (**b**) bending, and (**c**) underneath surface

References

1. Dadkhah, M., Tulliani, J. M., Saboori, A., & Iuliano, L. (2023). Additive manufacturing of ceramics: Advances, challenges, and outlook. *Journal of the European Ceramic Society, 43*(15), 6635–6664.
2. Noronha, J., Dash, J., Downing, D., et al. (2025). Thin-plate lattices in AlSi10Mg alloy via laser additive manufacturing: Highly enhanced specific strength and recovery. *Additive Manufacturing, 99*, 104664.
3. Chou, Y. S., & Cooper, K. (2017). Systems and methods for designing and fabricating contact-free support structures for overhang geometries of parts in powder-bed metal additive manufacturing. US patent 9767224.
4. Calignano, F. (2014). Design optimization of supports for overhanging structures in aluminum and titanium alloys by selective laser melting. *Materials & Design, 64*, 203–213.
5. Cheng, B., & Chou, K. (2015). Geometric consideration of support structures in part overhang fabrications by electron beam additive manufacturing. *Computer-Aided Design, 69*, 102–111.
6. Zhao, G., Zhou, C., & Das, S. (2015). Solid mechanics based design and optimization for support structure generation in Stereolithography based additive manufacturing. In *Volume 1A: 35th computers and information in engineering conference*.

Chapter 143
Support Structure: Effect on Fabrication Time

The effects of various types of support structures (SS) on fabrication time are given as follows:

- Different materials [1]—if SS and the main structure are of different materials, both materials can mix during their fabrication and the removal of SS, which causes them not to be recycled. Consequently, fabrication with different materials will require attention, increasing the fabrication time.
- Optimizing design: If lightweight SS is made, it will decrease the fabrication time [2].
- Weak material [3]—SS is made from a weak material, as SS has no use after fabrication. Weak material is less expensive than the material of the main structure, causing the fabrication not to become expensive when many SS are required. Weak SS will take less time to be removed.
- Removable or nonremovable [4]—SS made from water-soluble materials (sugar, polyvinyl alcohol, etc.) are removed by dipping them in water [5]. Though these materials cannot be used in all AM processes, especially in powder bed fusion (PBF), where using another material to create weak SS requires system modification [6]. Using a removable SS will save time for its removal.
- Contactless support [7]—in PBF, when fabrication starts not exactly on SS, but after spreading one or two additional layers on SS, then the SS does not contact the part fabricated on it. It is easy to remove parts from this type of SS because no contact needs to be broken to separate SS, saving time to remove it.
- Using layerless AM—layerless AM does not use SS and therefore does not consume time in fabricating SS. However, this AM is for small photopolymer parts [8].

S. Kumar, *A Concise Encyclopedia of Additive Manufacturing*,
https://doi.org/10.1007/978-3-031-98394-8_143

References

1. Lušić, M., Feuerstein, F., Morina, D., & Hornfeck, R. (2016). Fluid-based removal of inner support structures manufactured by fused deposition modeling: An investigation on factors of influence. *Procedia CIRP, 41*, 1033–1038.
2. Kuo, Y. H., Cheng, C. C., Lin, Y. S., & San, C. H. (2018). Support structure design in additive manufacturing based on topology optimization. *Structural and Multidisciplinary Optimization, 57*(1), 183–195.
3. Bobbio, L. D., Qin, S., Dunbar, A., et al. (2017). Characterization of the strength of support structures used in powder bed fusion additive manufacturing of Ti-6Al-4V. *Additive Manufacturing, 14*, 60–68.
4. McConaha, M., Venugopal, V., & Anand, S. (2020). Integration of machine tool accessibility of support structures with topology optimization for additive manufacturing. *Procedia Manufacturing, 48*, 634–642.
5. Ni, F., Wang, G., & Zhao, H. (2017). Fabrication of water-soluble poly(vinyl alcohol)-based composites with improved thermal behavior for potential three-dimensional printing application. *Journal of Applied Polymer Science, 134*, 44966.
6. Wei, C., Chueh, Y., Zhang, X., et al. (2019). Easy-to-remove composite support material and procedure in additive manufacturing of metallic components using multiple material laser-based powder bed fusion. *ASME Journal of Manufacturing Science and Engineering, 141*(7), 071002.
7. Ma, R., Liu, R., & Lu, B. (2024). Approaches used to design support structures for ceramic additive manufacturing: A review. *Additive Manufacturing Frontiers, 3*, 200136.
8. Peng, S., Xu, J., Li, D., et al. (2023). Ultra-fast 3D printing of assembly—Free complex optics with sub-nanometer surface quality at mesoscale. *International Journal of Extreme Manufacturing, 5*(3), 035007.

Chapter 144
Two-Photon Photopolymerization

Making a layer can be a problem if 100 nm resolution is required. Two-photon photopolymerization (2PP) (Fig. 144.1) makes sub-100 μm size parts having sub-100 nm resolution for medical, optical, and acoustical applications. With the help of a micro-mold created by 2PP, a complex structure of 2 μm resolution can be made by infiltration [1].

L-beam consisted of femtosecond pulse required to excite photopolymer and solidify it. In order to excite, two photons within an interval of 10^{-15} s are required. One photon excites the molecule to an intermediate virtual state, while another photon excites it from the intermediate state to the final state. If another photon is not available before the molecule loses its intermediate state and comes back to the ground state, the molecule will not be able to reach a higher excited state.

The absorption of two photons present in space and time made available through an l-beam leads to excitation of polymers that causes polymers to form either free radicals or cations for polymerization. Energy contributed by two photons is necessary to excite a molecule. If the same energy is contributed by a single photon, the molecule will not excite because of the minor difference in quantum states in two cases [2]. Therefore, 2PP is radically different from single-photon polymerization, which is used in stereolithography.

The probability of finding two photons within such a small time is possible only in an environment of high density of photons; such an environment is found at the focal point of an l-beam. Therefore, when the beam irradiates the liquid, only a small volume of photopolymer present near the focal point gets cured, while the vast amount of the liquid facing the beam remains unchanged.

This small amount of solidification makes the process slow, and even a sub-mm-sized part is too big to be made. But the same amount gives the process a capability to make small features and have a high resolution. The smallest feature made in a beam-based process is limited by the diffraction limit imposed by the beam, while 2PP is able to make far smaller features than the wavelength of the laser used, showing the particle characteristic of light is dominating over its wave characteristic.

S. Kumar, *A Concise Encyclopedia of Additive Manufacturing*,
https://doi.org/10.1007/978-3-031-98394-8_144

Fig. 144.1 Schematic diagram of 2PP

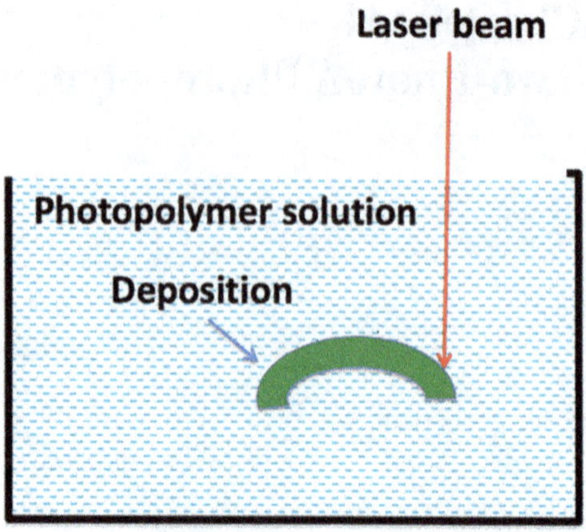

Tracing the beam in liquid will leave a trail of solidified material, and this is how a 3D feature is made by scanning the beam as per geometry without going through layers. The features can be made on the surface or inside the liquid. Making it inside the liquid will free the process from availing feedstock (liquid) at reaction sites, as the feature made inside the liquid remains self-supported, which neither sinks nor buoys.

References

1. Park, J., Brugger, J., & Bertsch, A. (2025). Additive manufacturing of water-soluble 3D micro molds for complex-shaped lipid microparticles. *Nature Communications, 16*(1), 1734.
2. Fourkas, J. T. (2016). *Fundamentals of two photon fabrication. 3D fabrication using two photon polymerization* (pp. 45–61). Elsevier Inc.

Chapter 145
Ultrasonic Consolidation

Ultrasonic consolidation, also known as ultrasonic additive manufacturing, is a layer-upon-layer process in which a 3D object is fabricated by joining metal foils using ultrasonic bonding, followed by shaping using a milling machine [1].

Ultrasonic bonding between foils is accomplished with the help of a solid metallic probe called a sonotrode that is pressed onto these foils. After pressing, the sonotrode is given ultrasonic energy, causing it to vibrate with a certain amplitude and frequency across foils (Fig. 145.1). A forced and vibrating sonotrode is enough to join foils. Its vibration smoothens the interface between two foils, destroys the oxide layer of the interface, and exposes the clean surface of one foil to that of another foil, causing plastic flow at the interface—these results in the creation of a metallurgical bond between two foils. To extend this bond along the lengths of the foils, the sonotrode travels from one end of the foil to its other end while being pressed and vibrated. The vibration happens perpendicular to the direction of the travel. If the sonotrode is moving in an x-direction and the thickness of two foils is measured in a z-direction, the vibration is planned to occur in a y-direction (Fig. 145.1a). To prevent the foils from getting displaced, they are clamped. The combined action of the force and the movement does not raise the foil temperature to its melting point. Thus, the process is suitable to fabricate those parts that require being fabricated at low temperatures, e.g., a part having sensors. The most important parameters for ultrasonic bonding are travel speed, vibration amplitude, and applied normal force on the sonotrode. Post-process heat treatment is done to improve bonding [2].

A milling machine is used for the following: to separate foils from a pool of foils (Fig. 145.1b), to trim foils at boundaries, to remove periodically excess material from bonded foils as per a CAD model, to make pockets on the ongoing build to accommodate fibers, antennas, sensors, or electronic circuits, to smooth the ongoing build for removing building defects, and to control z-height.

S. Kumar, *A Concise Encyclopedia of Additive Manufacturing*,
https://doi.org/10.1007/978-3-031-98394-8_145

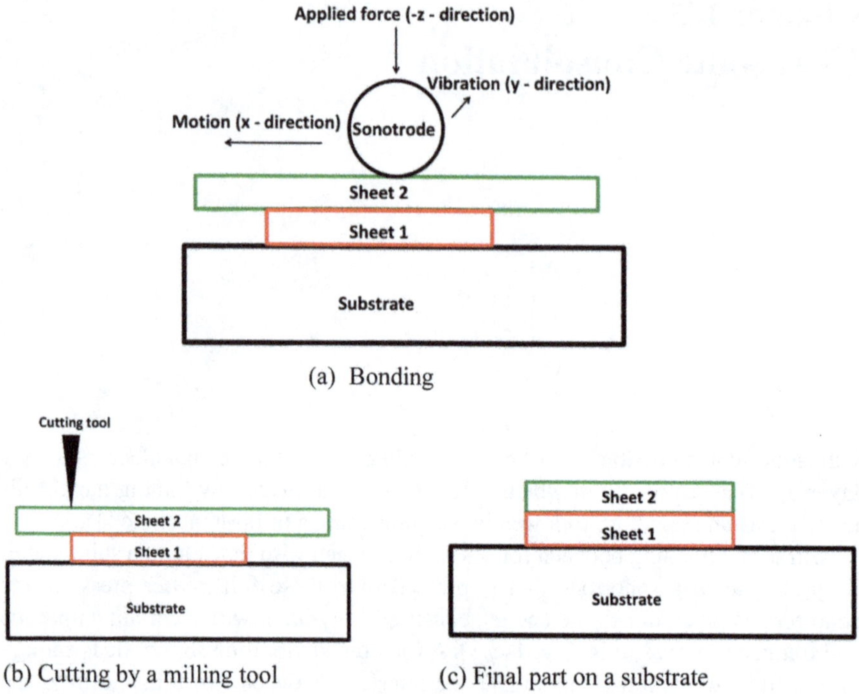

Fig. 145.1 Part made in ultrasonic consolidation: (**a**) sheet 2 is bonded on sheet 1, (**b**) excess material of sheet 2 is removed, (**c**) final part

References

1. Li, J., Yang, T., Liu, S., et al. (2025). A review on recent progress in light metal solid-phase additive manufacturing. *Materials Science and Technology*, 02670836251314968.
2. Ishfaq, K., Abas, Z., Saravana Kumar, M., & Mahmood, M. A. (2023). Review of recent trends in ultrasonic additive manufacturing: Current challenges and future prospects. *Rapid Prototyping Journal, 29*(6), 1195–1211.

Chapter 146
Water Deposition Process

The process, known as rapid freeze prototyping [1], makes a part, e.g., an ice structure, by depositing water (Fig. 146.1).

The deposition needs to be controlled so water solidifies only when it reaches a substrate. For the solidification to take place, the substrate needs to be kept at a subzero temperature. When the structure grows, it starts to melt because the substrate, though maintained at liquid nitrogen temperature (−140 °C), will not be able to cool it. This requires the whole setup of a nozzle and the substrate to be kept inside a cold chamber (−20 °C) to prevent the structure from melting. Keeping the setup inside affects the water present either within the nozzle or in transit to the substrate, causing the water to start solidifying. Therefore, the nozzle is kept near the substrate to facilitate the reaching of water to the substrate before the water starts to lose its form due to solidification. Besides, the water is ejected from the nozzle using more than a critical pressure, so the flow of water is not obstructed by icing on the nozzle. The flow breaks the icing if obstructed [2].

Two nozzles are required to make a structure that has an overhang: one nozzle for the main material and the other for the supporting material. Although one nozzle can do the job as well—first, it will deposit the main material, and the next time, it will deposit the supporting material as per need—but then the same nozzle needs to be fitted with different liquid sources periodically, which will delay the fabrication, contaminate the liquid, change the heat transfer time, and compromise the accuracy.

When two materials are used—one is water and the second is the supporting material of a lower freezing point than water (e.g., NaCl or brine solution), then after fabrication, the supporting material structure is removed by placing the part in a chamber having a temperature more than the freezing point of brine and less than the freezing point of water. This will melt the supporting structure and not the main structure.

In the absence of two materials, what will happen if water does the job of two materials? Water can do the job of a supporting material—the supporting structure is made weaker so it is removed by the application of force—this method is used in

S. Kumar, *A Concise Encyclopedia of Additive Manufacturing*, https://doi.org/10.1007/978-3-031-98394-8_146

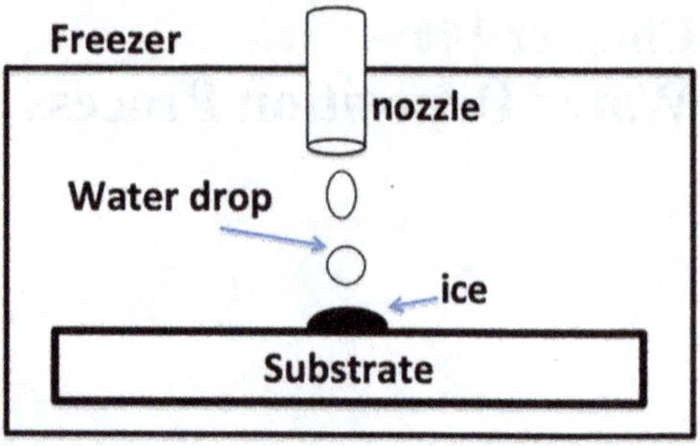

Fig. 146.1 Schematic diagram of rapid freeze prototyping

other AM processes without damaging the main structure [3], but this may not be as convenient since the removal is done inside a cold chamber.

The ice structure thus fabricated will melt away if removed from the cold chamber, and therefore, its use needs to be worked out before it is removed. If the ice structure after fabrication is used to make a non-ice structure, the non-ice structure can be removed from the cold chamber for further application. The example of such a non-ice structure is a ceramic shell made from an ice structure, i.e., an ice pattern, while the application is investment casting.

In comparison to a polymer pattern [4], the ice pattern gives an advantage as it does not need to be burnt to remove it from a ceramic shell, but it can be removed by bringing the pattern (covered with the shell) outside the chamber.

To create a ceramic shell, the ice pattern is dipped in a ceramic slurry. The slurry should not be frozen, so it needs to be free from water. Therefore, a special slurry free from water is required. The slurry coated on the pattern dries with the help of a catalyst, and the shell thus formed from the slurry is further strengthened by heat treatment [5].

Ice parts made from this process do not have many applications if possible, applications in the area of making ice sculptures are not counted. But the method learned is utilized in the area of tissue engineering, where a scaffold, if not printed in a subzero environment, is affected by porogens. In cryogenic prototyping, scaffolds are printed using this method by replacing water with a biological solution [6].

References

1. Leu, M., Isanaka, S. P., & Richards, V. L. (2009). Increase of heat transfer to reduce build time in rapid freeze prototyping. In *SFF sym proc* (pp. 219–230).
2. Barnett, E., Angeles, J., Pasini, D., & Sijpkes, P. (2009). Robot-assisted rapid prototyping for ice scluptures. In *2009 IEEE international conference on robotics and automation, ICRA 2009, Kobe, Japan*.
3. Kumar, S. (2022). Fabrication strategy. In *Additive manufacturing solutions* (pp. 111–144). Springer.
4. Kumar, S. (2022). Application. In *Additive manufacturing solutions* (pp. 93–110). Springer.

5. Zhang, W., & Leu, M. C. (2000). Investment casting with ice patterns made by rapid freeze prototyping. In *Proceedings of the 11th annual solid freeform fabrication symposium* (pp. 66–72).
6. Pham, C. B., Leong, K. F., Lim, T. C., & Chian, K. S. (2008). Rapid freeze prototyping technique in bio-plotters for tissue scaffold fabrication. *Rapid Prototyping Journal, 14*(4), 246–253.

Index

S. Kumar, *A Concise Encyclopedia of Additive Manufacturing*,
https://doi.org/10.1007/978-3-031-98394-8

J

K

L

M

www.ingramcontent.com/pod-product-compliance
Lightning Source LLC
Chambersburg PA
CBHW051426031125
34911CB00001B/30
* 9 7 8 3 0 3 1 9 8 3 9 3 1 *